Total Engineering Project Management

McGraw-Hill Engineering and Technology Management Series
Michael K. Badawy, Ph.D., Editor in Chief

RITZ · *Total Engineering Project Management*

STEELE · *Managing Technology*

WHEELER · *Computers and Engineering Management*

Total Engineering Project Management

George J. Ritz

McGraw-Hill Publishing Company
New York St. Louis San Francisco Auckland Bogotá
Caracas Hamburg Lisbon London Madrid Mexico
Milan Montreal New Delhi Oklahoma City
Paris San Juan São Paulo Singapore
Sydney Tokyo Toronto

Library of Congress Cataloging-in-Publication Data

Ritz, George J.
Total engineering project management / George J. Ritz.
p. cm. — (McGraw-Hill engineering and technology management series)
ISBN 0-07-052966-3
1. Engineering—Management. 2. Industrial project management.
I. Title. II. Series.
TA190.R47 1990
658.4'04—dc20 89-13200
CIP

234567890 DOC/DOC 9543210

ISBN 0-07-052966-3

The editors for this book were Robert W. Hauserman and Galen H. Fleck, the designer was Naomi Auerbach, and the production supervisor was Dianne L. Walber. This book was set in Century Schoolbook. It was composed by the McGraw-Hill Publishing Company Professional and Reference Division composition unit.

Printed and bound by R. R. Donnelley & Sons Company.

To all the business associates who helped me gain the experience to write this book.

To my wife, Helen; my children, Don, Barry, and Trish, for supporting my career moves.

Contents

Preface xi

Series Introduction xiii

Chapter 1. The Project Environment 1

Some History of Project Management 1
Defining Project Management 4
Project Goals 8
Basic Project Management Philosophy 12
What Are We Going to Plan, Organize, and Control? 13
Project Manager's Job Description 15
Project Size 16
References 22
Introduction to Case Studies 22
Case Study Instructions 22

Chapter 2. Proposals and Contracts 25

Project Execution Approach 25
The Contractor Selection Process 27
Managing the Proposal Effort 36
Capital Project Contracting 37
The Role of the Project Manager in Contracting 46
Contract Technical Terms 46
Contract Commercial Terms 50
Contract Administration 59
References 61
Case Study Instructions 62

Chapter 3. Project Planning 63

Planning Definitions 63
Planning Philosophy 63
Types of Planning 65
Operational Planning Questions 66

The Project Master Plan 66
Project Execution Format for Capital Projects 68
Instrumentation 76
A Typical Nonprocess Capital Project Format 78
Summary 84
References 84
Case Study Instructions 84

Chapter 4. Project Scheduling 87

Scheduling a Nonprocess Project 88
Procurement 91
Scheduling a Process-Type Project 94
Major Activities to Be Scheduled 94
The Design Phase 95
Engineering Schedule 95
Procurement Activities 99
Construction Schedule 100
Overall Schedule 101
Scheduling Systems 103
Bar Charts 103
Basic Network Diagramming 106
Advantages and Disadvantages of CPM and Bar Chart Methods 00
Scheduling System Selection 118
References 120
Case Study Instructions 121

Chapter 5. The Project Money Plan 123

Project Cost Estimating 123
How to Make the Four Classes of Estimates 127
Estimating Methods 133
Estimating Tools 134
Estimating Escalation and Contingency 135
Estimating Home Office Services 138
Requirements for a Good Estimate 144
Project Budgeting 145
Budgeting Escalation and Contingency 148
Computerized vs. Manual Budgets 150
The Project Cash Flow Plan 151
Summary 152
References 153
Case Study Instructions 153

Chapter 6. Project Resources Planning 155

Human Resources Planning 155
Subcontracting Design and Construction 163

Design Subcontracting 164
Construction Subcontracting 167
Material Resources Planning 167
Summary 171
References 172
Case Study Instructions 172

Chapter 7. Project Organization 173

Organizational Overview 173
Constructing the Organization Chart 176
Selecting and Motivating the Project Team 186
Project Modes: Matrix vs. Task Force 191
Some Advantages and Disadvantages of the Task Force vs. Matrix Systems 197
Organizational Procedures 200
The Project Procedure Manual 200
Review of the Main Sections 201
Design Procedures 205
Planning and Scheduling 206
Project Control and Reporting 207
Change Order Procedure 208
Computer Services 208
Project Procedure Manual Publication 209
Satellite Office Procedures 210
Summary 210
References 210
Case Study Instructions 211

Chapter 8. Project Control 213

The Control Process 213
Controlling the Money Plan 214
A Typical Cost Control System 218
How a Cost Control System Really Works 220
Controlling Material Resource Costs 228
Specific Areas for Cost Control 236
Controlling Cash Flow 245
Schedule Control 250
Equipment and Material Control 260
Controlling Quality 262
Operating the Controls 268
References 271
Case Study Instructions 271

Chapter 9. Project Engineering 273

The Project Engineering Environment 274
The Administrative Side of Project Engineering 275

The Technical Side of Project Engineering 280
Human Factors in Project Engineering 284
Leadership 286
References 286
Case Study Instructions 286

Chapter 10. Project Communications 289
Project Communications 291
Summary 314
References 314
Case Study Instructions 314

Chapter 11. Human Factors in Project Management 317
Qualities for a Successful Project Manager 318
Human Relations 322
Personal Human Factors 333
References 335
Case Study Instructions 336

Appendixes

A. Project Manager's Job Description 337
1.0 Concept 337
2.0 Foreword 337
3.0 Duties and Responsibilities 338
4.0 Authority 340
5.0 Working Relationships 341
6.0 Leadership Qualities 341
B. Project Services Checklist 343
C. Case Studies 357
Project 1 357
Project 2 358
Project 3 358
Project 4 359
Project 5 359
D. Typical Project File Index 361

Index 365

Preface

This book presents my personal thoughts on the practical aspects of effective total project management. It stresses a practical approach to avoiding the problems normally encountered in executing projects. However, no single book can meet all of your needs in the practice of effective project management. I hope this one will expand your interests in the right direction and thereby improve your project performance.

The goal of the book is to give readers an understanding of how important total project management is for the successful execution of capital projects. It also emphasizes the key role played by the project manager in making the system work. The book uses a hands-on approach to managing capital projects from start to finish. It includes practical suggestions and examples that are adaptable to the reader's project environment.

This book evolved from a series of project management training seminars which I developed over the years. It uses a training format, so it can double as a textbook or manual. A slate of case studies appears at the end of each chapter to lead readers through the execution of a total project. Workshop sessions using group participation and role playing to prepare the case studies work well in a training environment.

The total project management concept applies to the execution of all kinds of capital projects. However, the special requirements of each project field have led to systems, practices, and a jargon peculiar to the field's own needs. It was hard to bridge the specialized differences completely, so I concentrated on the principles that are common to all fields.

The term "project manager" as used throughout this book means the person in charge of the project regardless of his or her actual title. The project leader may have the title "project engineer," "project director," "job captain," "lead architect," "construction manager," or any other that indicates responsibility for the work. Anyone who is responsible for and in charge of a capital project is practicing project management.

One problem which often arose in writing this book was that of addressing areas each of which in itself was broad enough to be the subject of a book. Project management involves many areas such as human relations, legal matters, accounting, finance, estimating, budgeting, cost control, scheduling, organization, and management science. Since there was not room to cover all of those and a host of others in detail, I have addressed them from the project manager's viewpoint. That approach gives the essential points in the areas which will permit a project manager to survive and prosper. If you need more details, by all means consult your library and the references listed at the ends of the chapters.

I hope the information passed along in this book will help you enjoy managing projects as much as I have. Remember, none of us will ever know all there is about project management, so keep an open mind and good luck!

George J. Ritz

Series Introduction

Technology is a key resource of profound importance for corporate profitability and growth. It also has enormous significance for the well-being of national economies as well as international competitiveness. Effective management of technology links engineering, science, and management disciplines to address the issues involved in the planning, development, and implementation of technological capabilities to shape and accomplish the strategic and operational objectives of an organization.

Management of technology involves the handling of technical activities in a broad spectrum of functional areas including basic research; applied research; development; design; construction, manufacturing, or operations; testing; maintenance; and technology transfer. In this sense, the concept of technology management is quite broad, since it covers not only R&D but also the management of product and process technologies. Viewed from that perspective, the management of technology is actually the practice of integrating technology strategy with business strategy in the company. This integration requires the deliberate coordination of the research, production, and service functions with the marketing, finance, and human resource functions of the firm.

That task calls for new managerial skills, techniques, styles, and ways of thinking. Providing executives, managers, and technical professionals with a systematic source of information to enable them to develop their knowledge and skills in managing technology is the challenge undertaken by this book series. The series will embody concise and practical treatments of specific topics within the broad area of engineering and technology management. The primary aim of the series is to provide a set of principles, concepts, tools, and techniques for those who wish to enhance their managerial skills and realize their potentials.

The series will provide readers with the information they must have and the skills they must acquire in order to sharpen their managerial performance and advance their careers. Authors contributing to the series are carefully selected for their expertise and experience. Although the series books will vary in subject matter as well as approach, one major feature will be common to all of them: a blend of practical applications and hands-on techniques supported by sound research and relevant theory.

The target audience for the series is quite broad. It includes engineers, scientists, and other technical professionals making the transition to management; entrepeneurs; technical managers and supervisors; upper-level executives; directors of engineering; people in R&D and other technology-related activities; corporate technical development managers and executives; continuing management education specialists; and students in engineering and technology management programs and related fields.

We hope that this series will become a primary source of information on the management of technology for practitioners, researchers, consultants, and students, and that it will help them become better managers and pursue the most rewarding professional careers.

MICAHEL K. BADAWY
Professor of Management of Technology
The R. B. Pamplin College of Business
Virginia Polytechnic Institute and State University
Falls Church, Virginia

Chapter

1

The Project Environment

Many recent authors of project management books would have us believe that project management is a relatively new technique developed since the arrival of the computer. Anyone in the field of capital projects should be aware that is simply not the case. Upon entering the engineering and construction field in the late 1940s, I discovered project management systems operating in much the same context as we practice them today. The only real difference between then and now is the plethora of data generated by modern scheduling methods and the computer. Another plus is the improved communication systems available to today's project manager.

Some History of Project Management

I even traced the system's origins a few years further back to the synthetic rubber plants built during World War II. That shows that project management has been in active use on industrial capital projects for almost 50 years. The project management system has broadened its base in recent years because of its successful results in the capital projects arena.

One of the more publicized successes was the famous Manhattan Project. It started in the early 1940s under the direction of General Leslie Groves to develop the atom bomb. Groves certainly qualifies as the first bona fide project director. Being a military man with a blank check during a major war, we can assume that he operated without today's strict budgetary restraints.

Many centuries earlier, Jesus Christ actually recognized the need for project management when he said, in Luke 14:28–29:

> If one of you is planning to build a tower, he sits down first and figures out what it will cost, to see if he has enough money to finish the job. If he

doesn't, he will not be able to finish the tower after laying the foundation; and all who see what happened will make fun of him.

Present-day retribution befalling project managers failing to complete their towers because of lack of money leads to stronger censure than public ridicule. I hope the material presented in this book will help readers avoid censure and ridicule through improved project execution.

After World War II, capital project development enjoyed an unprecedented growth in both size and dollar value. The market resulted from the demands of a huge post-war economic boom throughout the world. The rebuilding of Europe and satisfying pent-up demand because of wartime restrictions created a climate for economic expansion and capital spending. The practice of project management had to respond to the explosive demand for new facilities.

In the sixties, projects became larger as the advantage of economy of scale came into vogue. Chemical plants, steel mills, office buildings, shopping centers, airports, power plants, nuclear projects, and resorts became larger, more complex, and more costly. Horror stories of project cost and schedule overruns started to surface throughout the capital projects industry.[1] Existing project management systems began to show the strain of the increased demands placed on them. There were not enough high-quality project management practitioners using effective project procedures to go around.

Perhaps the greatest benefit derived from that explosive period was top-management's recognition of the need for strong, effective project management. That was the only logical solution for correcting the problems and improving results. To effectively meet their responsibilities, project managers received more authority to discharge their project goals. Use of improved procedures like CPM scheduling, task forcing, and computerized data banks strengthened project management techniques and systems.

The seventies saw continued expansion, runaway inflation, and heavy growth in overseas projects. That created an additional need for project managers to learn how to operate in a foreign environment. Again there was increased demand for project management systems and training to satisfy that latest requirement of the marketplace.

The late seventies saw the long postwar capital projects boom culminate with the impending need for synthetic fuels megaprojects and their associated infrastructure. Industry buzzed with plans to serve the huge new market for megaprojects. Each one would require millions of labor-hours and billions of dollars to design and construct. The huge plants needed new cities and towns to service them.

There did not seem to be enough human resources in the total capital projects system of owners, labor unions, or contractors to meet the massive personnel requirement. With good project managers in the shortest supply, it would have been interesting to see how that scenario played out. Unfortunately, the major test of the capital projects industry never came to pass. The underlying economics of world energy supply did not support the need for synthetic fuels at that particular time.

In fact, the oil glut of the early eighties, coupled with the 1981–1982 recession, drastically curtailed the capital projects market. The era of the super- and megaprojects waned. Industrial plant usage dropped below 70 percent from the normal 85 percent rate. Industrial plants were mothballed; office buildings were standing vacant; nuclear plant construction ground to a halt; and new fossil-fueled power plants were delayed. The capital projects market became severely restricted across the board.

The only expanding marketplace was that of the high-tech industries. That meant facilities for fine and specialty chemicals, electronics, foreign autos, foods, biotechnology, specialty structures, and so on, were in demand. The capital projects marketplace became smaller and more specialized and required more sophisticated design, construction, and project execution methods.

The demands on the project management system were again changed by the needs of the marketplace. Fewer project managers were required, and they had to meet the specific needs of the clientele. Project size declined but complexity increased. Specialized manpower became scarce; budgets became smaller; schedules became tighter; and the demands of new technology grew. It is difficult to predict the future, but it looks as though the marketplace will remain in the same mode for the next 5 years. By then the next macroeconomic cycle may gather the momentum to create a new series of requirements in the capital projects marketplace.

The history of project management has proved that project managers must be ever alert to the changes required to meet the demands of the marketplace. They must be flexible enough to adapt to the ever-changing challenges they are bound to face during their careers. Continual changes in the marketplace, coupled with the normal changes inherent in managing projects, ensures that the project manager's life will never be dull! Those who wish to practice project management must feel comfortable with managing change if they are to be successful.

Defining Project Management

Since there is no clear concise definition of the term "project management," we will have to dissect the term and define its parts.

What is a project?

"Capital project" means different things to different people. It can mean building a new oil refinery, a high-rise building, a dam, a chemical plant, or an airport or even remodeling or upgrading a facility. To summarize the various major types of capital projects, I have prepared Figure 1.1. This chart sorts capital projects into three process and six nonprocess groups. It lists the type of documentation normally prepared for each type. I have included a line showing the relative design and engineering cost as a percent of total installed cost (TIC) for each type of project. My intent was to make the list representative rather than exhaustive. You should be able to fit your type of project into a suitable class.

Although the types of projects shown in Figure 1.1 differ, they do have some features in common:

- Each project is unique and not repetitious.
- A project works against schedules and budgets to produce a specific result.
- The project team cuts across many organizational and functional lines that involve virtually every department in the company.
- Projects come in various shapes, sizes, and complexity.

Figure 1.2 is a visual presentation of the organizational lines that project people may cross on any given day. Not all these departments have the same goals as the project team. Therefore, we must handle each contact with the utmost care, as we will learn later.

What are the project variables?

Some unique project variables such as size, complexity, and life cycle also occur in the project environment. Usually, large projects become complex just by problems of size and the scope of work. You may want to refer to Figure 1.5 to get a feel for the effects of project size. We will discuss Figure 1.5 in more detail later.

Small projects also can be made complex by new technologies, remote locations, tight schedules, or other unusual factors in the project. Small projects are often more difficult to execute than large ones, so do not take them lightly.

All projects go through a typical life cycle curve as shown in Figure

Figure 1.1 A matrix of capital project characteristics

Process-Type Projects			Nonprocess-Type Projects					
Liquid Chemical Plants (1)	Liquid/ Solid Plants (2)	Solids Process Plants (3)	Power Plants (4)	Manu-facturing Plants (5)	Civil Works Projects (6)	Support Facili-ties (7)	Commer./ A&E Projects (8)	Miscel-laneous Projects (9)
Petrochem Refineries Organics Basic chemicals Monomers Etc.	Pulp and pa-per Mineral dressing Paints & poly-mers Syn-thetic fibers Foods Pharma-ceuticals Soaps, etc. Films Etc.	Cement Clays Mining Steel & non-ferrous metals Foods Fertiliz-ers Etc.	Fossil fuels Hydro Nuclear* Power lines Co-genera-tion Etc.	Autos Textiles Light assem-bly Fabrica-tion Elec-tronics Aero-space Etc.	Dams High-ways Trans-portation Infra-struc-ture Water & sew-age Ports & marine Etc.	Labora-tories Testing facili-ties Aero-space facili-ties Etc.	Offices Shopping malls Housing schemes Hospitals Institu-tions Etc.	R&D work Studies Feasi-bility analy-ses Project control Auto-mation Loca-tion studies Etc.
% of TIC† 12%	10%	9%	7–8%	7%	6%	6%	4–6%	0.5–1.5%
Types of documentation: Process flow diagrams Heat & material balances Process design calculations Piping & instrument diagrams Process equip. data sheets Plant and equip. layouts Piping design and/or models Piping isometric drawings Piping material takeoff lists Instrument lists and specs. Instrument material takeoff Equipment specifications Procurement activities: Buying, expediting, and inspecting Civil, arch., struct. (CAS) design Electrical design Instrumentation design Process equipment design Effluent treatment systems Construction management			Types of documentation: Some material flow diagrams Some P&I diagrams Equipment layouts Block models Equipment specs Instrumentation Procurement: Buying, expediting, and inspecting CAS design Electrical design Effluent treatment Contract bid packages Construction management *More than for process plants		These are considered architect-engineer (A&E) type projects with only minor process design involvement except some in column (9). Plot plans and general layouts Equipment or floor plan layouts Interior design Block or scale models Architectural design Civil/structural design Electrical design power/lighting Plumbing and HVAC systems General specifications Contract bid packages Contractor selection Construction inspection Construction management			

*Nuclear plants are in a class by themselves because of safety and regulatory requirements.
†Engineering cost as a percent of total installed cost (TIC).

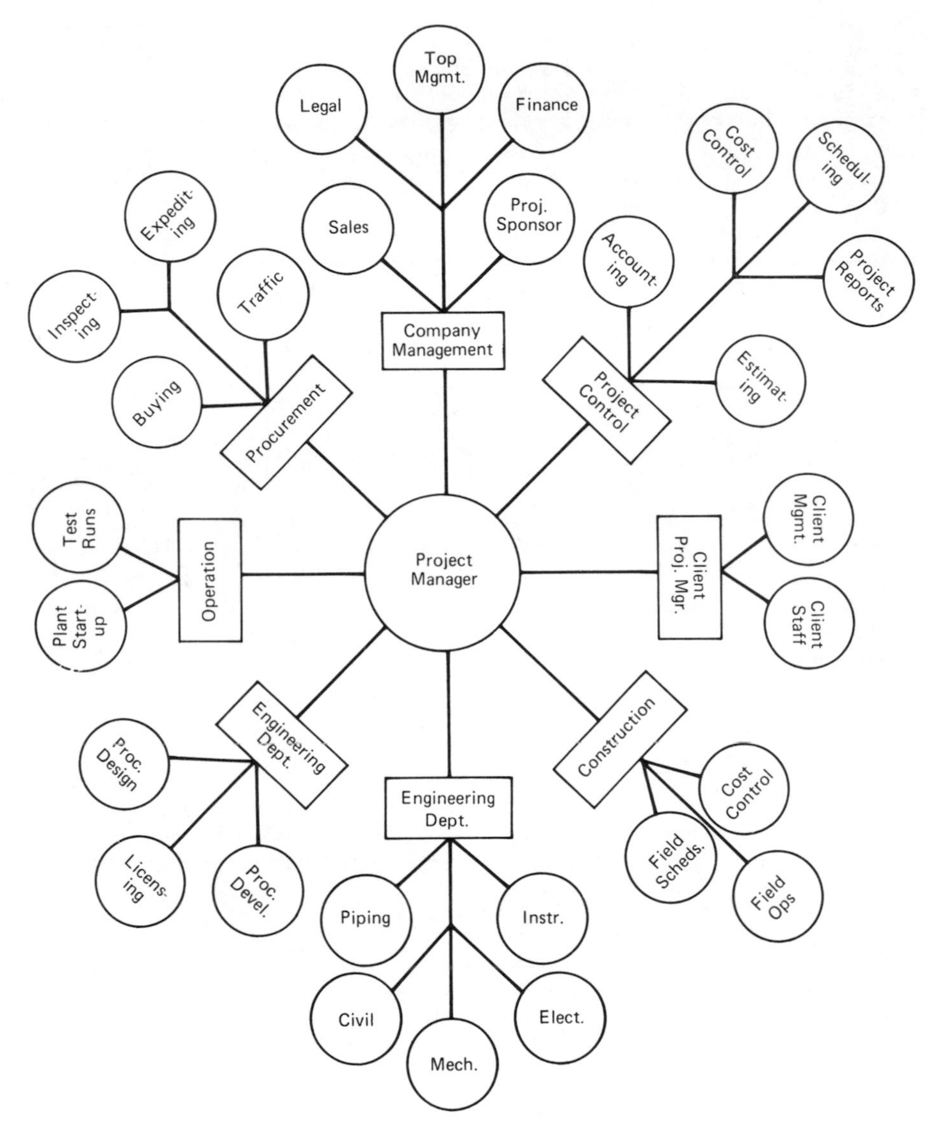

Figure 1.2 Project coordination interfaces.

1.3. The project starts with a conceptual phase and passes through a definition phase into an execution phase. Finally, the work tapers off into a turnover stage and the owner accepts the project. The project execution phase consumes most of the project resources so it becomes the focal point of the life cycle curve.

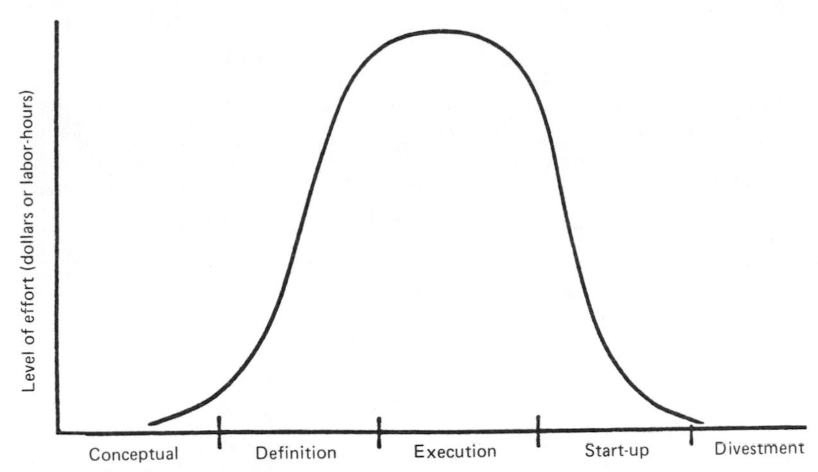

Figure 1.3 Project life cycle.

In most cases, a single project manager does not handle a project completely through the full life cycle. Most owners have specialists who develop the conceptual phase and other specialists who handle the execution phase. The conceptual phase has a heavy accent on R&D, market analysis, licensing, and finance and economics. Each area requires a set of skills different from those needed in the project execution phase. Many projects never make it out of the conceptual phase so one project manager may handle several conceptual phases simultaneously.

The life cycle curve gains most of its vertical growth during the engineering, procurement, and construction phases (the peak of the curve). The major commitment of financial and human resources occurs during that part of the project. This is also the time when the master plan must be in place. Pressure on budgets and schedules occurs during this period and makes life interesting for the project manager.

We have defined a project; now let us define a manager. The dictionary tells us that "a manager is a person charged with the direction of an institution, business or the like." Synonyms listed are "administrator," "executive," "supervisor," or "boss." At one time or another during the project, the project manager will perform under each of those titles as it bears on the work in progress.

Going one step further, we can define the project management system. It is a centralized system of planning and control of a project (or group of projects) under a project manager responsible for *time, cost, and end product.*

Figure 1.4 graphically illustrates the definition by showing the key functions of the project management system. The project team uses the inputs of people, money, and materials and organizes them to deliver the

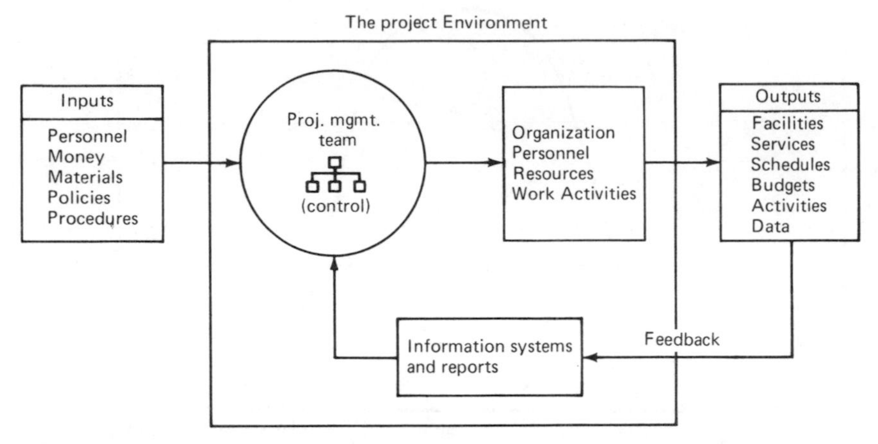

Figure 1.4 The total project management process. (*Reproduced with permission from* Project Management Handbook, *Van Nostrand Reinhold, New York, 1983.*)

outputs of products, and services. Figure 1.4 also shows the feedback and control system that ensures the project goals will be met.[2]

The Project Management Institute has published an excellent treatise on project management theory and practice entitled *The Implementation of Project Management.*[3] It does an excellent job of describing the concepts for installing a project management system. I recommend that you read this book and consider adding it to your technical library.

Project Goals

The primary goal of the project team is to finish projects *as specified, on schedule and within budget.* I have dubbed this message the Project Manager's Creed. Attendees at my project management seminars get copies of the creed as certificates. I recommend they frame the creeds for hanging above their desks to remind them of the primary goal of project management.

That goal is so important that I am repeating it here since I am unable to hand each reader a certificate. I hope you will enlarge it and hang it on your wall.

The whole system of project management exists to ensure our meeting that goal! Obviously, the project management system has not met the goal every time. If it had, we wouldn't continue to hear about projects which have failed because they were ill-conceived, late, over budget, or abandoned.[1]

I designed this book to present the fundamental methods that project management practitioners must know and use for effective *total project management.* Fortunately, no particular mystique is in-

The Project Manager's Creed

Finish the Project

- As specified
- On schedule
- Within budget

volved, just good common sense. When projects fail, it is not from the lack of a complex computerized control system. More likely, it is from overlooking one of the basic principles that we will be discussing here.

Goal-oriented project groups

Let us look for a moment at the three key goal-oriented groups involved in the execution of any capital project:

Owner-client	The one who commissions and owns the completed facility
A/E firm or central engineering group	The entity accountable for delivery of the completed facility
The project team	The project manager(s) and key staff members

The underlying thought for all capital project management people must be the creation of facilities to earn a profit or deliver a service at a reasonable cost. The former applies to the business sector and the latter to government. In addition to the functionality of the facilities, we must consider the aesthetics, safety, and public welfare. The final trade-off decisions on these two groups of project requirements often center around the project manager.

Owner-client goals

The owners create both the need and the financial resources necessary for the creation of the facilities, so they certainly rate top billing in the goals department. Actually, the owners' needs are quite simple:

The best facility for the money	To maximize profits or services at a reasonable cost
On-time completion	To meet production or services schedules
Completion within budget	To meet financial plans for the facility

In their search for a tool to maximize profits or services at a reasonable cost, owners want reliability, efficiency, safety, and good onstream time. We hope they do not want gold plating and overdesign, which add unnecessary costs. All those conditions should add up to either an edge over the competition or acceptance by the public.

On-time completion allows owners to meet production quotas and schedules while avoiding the expense of added interest and high startup costs. On-time completion usually results in the added bonus of a smooth start-up and rapid acceptance of the facility.

Finishing within budget avoids nasty surprises which can upset the owner's financial plan for the facility. Project overruns lead to slower payout and reduce the chance of early return on the investment. On very large projects, the financial jolts are often felt right down to the owner's bottom line for a long time. SOHIO's participation in the Trans-Alaskan Pipeline in the 1970s is an excellent example of such a situation.

Architect/engineer or central engineering group goals

The A/E or central engineering group (CEG) is the entity responsible for delivering the completed facility. It can be an outside contracting group such as an A&E firm or the owner's captive central engineering department. In the latter case, the client is the division or service group owning the facility. It is desirable to have a definite *contractual environment* to ensure meeting the operating division's goals. The project execution entity's goals are to:

Make a profit on each project	This applies only to A/E entities. CEGs must minimize cost only.
Finish on time	To satisfy the owner/client and meet contractual requirements.
Finish within budget	Ensure goal 1 and satisfy owner.
Furnish quality per contract	Ensure goal 1, satisfy owner; meet contractual requirements.
Get repeat business	Maintain company reputation and reduce selling expense.

The contracting firm must strive to make a profit on every project performed or it will not be able to stay in business. Because of the fixed duration of the project life cycle, the contracting entity has only one chance to make its profit. Owners may suffer some financial setback on projects with unmet goals, but they can often make up the losses later than originally anticipated. That one-time profit requirement places additional pressure on contracting project managers to meet their company and personal project goals.

The owner-client may not want to see the contracting entity lose money but does not have the contractor's profit goal strongly in mind. That is the only major point of difference among the goal-oriented groups involved in most capital projects.

Finishing projects on time and within budget has several attractions for the contracting entity. That is especially true when the contract involves an incentive clause or lump sum bid. In that case, any budget overruns are net losses to the contractor. The contractor's general management considers that a catastrophic event! On the other hand, finishing early and under budget increases profits or reduces costs, which are happy events for everyone in the contractor's organization.

Finishing on-time and under budget are important considerations when the performance of the contractor is evaluated. They are the key to getting repeat business from the owner. The possibility of withholding repeat business is the best leverage that clients have to ensure good contractor performance. They should use it to the fullest extent when contractor performance falls below par.

Building the facility to specifications is an important goal because it affects the overall cost of the facility and thus the budget. It also affects the overall quality of the facility and its capacity to carry out its mission. To keep that goal in perspective, it is critical to develop a thorough and sound project definition before starting any work. Many project failures result directly from inadequate project definition before the project is started.[1]

The project team's goals

The last goal-oriented group to consider is the project team. Here we are talking about people much like ourselves, so the matter becomes even more personal. The project teams goals are to:

Goal	Reason
Earn a financial reward	This can be in the form of a bonus or promotion.
Identify with a successful project	Build a reputation and personal satisfaction.
Finish on time and within budget	Attain goals 1 and 2 and satisfy client and company management.
Maintain one's reputation in the business	Personal satisfaction and professionalism.

It should come as no surprise that personal satisfaction and financial reward top the list of our human needs. Behavioral scientists have been telling us that for years.[4] With life cycles of most projects running from 6 months to 2 years, our personal satisfaction comes quite soon compared to a manufacturing environment. Since the results of our labors are usually visible and permanent, we remember them whenever we see them or hear about them. People who do not take pride or personal satisfaction in their work will never become successful project managers.

Completing a difficult project on schedule and under budget gives a PM additional satisfaction from having met a challenge. Having both client and management satisfied at the end of a project is indeed much more satisfying than the reverse would be.

Financial rewards can come in either of two forms: a bonus or a promotion. Many service firms set their project teams up on an incentive program in which the rewards are tied to successful performance. Commercial and industrial clients also favor incentives because they ensure meeting their project goals at minor additional cost.

It is extremely important for a project manager to build and maintain a reputation in the field of capital project execution. The number of capital project management practitioners is relatively small, particularly within each specialized area of practice. News of any unsuccessful projects will travel through the industry quite rapidly and detract from one's reputation. It takes several successful project performances to overcome the bad press from one bad project.

Looking at all the goals for the three groups, we see that they are almost identical. The only goal not common to the key players is that of the contracting entity making a profit on the project. It is difficult to understand how that single minor difference in goals can cause so many client-contractor relations problems. We will explore that problem further in later chapters.

Assuming the project has gone through a sound conceptual stage, it is now ready for implementation. The project manager and his or her team are the keys to delivering a successful facility satisfying all the goal-oriented groups. Using the total project management principles presented in this book should allow all parties to meet their project goals.

Basic Project Management Philosophy

My basic project management philosophy is simply stated in three words: *plan, organize, and control.* I call that my Golden Rule of Project Management. It should hang right alongside the Project Manager's Creed. Practicing the Golden Rule will deliver the goals in the creed!

Many of you may recognize my Golden Rule as a distillation of a general management philosophy of plan, organize, execute, coordinate, and control. To ensure its easy application by busy project managers, I boiled it down to the three essentials.

Total project management requires that you practice the Golden

Rule on your projects in general as well as on each project activity. By that I mean that you must plan, organize, and control every activity on the project. A good example of applying the rule to part of the project is your approach to project meetings. Plan your meeting (or your part in it); organize the logistics for the meeting; and control the meeting if you are its leader. I will cover handling meetings in a later chapter on project communications.

As a corollary to the Golden Rule, I want to add the KISS principle to further stress the need for a simple uncluttered approach to project management. KISS is the acronym for the phrase "Keep it simple, stupid," which turned up in a joke some years ago. Some people (including my RightWriter style program) are a little stuffy about using the "term" stupid in connection with executives. However, I believe that any unintended derogation is well worth the benefits gained by remembering to keep it simple! At the risk of spoiling your office decor, I recommend hanging a KISS sign. It is a valuable reminder for a serious project management practitioner.

My aim in introducing simplicity is to point out that overdetailed plans, budgets, schedules, and control systems do not automatically equate to more successful projects. In fact, the reverse is true: Overkill systems have hurt more small projects than underpowered ones have hurt large projects. There will be more on this subject as we progress through the book.

What Are We Going to Plan, Organize, and Control?

The key project activities that we need to *plan, organize, and control* lie at the very heart of the total project management system. They also establish the format of the project execution portion of this book. They are the basic building blocks of the *total project management approach.*

Project planning activities

The planning activities for a typical capital project are:

- Project execution plan
- Time plan—schedules
- Money plans—budget and cash flow
- Resources—people, materials, and money

The project execution plan is the master plan for executing the project from conception to completion and turnover to the owner. The master plan must operate within the restraints of overall project financing, strategic schedule dates, allocation of project resources, and contracting procedures.

The time plan is the itemized working plan for the project execution, and it results in the detailed project schedule. The project team makes a work breakdown structure by dividing the project scope into major work activities. It then assigns a completion date for each operation.

The money plans consist of a detailed project budget based on a sound project cost estimate. The cash flow plan results from the budget and the schedule. It forecasts how the budget will be spent.

The last major planning effort is forecasting the human, material, and systems resources required to execute a project according to the master plan and the schedule.

Organizational activities

The activities in the organizational activities area cover deployment of the human resources and systems required to meet the master plan and the project schedule.

- Prepare organization charts and personnel loading curves
- Write key position descriptions
- Issue project procedures
- Mobilize and motivate project staff
- Arrange facilities and systems
- Start control procedures

Base the project organization on the scope of work and the personnel plan. Issue detailed project procedures promptly to instruct and guide the organization. Mobilize the staff and suitable facilities to kick off the project on schedule. Start the cost controls immediately because you are spending money!

Control activities

Controlling the project plans and activities extends over the life of the project. Control is the most vital of the three total project management steps. The major areas of control are:

- Quality—engineering, materials, and construction
- Time—by the project schedule
- Money—by budget and cash flow plan

- Measuring physical progress
- Project reporting

The controlling function must monitor the quality of all phases of the work to meet the universal goal of building the project as specified. Monitor time by checking physical progress (not labor-hours) against the schedule. Monitor costs through a cost control system based on the project budget. The project reporting system regularly informs the key players on the project of the status of project activities and the results of the project control systems in detail.

The above lists present the skeleton of a proven route to the successful execution of capital projects. In later chapters we will fill in the flesh, sinew, and blood to give you a complete body of knowledge for *total project management!*

Project Manager's Job Description

A discussion of the project environment would not be complete without addressing the duties of the project manager. I have included a comprehensive project manager's job description in Appendix A. The job description is actually a hybrid of several versions that I had the opportunity of contributing to over the years.

Section 1.0, Concept, sets the stage for granting the project manager's charter and delegates responsibility to complete the project *as specified, on time, and within budget.* That is also where top management makes its commitment to the project management system. If the charter statement is weak at the outset, little can be done later to strengthen it.

Section 2.0, Foreword, suggests some good management techniques for successful project management: goal setting, effective delegation, planning, organizing, and controlling the work, and leadership. They are vital to successful project management.

Sections 3.0, Duties and Responsibilities, gets down to defining the on-the-job operating techniques required to execute a project. It gives a step-by-step approach on how to plan, organize, and control a capital project. It serves as a good checklist to follow as you proceed through the project life cycle even if you do not use it as a job description.

A key feature in Section 3.0 is the number shown in parentheses at the end of each paragraph. The number shows the degree of autonomy given to the project manager in decision making: (1) delegates complete authority; (2) requires notification to management; and (3) indicates prior management approval before acting. Obviously, a strong project charter will have more (1)s and (2)s than (3)s.

Section 4.0, Authority, expands on the project manager's authority in relation to other company departments. Strength in that respect is important to set up the project manager's power base in the company hierarchy. Weakness here can seriously hobble the PM in staffing projects and marshalling the firm's resources.

Section 5.0, Working Relationships, gives a few suggestions about the human relations involved in using the power granted under Section 4.0 when dealing across departmental lines. You can expand this list with other interfaces in your particular environment.

Section 6.0, Leadership Qualities, lists some leadership qualities which are most useful in motivating the project team and others involved in the project. We will expand on those valuable qualities in later chapters.

The job description applies within the context of a contracting environment, which usually grants a project management charter that is stronger than that of a typical owner. That is primarily due to the competitive nature of the contracting business in which the profit motive is much more critical. If you are an owner's organization, you may have difficulty getting a strong version of your job description approved. In any event, it's worth a try, since your management's volunteering a strong project management charter is doubtful.

If you do use the job description in Appendix A as a model for your own position, you can adapt the parts that apply to your work. Also, add any new material describing your particular situation. It makes good sense to write your charter as broadly as possible and let your management cut it back. You might as well see how far management is willing to go. But remember this: If you are given total project responsibility to meet the project goals, you must have the authority necessary to do the job!

I hope that you noticed the application of my Golden Rule of Project Management in the layout of Section 3.0. The duties and responsibilities are divided into subsections for the planning, organizing and controlling functions that are so essential in effective project management.

The job description is very general, as it must be when there is no specific project assignment. Therefore, you should prepare a written list of *project-specific* goals to measure your performance and that of your staff. You should in turn incorporate these specific project goals into your staff's job descriptions. That will allow you to take a management by objectives (MBO) approach to the execution of the project. We will go into more detail on MBO systems in later chapters.

Project Size

The one project variable that needs more detailed discussion is project size. Project management people seem to have more difficulty with it than with any other subject in project management discussions.

The matrix of project size characteristics chart shown in Figure 1.5 shows the anatomy of projects of various sizes. The size divisions in the left-hand column represent my opinion of project size categories. Opinions may vary somewhat among the different types of capital projects listed in Figure 1.1. The common denominators for judging project size are scope of services, numbers of drawings, major equipment items, and the home office and field labor-hours. It behooves project managers to give those size factors some careful consideration when reviewing their project assignments for the first time.

Since a high percentage of capital projects fall into the small-project group, many readers may feel overwhelmed at the numbers for the larger projects. However, larger projects do come along occasionally, so you should know how to deal with them. I have found that doing a single large project is much easier than doing several small projects simultaneously.

The important thing to remember when reading the chart in Figure 1.5 is that the same principles of total project management covered in this book apply to projects of any size. Only the systems, tools and procedures vary to suit the size. Smaller projects require keeping the KISS principle foremost in your mind. Use simple uncomplicated methods that do not get in the way of the work. Larger projects demand more complicated methods permitting use of computerized *management-by-exception* techniques to control the larger number of work activities.

Notice in the right-hand column of Figure 1.5 that staffing and systems become more sophisticated as the projects increase in size. As projects become larger, they are broken down into smaller, more manageable units (small projects). A project engineer or project manager is responsible for each unit and is responsible to the project director for that piece of the project. On large projects, we often delegate complete units to subcontractors specializing in that type of work.

The most important differences in handling small vs. large projects are the systems and tools used for project control. We can effectively control small projects (under $10 million) by using manual systems to control schedule and cost. Now we have personal computers (PCs) and commercial software to assist with those chores. Notice that I said *commercially available* software. Creating custom software for small projects becomes too involved, expensive, and time-consuming, so be wary of it. Spreadsheets and data base programs can do the job very well.

As projects become larger, computers become productive in processing the more complex budgets, material lists, schedules, cost reports, and so on. Remember, computers are tools capable of sorting and crunching large volumes of data. They do not add any intelligence to your input data.

Size TIC	No. of Units	Off-Sites	Tagged Equip. Items	Pipe-lines	Instr. Loops	HO Hours	Field Hours	Staffing			Sub-contracts Const.	Systems		Remarks
								PM	Area Engs.	Proj. Cont.		Cost	Sched.	
Small $1MM to $10MM	1 or 2	None	10 to 50	40 to 175	25 to 80	4000 to 40,000	24M to 240M	0 to 1	1 or 2	1	3 to 5	Man. or PC	Bar chart	Use manual methods or PCs
Medium $11MM to $75MM	3 to 6	Minor to med-ium	60 to 500	200 to 1800	100 to 250	40,000 to 200M	240M to 1.2MM	1	2 to 3	1 to 2	6 to 10	Man., main-frame, or PC	Bar chart or CPM	Could use mini-computer or PC
Large $80MM to $200MM	7 to 12	Me-dium to heavy	550 to 1500	2000 to 4500	260 to 450	225M to 480M	1.3MM to 3.0MM	1	3 to 5	2 to 5	10 to 15	Mini or main-frame.	CPM comp-uterized	Fully mech. split OSBL/ISBL
Super $250MM to $600MM	15 to 20	Major	1500 to 3000	5000 to 8000	460 to 750	520M to 900M	3.2MM to 6.0MM	1 to 3	5 to 9	4 to 8	12 to 20	Main-frame comp.	Main-frame CPM	S/C where poss. max. mechanize
Super* Energy $1.0BB to $3BB+	25 to 60	Heavy major	3500 and up	8500 to 20,000	2000 to 5000	1.6MM to 4.0MM	10MM to 24MM	4 to 7	15 to 30	10 to 20	3 to 5 major; 25 to 30 mi-nor	All mech.	CPM for ea S/C+ a master project schedule	Many dupli-cate units. Maximize S/Cs; unitize to subprojects

*Also known as *Megaprojects:* 1 MM + engineering hours or 10MM + field labor-hours.

Figure 1.5 THE ANATOMY OF A PROJECT: A matrix of project characteristics.

Project size ratios

There are several good rule-of-thumb ratios between labor-hours and total project cost which are useful for rough cost estimates early in the project. The simplest one is the ratio between home office labor-hours and field labor-hours. On complex projects, field labor-hours are from 5 to 7 times the home office hours. Thus each design labor-hour generates 5 to 7 construction labor-hours.

In Figure 1.1, we show ratios of home office costs to total installed cost (TIC) ranging from 4 to 12 percent depending on the type of facility. By using the applicable percentage and an average home office labor-hour cost, one can readily get from the home office hours to the total project cost. The same statement applies to the field labor-hours.

Today, average overall home office hourly rates run from \$30 to \$50 per hour inclusive of overhead and fees. For example, by using a medium-size project with 250,000 home office hours and a \$38.00 per average hourly rate, we can estimate the total project cost as follows:

$$250{,}000 \text{ hours} \times \$38/\text{hour} = \$9{,}500{,}000 \qquad \text{design cost}$$

$$\frac{\$9{,}500{,}000}{0.13\ (\%\ \text{TIC})} = \$73{,}000{,}000 \qquad \text{approx. total project cost}$$

One should only use that method of factoring costs in the early phases of a project before more accurate and reliable estimating data become available. Owners use the cost and labor-hour relationships in checking contractors' proposals. A&E firms and contractors also use them when preparing proposals. You can develop similar numbers for your project from the cost reports and average them to get your own rule-of-thumb ratios.

Handling megaprojects

The largest projects on the list are the megaprojects, which are the World Series and Super Bowls of project management. Some of us secretly dream of being at the controls of such a major undertaking.

Let's look at one such project which is now looming on the horizon: the newly proposed Linear Particle Accelerator project. The project will further extend America's frontiers in that most recent high-tech discovery, superconductivity. It is about a 10 times scale-up of the already impressive Enrico Fermi linear accelerator facility located outside Chicago.

Preliminary budget forecast for that project is over \$4.4 billion, which is mega in anybody's book. The new research center will include about 53 miles of accelerator tunnel, an operations center, research buildings, and support facilities. The facility is to be located in a small town in Texas and will require major additions to the local infrastructure. The \$4.4 billion budget figure does not include any infrastructure cost.

Assume you are the project director on this project. How would you get such a mind boggling operation organized? Remember, your mission is to bring this venture to a successful conclusion as specified, on time, and within budget!

After the initial panic has passed, I expect you to remember to divide the monster into manageable packages. That will allow you to bring additional resources to bear on the project. Because of the specialized nature of the process and equipment, that is not your everyday type of facility. There is probably not a firm in the world experienced in all the diverse technical details of the facility.

What are some of the natural dividing lines we could use to subdivide this megaproject into workable units? Some major areas that quickly come to mind are:

1. The accelerator tunnel
2. The accelerating equipment and associated electrical work
3. The control building and operations center
4. The research center and laboratories
5. The support facilities—engineering building, shops, power plant, food services, and so on
6. Site development—site selection, site preparation, drainage, roads, utilities, security, and so on
7. Supporting infrastructure, offsite facilities, and so on

With our present limited view of the project, those are seven readily identifiable areas available to start a project breakdown. Let us also assume that a preliminary design study has been completed to set the overall parameters and budget for the project. We are also early in the project life cycle and are just starting to commit major funding.

In reviewing the list of major areas, we find the items readily fall into several areas of design and construction specialization. Tunneling, for example, is a highly specialized skill handled by only a few firms with that expertise. It is a large dollar value piece of the work that we can lay off onto a qualified specialist. Only our moderate supervision should be required to get a good performance.

The accelerator equipment design and build is highly specialized and limited to the capability of only a few firms in the world. With our staff specialists overseeing the work, we can effectively delegate a design, build, and install contract to the low bidder. By prequalifying the bidders, we can expect to get a reliable performance.

Site selection is a specialty area where we could use an experienced consultant to help in the technical and political aspects of this emotion-charged task. The public relations part of our project management skills arsenal will play a major role in this area.

Site development is basically civil-oriented work, so we will need a qualified civil engineering project manager to supervise it. In turn, the site development manager can subdivide this major piece of the budget into work packages bringing qualified human and physical resources to bear on it.

Other areas of the project involving buildings can be broken down into related packages and let out to several qualified A&E firms. In that way we can bring in firms who specialize in the various types of buildings involved. Most of these building projects will be large enough to warrant a project manager to supervise and coordinate their integration into the overall project.

We have farmed out most of the technical work to a group of highly qualified specialists; all that remains is the overall coordination of the project. For that we will need a staff of specialist engineers, planners, cost engineers, contract administrators, accountants, legal advisers, procurement people, inspectors, data processing people, and a clerical staff to handle the paperwork. The staff will sort all of the exceptional items which require action by the project director.

Aside from the sheer numbers involved, that is not much different from the work of a project manager on a medium-size job. The only other difference is that the stakes are much higher and the risk for failure much greater. A steely set of nerves and a cool head are also essential for survival in the megaproject arena. We will get into that part of the business later under human factors.

This chapter has been a discussion of some major points that I have found to be important in practicing total project management. There are many more factors you should know about. Therefore, I recommend that you expand your viewpoint through an outside reading program. Select applicable material from the references listed at the end of each chapter and the bibliographies available in the literature.

I think you will find it particularly valuable to expand your interest into general management books such as those by Peter Drucker and other management gurus. That will give you an insight into how project management fits into the global management picture. Project managers usually do a lot of traveling. The time you otherwise waste in airplanes and airports can be put to good use by reading some of these books. Remember, project management is a possible path into general management.

References

1. Peter Hall, *Great Planning Disasters,* Weidenfield and Nicholson, London, 1980.
2. David I. Cleland and William R. King, *Project Management Handbook,* Van Nostrand Reinhold, New York, 1983.
3. Linn C. Stuckenbruck, *The Implementation of Project Management: The Professional's Handbook,* Addison-Wesley, Reading, Mass., 1982.
4. F. I. Herzberg, *Work and the Nature of Man,* World, New York, 1966.

Introduction to Case Studies

One of the purposes of this book is to serve as a textbook for training programs in project management. To support that feature, a set of case study instructions is given at the end of each chapter. That gives readers an opportunity to apply some of the ideas developed in that chapter.

Appendix C contains a broad selection of typical case study projects. Included is a generic project format which allows readers to tailor a project to a specific area of interest. Use of the original project selection throughout the book will result in a complete project execution file as the case studies evolve.

In a group-training environment, use a team approach to bring more minds to bear on each aspect of the case study. There are no set solutions to the case studies, which encourages readers to develop new ideas for handling the situation. Critiquing each group's presentation solution by the others is also an excellent character-building device.

A goal of the case study approach is to stimulate the flow of your creative juices to develop innovative solutions to some everyday project situations. If you are fortunate enough to be in a group, you can try some of your ideas out on your peers before using them in actual combat.

Please refer to Appendix C to make your project selections for use with the instructions listed at the end of each chapter. Set up a project file or notebook as a record of the solutions you develop as we proceed through the project life cycle. That will give you a complete project profile for future reference.

Case Study Instructions

1. Assume that your company has not previously operated in a project management mode. Your top management has asked you to formulate and present a plan for installing an effective project management system in the firm. Develop an outline of the main features of your recommended plan to install such a system.
2. Prepare a job description for your assignment as project manager

(director) on the project(s) which you selected from the list in Appendix C.

3. Based on the general project goals discussed in this chapter, prepare a list of specific project goals for the goal-oriented groups involved in your selected project(s).
4. Use the chart in Figure 1.5 to classify your project as to size and complexity. Prepare a project execution master plan for presentation to your management. Include your recommendations for contracting, staffing, scheduling system, and standard company procedures for use on the project. Base them on the expected company resources required vs. those available.

Chapter

2

Proposals and Contracts

There are various ways to bring human resources to capital projects. Owners may execute their project with their own forces, or they may choose to engage outside services. They often use the services of an engineering-construction, consulting engineering or architectural firm. To simplify the terminology for this chapter, I will refer to that group of professional service suppliers as *contractors.* They do, in fact, enter into contracts with the owner to supply a broad spectrum of technical services.

Project Execution Approach

The project execution approach selected depends largely on the owner's approach to developing its new facilities and the amount of human resources on its captive staff. It is logical that an owner will want to use its own people, since they have experience in the owner's business. Also, it will want to avoid bringing outsiders into confidential projects. Sometimes, however, owners bring in outsiders because they do not have in-house capability in specialized areas. In that case, an owner may buy a process license or a design package for a certain process to meet its needs. Buying a process design can save an owner a good deal of R&D expense and time in bringing a new product to market.

Most government agencies, for example, do not maintain large staffs of engineers and architects because of variations in workload. The heavy overhead charges of such organizations during periods of low project activity are hard to justify. Therefore, government agencies contract with outside firms for most of their engineering, construction, and project management services.

Most government agencies prefer to have a capital project designed by an A&E firm or consulting engineer and take lump sum bids for the construction work. On some highly specialized projects requiring a *fast-track* schedule, they may let a total engineering, procurement, and construction (EPC) contract.[2]

Some large industrial companies can support central engineering departments to design their projects in-house. When business falls off however, they too face the same overhead cost crunch with their central engineering departments. Therefore, many large companies build and disband their central engineering groups depending on the business cycle. Since 1980, the business environment has become extremely competitive, and that has forced many owners to reduce their central engineering groups to lower overall overhead expenses.

Owners use their leaner central engineering groups for the critical project development and conceptual phases of larger projects. The same project team can then follow through to supervise a service contractor for the larger human resources effort needed to complete the project. The arrangement gives the owner the best of both worlds, and it can result in a highly efficient project execution mode. With both organizations combining in a team effort to meet the project goals, total project management can deliver a successful project.

Most small projects must still be done by plant and central engineering groups because it is not efficient to farm them out. Today's competitive economic climate has even led owners to import outside service teams for peak loads for smaller in-house projects. In that case the contractor's personnel work right in the client's plant or engineering department. The situation calls for a mode of project management different from that used for doing the project off the owner's premises.

In-house operating mode

Owners using the in-house mode do not require a formal proposal and contract because both client and service organizations belong to the same company. Although the arrangement does not require an arms-length agreement, it is necessary to define the scope of services. Defining other key project variables such as scope, schedule, budget, and quality in writing is also important. Such an approach is critical to ensure that all partners in the project clearly understand each other's intentions. That is the only way to assure meeting the project goals.

A quasi client-contractor relationship should exist during the project execution even though the owner and designer have a common owner. Except for the missing contractor's profit motive, both groups will operate within the normal project goals format as discussed in Chapter 1.

Owner-contractor operating mode

Once the owner decides to bring in outside help for a project, it introduces a third party into the project proceedings. That complicates matters slightly. We must now have an arm's-length agreement between the owner and the third party, which brings a proposal and a contractor selection phase into the picture.

The owner's decision to enter a proposal phase brings a number of project managers into the action. The first one is the owner's project manager, who heads up the owner's contractor selection team. The owner's request for proposal (RFP) will in turn activate several contractors project managers who are responsible for much of the proposal effort. Remember, one of the precepts of project management is that the project manager should have responsibility for the project from initial proposal to project closeout.

The Contractor Selection Process

The process for contractor selection that I am going to discuss here will apply across the board from major prime contractors to minor subcontractors. Naturally, the process is less detailed when it is used for proposals on a smaller project with a smaller work scope. I hope that you will remember to use the KISS principle when you find yourselves engaging in that activity!

The selection procedure discussed here comes from a paper given at the 10th Annual Seminar/Symposium of the Project Management Institute in Anaheim, California, in 1978.[1] I recommend that you read that paper if you become responsible for a contractor selection. The goal of the contractor selection process is to arrive at a mutually agreeable contract with an excellent supplier of the desired services. That should result in meeting all the owner's project goals.

Figure 2.1 shows the part of the project life cycle in which we are now working. The contracting phase fits neatly between the project development and execution phases. The bottom line of the graphic shows the party with the primary responsibility for the major activity taking place in the block above.

The owner's input

The owner must do a thorough job in the development phase to allow the following two phases to proceed smoothly and efficiently. The design and construction phases are critical because they involve the largest commitment of human and financial resources. Poor definition during project conception often introduces delay and indecision, which are costly and goal-wrenching in the phases that follow.

DEVELOPMENT PHASE	CONTRACTING PHASE	EXECUTION PHASE
ACTIVITIES	ACTIVITIES	ACTIVITIES
Project Planning	Contracting Plan	Detailed Engineering
Market Development	Contractor Screening	Procurement
Process Planning	Selection of Bidders	Construction
Cost Estimating	Invitation for Proposals	
Basic Design	Contractor's Proposals	
	Bid Review	
	Contract Award	
. . . By Owner	. . . By Owner and Contractor	. . . By Contractor

Figure 2.1 Phases of a project. (*Reproduced with permission from* Proceedings of the 10th Annual Seminar/Symposium, *Project Management Institute, October 1978, p. II-L1.*)

Contractor selection

Successful contractor selection can be best accomplished through an open and honest approach by all parties. Using a list of prequalified contractors can further improve the owner's selection process. There is no point in working with contractors with whom you are not willing to enter into a contract.

Selecting a contractor for any project requires a competent project team that includes a broad spectrum of technical and business skills. A project manager heads up the selection team to coordinate the other departmental inputs to the selection process. In addition to the technical people on the team, tax, risk management, procurement, legal, and accounting specialists also are necessary. The selection team should participate for the duration of the selection process to minimize the adverse effects of repeated learning curves.

The owner's contracting plan is the foundation of the selection process. A well-conceived and management-approved plan is necessary to avoid revising it later. Management should tailor the contracting plan toward the best arrangement to meet the goals for this project. The plan should not necessarily conform to long-standing corporate policy.

It is wise to make a complete analysis and evaluation of all contracting alternatives and project factors to arrive at a sound project strategy. Major items to consider are:

- Project needs
- Requirements of the project execution plan
- Key schedule milestone dates

- The scope of contractor's services
- Possible contracting alternatives
- Local project conditions
- Contracting market conditions

The owner is looking for the contractor who will perform best by delivering a quality facility on time at the lowest overall cost. In striking the ideal owner-contractor match, the owner should select a contractor from a list of firms that really want the job. Factors affecting a contractor's desires are present workload, prestige, repeat business, and so on.

Incentives are necessary on both sides of the contracting equation to get outstanding performance and results. It is the merging of the owner's goals with the contractor's goals that leads to a successful contract award.

Contractor selection is much like a miniproject; it needs a project management approach. That means a total project management approach to plan, organize, and control the selection effort is necessary.

The bar chart in Figure 2.2 is a typical schedule for major activities in the selection process. You will have to add the applicable time scale and any other activities to suit your particular project. In many cases, setting the proposal opening date controls the timing of the other activities. Be sure to allow enough time for the contractors to prepare their proposals.

The contracting basis used causes the time scale for the proposal to vary widely. A cost plus fixed fee (CPFF) based proposal requires less time than a lump sum (LS) proposal. Fixed price proposals require preparation of detailed cost estimates by the contractor and bidding documents by the owner. Documentation for CPFF proposals is much less complicated and time-consuming for both parties.

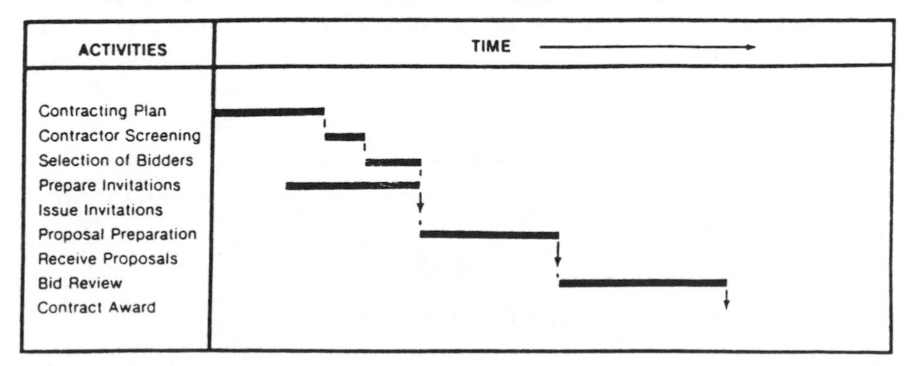

Figure 2.2 Sequence of contracting activities. (*Reproduced with permission from* Proceedings of the 10th Annual Seminar/Symposium, *Project Management Institute, October 1978, p. II-L.2.*)

Contractor screening

The screening process begins by issuing a request for a *statement of interest* from a prequalified list of about ten to twenty firms. For smaller projects the screening may even be a telephone canvas of five or six firms to determine their present workload and degree of interest. The goal of the screening activity is to arrive at a *short list* of four to six firms that are sure to respond with viable proposals.

The short-list approach has several goals. The first is reducing the selection team's work in reviewing the detailed proposals. The second is that the approach helps to control expenses for proposal preparation. Proposals for medium and large projects often require investing several hundred thousand dollars. Since there can be only one winner, the unsuccessful firms waste a total of 5 times several hundred thousand dollars. The proposing firms are profit-making entities, so they must charge the money spent on unsuccessful proposals to overhead. All owners eventually pay for unsuccessful proposal expenses when they employ contractors. Therefore, owners are wasting their own money when they request unnecessary proposals.

Selecting a short list of eager, qualified, and competitive contractors should assure owners of a good representative selection of proposals. The following are some of the major points to consider in the screening process.

- Screen as wide an area as is needed to satisfy your needs.
- Use a written screening document: letter, telex, fax, or mailgram.
- Prequalify the screening list.
- Use a simple screening request document.

Although concise, the screening document should cover all of the critical points needed to select a good short list of quality firms. The screening process really requests miniproposals to get some preliminary facts from the contractors. Here are some of the key points to cover:

- A brief project description
- A statement of the scope of services
- Key dates for proposals, contract award, project start, and desired completion date
- Tentative project plan and schedule
- A statement of contractor's interest
- Location of the work

- Request for status of contractor's existing work load
- List of contractor's current technical personnel capability
- List of recent similar projects
- Pertinent experience in area of project site
- Any other project-pertinent factors worthy of evaluation

The screening analysis requires a thorough review of the positive replies to the screening document. A major item to check is the contractor's capability to handle your project within the required time frame. Since you have only approached firms capable of handling your work, your main concern is their ability to staff your project properly. For that, the selection team needs to study the contractor's technical personnel loading curves submitted with the screening reply. Figure 2.3 is a suggested format for getting a contractor's current workload data.

The examples show personnel loading curves for two prospective contractors. Contractor A's curve shows the current backlog tapering off in time to add your expected project's workload with some capacity to spare. The curve for Contractor B shows that a backlog will not permit the commitment of enough personnel to your project to meet your proposed schedule. The only alternatives for Contractor B are to increase staff or subcontract some of the work. Neither of these options is acceptable to most owners except on very rare occasions.

At this stage you are only making a preliminary evaluation of workload and will do a more thorough investigation later on. Also at this stage, contractors will try to show themselves in the most favorable light, so the personnel loading curves will be somewhat optimistic. Furthermore, the contractor may book some additional work during your proposal-seeking efforts.

Check the remaining responses for other pertinent factors such as experience, location, dedication, interest, project execution plan, and worthwhile suggestions. Scoring the individual factors on a weighted scale of 1 to 5 points gives a total score for each proposal. The short list for proposals consists of the five or six highest scores.

Preparing the RFP

As shown on the proposal schedule in Figure 2.2, the work for preparing the RFP documents proceeds concurrently with the screening activity. Thus, there should be no delay in issuing the RFP as soon as the proposal slate is ready.

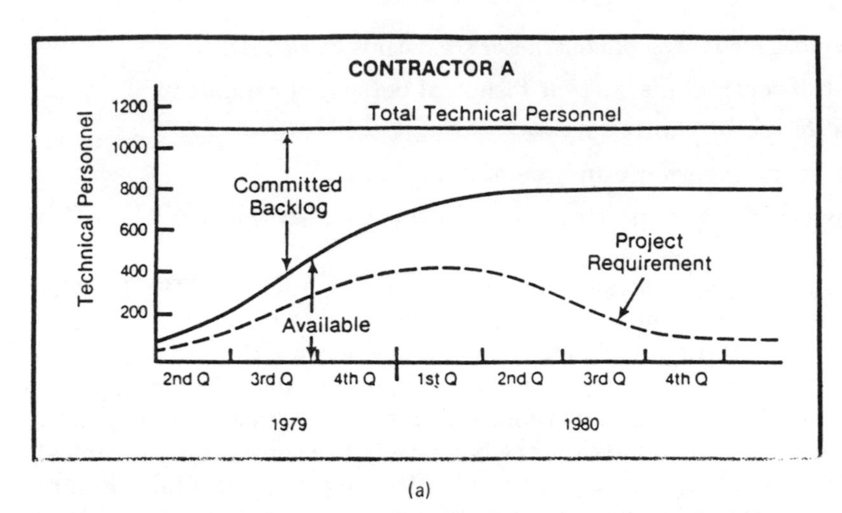

(a)

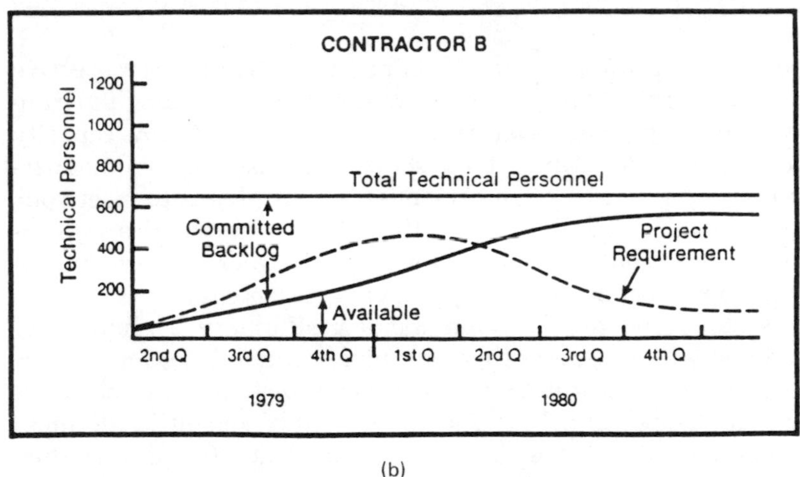

(b)

Figure 2.3 Contractor personnel loading curves. (*Reproduced with permission from* Proceeding of the 10th Annual Seminar/Symposium, *Project Management Institute, October 1978, p. II-L.3/*)

As a minimum, the RFP documents should include the following major sections:

- Proposal instructions
- Form of proposal
- Pro forma contract
- Coordination procedure and job standards

The size and complexity of RFP documents can range from a few pages to several books depending on type of contract and size of project. A small air-conditioning job, for example, might require several pages, whereas a grass roots paper mill might need one or more volumes.

The proposal instructions should include the following subjects to ensure receiving uniform proposals to permit a meaningful comparison. Receiving proposals in a variety of formats makes comparison difficult, and it can even lead to a poor selection resulting in unmet project goals.

Important areas to include in the instructions to contractors are:

- General project requirements: project description, proposal document index, scope of work, site location and description, and so on.
- Special project conditions: owner contacts, site visitation, proposal meetings, and so on.
- Technical proposal requirements: as needed for contractor's technical input.
- Project execution proposal requirements: could include a table of contents of key information that the owner desires (i.e., corporate organization chart, experience record, project execution plan, project organization chart, key personnel assignments, current workload charts, preliminary schedule, description of project control functions and any other pertinent data desired).

The proposal form is at the heart of the contractor's commercial proposal. It requests all of the contractor's business terms such as pay scales, fringe benefits, overheads, fees, out-of-pocket charges, travel expense policy, and computer charges. This section also asks contractors to state any exceptions to the terms of the proposal agreement, the pro forma contract, and any other terms of the proposal invitation.

The pro forma contract is a draft of the contract that the owner plans to execute with the successful firm. The number of exceptions taken to your pro forma contract will be directly proportional to the volume of the document. Standard short-form contracts, such as those developed by the American Institute of Architects (AIA) and the National Society of Professional Engineers (NSPE), can offer some particular advantages in this regard. At the other end of the scale are the long-form contracts generated by large corporate legal staffs and the federal government. The long-form contracts often generate more problems than they solve. Obviously, most legal people don't recognize the KISS principle.

The coordination procedure includes the basic standards and practices that owners expect contractors to follow in executing the project.

This section may include the owner's design practices and standards, local government requirements, safety regulations, machinery standards, procurement procedures, mechanical catalogs, operating manuals, accounting standards, and acceptance of the work. This section is probably the largest volume producer in the proposal documents, depending on how much of this material an owner normally uses. Owners with limited standards can select the standards needed from a menu provided by the contractor and still have a good project.

Most of the material contained in the RFP is an outline for the execution of the project. Design it to get the contractors more deeply involved in thinking about your project. One goal of the RFP is to get contractors' creative juices flowing in generating some input for the owners use in project planning. Good contractors are eager to prove their competence and improve their standing with owners. That is a good way to improve their chances of winning a contract.

Evaluation of the proposals

The purpose in having strict requirements in the RFP documents is to generate consistent proposals leading to easy and accurate comparison on a common basis. Nonconforming proposals are subject to rejection or revision which adversely reflects on the errant proposer. The three crucial sections of the proposal to check thoroughly are:

- The business proposal
- The technical proposal
- The project execution plan

Assign those sections to the separate specialty people on the proposal team for their expert input. Reviewing the business and technical portions separately avoids any influence of the commercial terms on the rest of the proposal. Finally, combine the best combination of commercial and technical offers to present to the selection committee for final approval.

The commercial evaluation will undoubtedly focus on the total cost of the various contractors' services. If the proposals are on a lump sum basis, the commercial evaluation is quite simple and involves comparing a few lump sum cost figures. The technical and performance evaluations are then more critical to ensure that the offerers are pricing the full scope of services. That is necessary to assure that the owner's project expectations will be satisfactorily met.

For a CPFF commercial proposal, the commercial evaluation is more complicated. In that case the evaluation team must reconstruct the total cost of the services from the estimated project labor-hours multiplied by the hourly rates of each proposal. Since the labor-hour estimate varies with the proposal, owners should compare the proposals against a standard set of labor-hours. The selection team then multiplies the estimated standard labor-hours by each offerer's total personnel costs, to arrive at the expected total cost for services. That puts the expected labor costs for all proposals on a common basis.

Before making the final selection, the selection team should visit the contractors' offices for an onsite examination of the firm's proposed personnel, systems, and facilities. Schedule your office visits well in advance so the contractor's top management and key project personnel are present for the personal interviews. The appraisal team should study the organization from top to bottom to get a feel for its ability to meet your project needs. Especially look for top-management commitment, capable personnel, modern facilities, workable systems and procedures, successful job history, and good references.

Now is the time to check out the workload curves to see that they are factual. Review the contractor's work in progress, as well as any outstanding proposals besides your own. Also, now is a good time to pursue the details of any technical proposal features that may need clarification or further discussions with the contractor's top technical people.

The contractor will supply much of the information you seek as part of the dog-and-pony show presented on your behalf. Be polite about that show but do not accept it as the final answer to your questions. Remember, the main theme of the presentation is to cover up any weaknesses in the contractor's proposal!

With the final office visits completed and evaluated, you are ready to make the final selection. The three main areas to rate are the technical, commercial, and project execution proposals. Combine the scores from the three major areas with the results of the office visits to get the final rating. You are then in a position to make a recommendation to top management for approval.

It is good practice to rate the three top proposals in 1, 2, 3 order of preference before starting final negotiations with number 1. If you cannot reach agreement with number 1, you can drop back to the number 2 contractor, and so on. It is usually not a problem to reach an agreement with your prime candidate if all matters were fairly presented by all parties. Using a well-organized selection process as that described above should assure a good selection and a happy project!

The final work of the selection team is to tell the unsuccessful candidates of the final decision as soon as possible. You may expect that the unsuccessful offerers will approach you for a debriefing as to why their proposal was unacceptable. If that is done in the spirit of improving their performance on the next proposal, a fair and open discussion of nonconfidential matters is in order. That is a partial satisfaction for the contractor's investment in the proposal.

Managing the Proposal Effort

A few words about the contractor's project management functions during proposal preparation process are now in order. However, having a better understanding of the selection process from the owner's side is probably more valuable than the how-to of this section.

Many project management practitioners regard proposal assignments as fill-in work to be done only when not assigned to an active project. That is indeed a shortsighted attitude toward a function which is the lifeblood of their organizations. Successful proposals are responsible for making everyone's job in the company possible.

Giving business development and top management a strong performance on a proposal is the most valuable contribution that project management practitioners can make to their companies. It also goes a long way in supporting their personal goals as well. A truly professional project manager cannot afford to take proposal assignments lightly!

The proposal goals

The proposal project managers have the goal of turning out winning proposals on a tight budget within the time allotted. As with owner project managers, they must treat the work as a miniproject using the same planning, organizing, and controlling functions. Miniproject schedules allow no room for false starts, missed targets, or mistakes along the way.

The business development group is usually responsible for the actual production of the proposal documents. However, the project manager is responsible for developing the crucial technical and project execution portions. The PM must be certain that the technical approach is sound and is presented to the best advantage of the proposal and the prospective client.

The PM must analyze the owner's goals and develop the all-important project execution proposal to meet those goals. The project manager prepares this key part of the proposal with the expert input of the applicable department heads. Editing and polishing the specialist inputs into a unified proposal format also falls to the project manager. That means that the PM must be an excellent communicator.

The PM also develops the proposed organization chart and key personnel assignments for the proposal with the approval of top management. The organization must present a can-do project execution team when the client selection committee makes its office visit. The PM should also prepare and rehearse the proposed project team for the live presentation. The training program and rehearsal culminate in a dry run before a devil's advocate tribunal. The tribunal includes top management, so everything had better be right! Make any final adjustments to the presentation from the comments received during the dry run.

The project manager should take the lead in reviewing and responding to any problem areas in the pro forma contract presented in the RFP. The legal people will handle some of that activity as it affects the liabilities and risks assumed by the company. The project manager is responsible for reviewing the day-to-day operating requirements called for in the contract.

The specific areas of interest in the contract for the PM are:

1. Scope of work
2. Payment terms
3. Change order procedure
4. Guarantees and warranties
5. Schedule requirements
6. Incentive clauses
7. Suspension and termination

Proposal work is more exciting if the assignment is approached as a contest of wits in trying to beat out the competition. You are probably spending good money on computer games which are not half as much fun as this real-life challenge!

Your key role on the proposal team puts you in position to come up with the hooks that make the selection committee pick your proposal. That is an opportunity to be creative, so don't pass it up. You must reduce any potential negative effects such as human resource shortages, office location, or costs which might hurt your chances of winning the proposal.

Let's assume that you have been good enough at your craft to win the proposal. You developed the most creative problem solutions, organized the strongest project team, orchestrated an outstanding dog-and-pony show, and made the fewest mistakes. On that happy note, you are now ready to play a key role on the contract negotiation team. That requires a solid background in the area of contract terms and conditions, which just happens to be our next subject.

Capital Project Contracting

The project managers on both sides of the contract equation play influential parts in the negotiation and acceptance of the contract documents. They are the people held responsible to fulfill the terms of the document and make it work. If all goes well, and it usually does, the lawyers and top management involved may never look at the contract again.

The reason for the contract is to clearly and fairly set out the responsibilities of each party involved in the project before the work begins. The contract also spells out the remedies for one party against the other for failure to perform or for defaulting on any of the terms. When that situation occurs, the contract becomes the primary document for settling any later arbitration or litigation claims.

Contracts for capital projects have assumed more importance in recent years because we have become a litigious society. Years ago the executed contract sometimes arrived closer to the end of the project than the beginning. The change to tighten up that procedure is not all bad, because it forces the project participants to finish their homework before starting the project.

Because of the complexity of capital projects, the contract cannot cover all possible situations that are likely to arise. That means there must be a good deal of give and take as well as reason on both sides. The cooperative approach is at least half the battle, with the other half being a sound and equitable contract document to start with.

This discussion of contracts is from the viewpoint of the project manager and not from the legal side. It covers what the project manager needs to know to arrive at a good contract, one that will allow all parties to meet their project goals.

Contract formats

The number of contract formats are virtually infinite. There are literally no two exactly alike, just as there are no two projects that are exactly alike. We can, however, classify the basic contract formats as listed in Table 2.1. Brief descriptions of the advantages and disadvantages and typical applications also are given. Although this chart first appeared about 20 years ago in *Chemical Engineering Magazine,*[3] I have never been able to find a more comprehensive summary of contracts in any later literature.

The contract format used on a given project depends on many factors. Some of them are project definition, owner preferences, public laws (government contracts), current market conditions, project location, project financing, schedule, and scope of work. Factors affecting format will vary over time and from job to job. That makes it difficult to standardize on any one contract format for all types of projects at any given time.

The length and complexity of contract formats can vary over a wide

TABLE 2.1 Types of Contracts

Primary advantages	Primary disadvantages	Typical applications	Comments
Cost-Plus PD: Minimal (Scope of work does not have to be clearly defined)			
1. Eliminates detailed scope definition and proposal preparation time. 2. Eliminates costly extra negotiations if many changes are contemplated. 3. Allows client complete flexibility to supervise design and/or construction.	1. Client must exercise tight cost control over project expenditures. 2. Project cost is usually not optimized.	1. Major revamping of existing facilities. 2. Development projects where technology is not well defined. 3. Confidential projects where minimum industry exposure is desired. 4. Projects where minimum time schedule is critical.	Cost-plus contracts should be used only when client has sufficient engineering staff to supervise work.
Cost-Plus with Guaranteed Maximum PD: General specifications and preliminary layout drawings			
1. Maximum price is established without preparation of detailed design drawings. 2. Client retains option to approve all major project decisions. 3. All savings under maximum price remain with client.	1. Contractor has little incentive to reduce cost. 2. Contractor's fee and contingency are relatively higher than for other fixed-price contracts because price is fixed on preliminary design data. 3. Client must exercise tight cost control over project expenditures.	When client desires fast time schedule with a guaranteed limit on maximum project cost.	—
Cost-Plus with Guaranteed Maximum and Incentive PD: General specifications and preliminary layout drawings			
1. Maximum price is established without preparation of detailed design drawings. 2. Client retains option to approve all major project decisions. 3. Contractor has incentive to improve performance, since he shares in savings.	Contractor's fee and contingency are relatively higher than for other fixed-price contracts because price is fixed on preliminary design data.	When client desires fast time schedule with a guaranteed limit on maximum cost and assurance that the contractor will be motivated to try for cost savings.	Incentive may be provided to optimize features other than capital cost, e.g., operating cost.

TABLE 2.1 Types of Contracts (*Continued*)

Primary advantages	Primary disadvantages	Typical applications	Comments
Cost-Plus with Guaranteed Maximum and Provision for Escalation PD: General specifications and preliminary layout drawings			
1. Maximum price is established without preparation of detailed design drawings. 2. Client retains option to approve all major project decisions. 3. Protects contractor against inflationary periods.	1. Contractor has little incentive to reduce cost. 2. Contractor's fee and contingency are relatively higher than for other fixed-price contracts because price is fixed on preliminary design data. 3. Client must exercise tight cost control over project expenditures.	1. Project involving financing in semiindustrialized countries. 2. Projects requiring long time schedules.	1. Escalation cost-reimbursement terms should be based on recognized industrial index. 2. Escalation clause should be negotiated prior to contract signing.
Bonus/Penalty, Time and Completion PD: Variable, depending on other aspects of contract			
1. Extreme pressure is exerted on contractor to complete project ahead of schedule. 2. Under carefully controlled conditions, will result in minimum design and construction time.	1. Defining the cause for delays during project execution may involve considerable discussion and disagreement between client and contractor. 2. Application of penalty under certain conditions may result in considerable loss to contractor. 3. Pressure for early completion may result in lower quality of work.	Usually applied to lump sum contracts when completion of project is absolute necessity to client in order to fulfill customer commitments.	1. Project execution should be carefully documented to minimize disagreements on reasons for delay. 2. The power to apply penalties should not be used lightly; maximum penalty should not exceed total expected contractor profit.
Bonus/Penalty, Operation and Performance PD: Variable, depending on other aspects of contract			
Directs contractor's peak performance toward area of particular importance to client.	1. Application of penalty under certain conditions may result in considerable loss to contractor. 2. Difficult to obtain exact operating conditions needed to verify performance guarantee.	When client desires maximum production of a particular byproduct in a new process plant to meet market requirements'	Power to apply penalties should not be used lightly.

Lump Sum, Based on Definitive Specifications PD: General specifications, design, drawings, and layout—all complete			
1. Usually results in maximum construction efficiency. 2. Detailed project definition assures client of desired quality.	1. Separate design and construction contracts increase overall project schedule. 2. Noncompetitive design may result in use of overly conservative design basis. 3. Responsibility is divided between designer and constructor.	When client solicits construction bids on a distinctive building designed by an architectural firm or when a federal government bureau solicits construction bids on a project designed by an outside firm.	Clients are cautioned against use of this type of contract if project is not well defined.
Lump Sum, Based on Preliminary Specifications PD: Complete general specifications, preliminary layout, and well-defined design			
1. Competitive engineering design often results in cost-reducing features. 2. Reduces overall project time by overlapping design and construction. 3. Single-party responsibility leads to efficient project execution. 4. Allows contractor to increase profit by superior performance.	1. Contractor's proposal cost is high. 2. Fixed price is based on preliminary drawings. 3. Contract and proposal require careful and lengthy client review.	1. Turnkey contract to design and construct fertilizer plant. 2. Turnkey contract to design and construct foreign power generation plant.	1. Bids should be solicited only from contractors experienced in particular field. 2. Client should review project team proposed by contractor.
Unit Price Contracts, Flat Rate PD: Scope of work well defined qualitatively, with approximate quantity known			
1. Construction work can commence without knowing exact quantities involved. 2. Reimbursement terms are clearly defined.	1. Large quantity estimate errors may result in client's paying unnecessarily high unit costs or contract extra. 2. Extensive client field supervision is required to measure installed quantities.	1. Gas transmission piping project. 2. Highway building. 3. Insulation work in process plants.	Contractor should define the methods of field measurement before the contract is awarded.

TABLE 2.1 Types of Contracts *(Continued)*

Unit Price Contracts, Sliding Rate PD: Scope of work well defined qualitatively			
1. Construction work can commence without knowing quantity requirements. 2. Reimbursement terms are clearly defined.	Extensive client field supervision is required to measure installed quantities.	1. Gas transmission piping project. 2. Highway building. 3. Insulation work in process plants.	Contractor should clearly define the methods of field measurement before the contract is awarded.
Convertible Contracts PD: Variable: depends on type of contract conversion			
1. Design work can commence without delay of soliciting competitive bids. 2. Construction price is fixed at time of contract conversion, when project is reasonably well defined. 3. Overall design and construction schedule is minimum, with reasonable cost.	1. Design may not be optimum. 2. Difficult to obtain competitive bids, since other contractors are reluctant to bid against contractor who performed design work.	1. When client has confidential project requiring a balance of minimum project time with reasonable cost. 2. When client selects particular contractor based on superior past performance.	Contractors selected on this basis should be well known to client.
Time and Materials PD: General scope of project			
1. Client may exercise close control over contractor's execution methods. 2. Contractor is assured a reasonable profit. 3. Reimbursement terms are clearly defined.	1. Project cost may not be minimized. 2. Extensive client supervision is required.	Management engineering services supplied by consulting engineering firm.	Eliminates lengthy scope definition and proposal preparation time.

SOURCE: Reproduced with permission from "A Fresh Look at Engineering and Construction Contracts," *Chemical Engineering*, Sept. 11, 1967, pp. 220–221.

range. Federal government contracts appear at the long-form end of the spectrum. Formats developed by associations for contractors and consultants such as the AIA, NSPE, and AGA are at the short end of the spectrum. Those developed within owner's organizations fall in the middle ground.

Lately there finally seems to be a trend in the industrial/commercial project area toward shorter contract formats. Even the legal people have come around to the opinion that a concise contract format makes a more effective agreement. Many state and local government bodies, along with some commercial developers, have embraced the shorter association formats. An example of a construction version is AIA Document *A201 General Conditions of the Contract for Construction.* Sidney Levy gives it a detailed review in his *Project Management in Construction.*[6] These printed contract forms are available direct from the various professional associations or through legal document dealers.

Let us review some of the key considerations involved in format selection by discussing the formats listed in Table 2.1. The twelve formats shown break down into four major categories as follows:

1. Cost-reimbursable contracts (cost-plus)
   - Cost-plus
   - Cost-plus with guaranteed maximum
   - Cost-plus with guaranteed maximum and incentive
   - Cost-plus with guaranteed maximum and provision for escalation
   - Time and materials
2. Bonus-penalty contracts
   - Based on time and completion
   - Based on operation and performance
3. Lump sum contracts
   - Based on definitive specifications
   - Based on preliminary specifications
4. Unit price contracts
   - Flat rate
   - Unit price—sliding rate

The twelfth format is a convertible contract which changes any of the above reimbursable contracts into a lump sum agreement. The change can be made at any previously agreed-upon point in the work. For example, we could start with a reimbursable-type contract when the project scope is not very firm and convert to a lump sum basis after finishing the conceptual design. We could then negotiate a lump sum price, including any contingency or incentives, based on the refined scope of work.

Selecting the contract form

Referring to Table 2.1, let's look at the factors that lead to the best contract form to suit our needs. The number 1 factor to consider is the degree of *project definition* that is available at time of contract award.

We said earlier that most public agencies split their contracts between the design and construction phases. Thus, when they are ready to commit about 85 percent of the total project cost for construction, the project scope is completely defined. That includes approved plans, specifications, and contract documents. It gives the agency a good handle on its costs by taking lump sum competitive bids for most of the project cost. The method offers definite advantages to an owner using public funds.

A major sacrifice the agency makes in getting the cost control advantage is prolonging the project schedule. It usually takes about 25 to 30 percent longer to get from concept to use of the facility by using the split contract mode. Many commercial and industrial project developers cannot afford the luxury of a delayed schedule in the fast-paced world of international competition.

The best alternative to shorten the completion date is to use a cost plus fixed fee (CPFF) approach. The approach allows design, procurement, and construction to proceed simultaneously. It is common for long-delivery equipment to delay project completion by months. With a CPFF contract, it is possible to specify that material for early delivery to meet an advanced construction schedule. We refer to that approach as fast-tracking the schedule. The faster schedule delivers the facility earlier and offers an early entry into the marketplace with potential for earlier return on investment.

With effective supervision of the CPFF contract, the difference in cost vs. the lump sum method is minor. Often it is well worth the additional investment. The extra expense results from the added owner's personnel required to control project costs. Lump sum contracts are *self-policing* from both the owner and the contractor sides. Both are working against the agreed-upon lump sum price for the work which controls both changes and costs. However, the owner can expect change order requests from the lump sum contractor at the slightest variation in the contract documents.

Owners can get the best of both worlds by converting from a CPFF to lump sum basis after finishing the project definition. At that time both parties can agree on a reliable open-book cost estimate to agree on a fair lump sum price for the work. The contract then switches into the lump sum format with the contractor assuming primary responsibility for the control of costs on the project.

I firmly believe in the philosophy that project teams and contractors perform better when they have a financial incentive. Linking the incentive to meeting project goals ensures a good return for the owner. An owner can inspire a good performance from both project teams and improve the odds of meeting the project goals.

By using that method, the owner and the contractor agree on the estimated cost of the project. They add a contingency or upset amount to the targeted cost and arrive at an *upset price*. The upset amount will be shared by the contractor and the owner on the basis of the savings actually made at the end of the project. This pot can be split (usually 60 percent owner 40 percent contractor) based on the contractor meeting or bettering the agreed-upon targets of cost and schedule. Industrial owners have used that arrangement for many years on critical projects with tight schedules. It actually offers a double-carrot approach instead of the carrot-and-stick approach of the bonus penalty contract format.

If a bonus-penalty contract format is necessary, use it only when one contractor has sole responsibility for the work. Otherwise, there will be constant claims for time extensions resulting from delays caused by other parties to the agreement, including the owner. That format is not common in industrial and commercial practice anymore. However, government contracts still use it in the form of liquidated damages, particularly on construction work. Liquidated damages claim to indemnify the owner against losses incurred by not having the facility on time. Actually, they are fines set against the contractors to force attention to the schedule and completion of the work. The damage clauses are difficult to enforce. When the project does get into serious trouble, they usually do more harm than good. John Gallagher lists additional disadvantages of bonus-penalty clauses in his article.[3]

Before leaving the subject of reimbursable and lump sum formats, I should warn against using the two simultaneously on one project with the same contractor. To do so runs the risk of mixing two different forms of payment, which can result in paying twice for the same work. Some minor time-and-materials items can be done on a lump sum job, but control them carefully.

A time-and-material basis should be considered only for small amounts of work when it is not convenient to use another form of contract. Open-ended orders are anathema to project managers. Avoid them whenever possible because they are fraught with surprises, particularly near the end of what you thought was a successful project.

Unit-price contracts are acceptable when it is not convenient to use one of the other methods. The key to controlling the cost is to get a unit price which gives the contractor an incentive to perform efficiently. The main disadvantage to that format is the added cost needed to count the work units. Prearranged unit prices are good to use when pricing additions to and deletions from the contract.

The Role of the Project Manager in Contracting

As project managers, we often think of contracts as being strictly the domain of the legal department. Nothing could be further from the truth when we look at just how the contract comes into being and how it is handled during project execution. The legal people are responsible for developing the format and properly protecting the company against possible legal actions and uninsured casualty losses.

Project managers are responsible for preparing and organizing the technical and commercial input for the contract. If the contractor prepares the contract, the owner's project manager has responsibility for reviewing it for the legal and top-management people. With that sort of responsibility resting on their shoulders, neither the owner nor the contractor PM can afford to be cavalier in their knowledge of contracting!

It is good business for both the owner and the contractor PMs to know the project contract in intimate detail. If they did not participate in the negotiation of the document as I recommended earlier, they must become conversant with it immediately upon arriving on the project. They are usually named as contract administrators for their respective firms, so they had better know what it says.

To discuss the most important areas of concern to the project manager, I have sorted them into technical and commercial terms. They will not appear in the contract in that order, but the grouping should help to organize our discussion. The project manager has *primary responsibility* for developing and defining the technical terms and a *secondary responsibility* for refining and coordinating the input for the commercial area.

Contract Technical Terms

The Technical Terms of the contracting procedure that we shall discuss are:

1. The scope of work
2. Licensing arrangements
3. Performance guarantees

4. Engineering warranties
5. Equipment guarantees
6. Schedule guarantees
7. Responsibilities of owner and contractor

The scope of work

That is the most critical section of any contract because it defines the services performed or the work to be done in specific terms. Defining the scope is especially critical in lump sum contracts, where there is very little room for error. If complete definition is lacking, contract problems will most surely occur.

The scope section directly specifies the geographical limits of the work. That includes buildings, offsite units, access roads, battery limits, remote units, and interconnections. There may be portions of the work, or even specific equipment furnished by the owner or third parties. Defining the interfaces is important to avoid misunderstanding by any of the parties. The problem is defining the scope in a direct and concise manner without writing volumes to do it.

One way to define a complex scope with minimum wording is by reference to other detailed documents. Include the documents in the contract by reference. The documents can be drawings, studies, design manuals, and any other descriptive documents prepared during the conceptual stage of the project. The only criterion for using such documents is that they must be reasonably accurate representations of the work.

The next most important item of scope definition is the scope of services performed by the contractor. On large projects that can include the design, procurement, construction, and start-up of the project under a single contract. The all-inclusive approach introduces a plethora of third-party interfaces not present in a simple design services contract. A scope of services that broad will probably require a long-form contract to cover the legal implications introduced by the procurement and construction activities.

A tool for organizing and recording the scope of work covered by the services portion of the contract is a project services checklist as shown in Appendix B. The list shown is for a process plant project, but it is easily convertible to any type of project by substituting items applicable to your work. Once you have developed a checklist for your type of project, you can apply it to similar projects. Even if the checklist does not become part of the contract, it is a useful tool for people who estimate the work.

Another useful format for defining the work on split responsibility projects is a division of responsibilities chart such as that shown in Table 2.1. The format is actually for a project with several contracting entities involved. Such a condition is common when an owner acts as the general contractor or hires a managing contractor to coordinate the project. You can tailor the chart to meet the particular contracting plan for any sort of project. Just make sure that you cover all possible participants and interfaces.

A sound definition of the scope of work on any project is critical. If possible, have other key project personnel review your final efforts to ensure completeness and clarity. Another approach is to place yourself in the role of a project outsider with only limited knowledge of the work. Now see if you can clearly understand the scope definition that you have written.

Licensing arrangements

Possibly your management has decided to use a third party license for a specific process or procedure on your project. That introduces a major third-party interface that affects all phases of the project.

An owner normally contracts directly with the licensor to supply a basic process information package. The owner then hires a third-party firm for the detailed design and construction. Now we have two scopes of work to define avoiding any over- or underlapping of the work. Any overlapping costs the owner double; any underlapped work falls into the owner's account. Therefore, that part of the contracting procedure requires special attention by the owner's project manager to prevent any losses.

Licensing usually involves executing secrecy agreements by all parties having access to confidential process data. The secrecy agreements include anyone working on the project who has a *need to know* the details of the process. The legal department develops the secrecy agreement format, but the project staff is responsible for conformance to the agreement.

Performance guarantees

Licensing contracts normally involve performance guarantees. They can also occur when the contractor is supplying the basic process design. PMs must understand the complexities involved in performance guarantees. Failure to meet a performance guarantee can be expensive in *make-good* work or lost royalties.

The performance guarantee spells out exactly how the facility will perform as to products, quality, utilities, on-stream time, and so on. Engineering/construction contractors rarely give a plant performance guarantee unless the contractor is furnishing basic process design as part of the services.

The term "performance guarantee" means that the licensor will modify the plant at its expense until the specified performance conditions are met. Even that type of guarantee has a limit of liability or a specific penalty payment. The limit of liability comes into play in cases when the actual performance does not meet the guarantee in all respects.

Engineering warranties

Design firms give this type engineering warranties to cover detailed design work. Using the term "warranty" instead of "guarantee" means that the design firm corrects any errors in its plans and specifications at its own cost. Although the contractor will furnish corrected design documents at its cost, it does not accept responsibility for associated construction costs caused by the errors.

Equipment guarantees

The contractor passes equipment guarantees through to the owner on any equipment and materials bought by the contractor. They are the usual one-year materials and workmanship guarantees required of the equipment vendors. The contractor merely serves as the clearing house for owner complaints if equipment problems should arise during the contractor's one-year warranty on the completed facility. All make-good costs under the warranty guarantees are borne by the vendor. They are never underwritten by the contractor if the vendor should fail to perform for any reason.

Schedule guarantees

Earlier, we talked about ensuring schedule performance through in centives or bonus-penalty clauses in a variety of contract formats. In the absence of incentive or penalty clauses, the contract should include a "time is of the essence" or a "best efforts to complete" clause. Such a clause does not invoke any financial obligations for either party. However, it does put contractors on notice of the strategic planning end date for the project. Thus, they cannot plead ignorance when the end date arrives and the project is not finished.

Responsibilities of owner and contractor

The responsibilities of owners and contractors are set forth in two sections of the contract which define any obligations in addition to those defined in the detailed scope of work. For example, they may involve the owner supplying the basic project design basis, property surveys, government approvals, owner-furnished equipment, or project financing by certain dates. Since any of these things can affect job progress, they must be clearly defined and coordinated in the overall project schedule.

An additional contractor responsibility might include the owner's request to furnish some additional items. These could be special reports, help in financing, certain types of schedules, project cost estimates, resident inspectors, or subsurface soils reports. These services may not specifically appear in the scope of work. Many of them can entail real added project costs. The project cost estimates and budgets should include these items.

All of the above technical terms require strong input from the project manager and the project technical staff. The respective project managers are often the only technical representatives at the contract negotiation. They must be ready to spot problem areas on either side of the table and bring the necessary technical expertise to bear on the problem.

Contract Commercial Terms

The commercial terms of the contract which we will discuss are:

1. The type of contract and fee structure
2. Payment terms, including the fee
3. Limits of liability
4. Project change notices
5. Cost-escalator clauses
6. Suspension and termination of the work
7. Force majeure
8. Insurance and risk management
9. Applicable law
10. Arbitration
11. Subcontracts

The type of contract and the fee

The owner determines the type of contract and the basis for the fee during the formulation of the original contracting plan. Changes to the plan may occur as a result of additional ideas developed during the proposal stage. The changes often occur during the contract negotiating stage. In any case, there must be a clear statement for the fee basis and payment terms in the agreement.

In lump sum work, make a clear statement that all costs, overheads, and profits for performing the scope of work are in the contract price. List any unit rates for optional additional services by the contractor in this section. They might include rates for inspection services, costs for soils investigations, and laboratory services. It is normal to handle these costs as reimbursable items in an otherwise lump sum contract.

A cost-plus contract should define in detail the allowable reimbursable costs such as overheads and labor rates. Make these documents attachments to the contract rather than including them in the text. Also, make the fee statement in this section. Some owners use percentage fees, but I do not recommend them. A percentage fee actually gives the contractor an incentive to increase project costs. To avoid that condition, convert the percentage fee to a fixed fee of a specific amount. The only ground for changing the fixed fee is a major change in project scope—upward or downward.

State clearly any incentive or penalty terms to the fee structure in this section. Spell out the formula for calculating the incentives or penalties to everyone's satisfaction. Set up the split now even though the parties may not yet know the amount of the upset price.

On cost-plus work, the owner reserves the right to audit the contractor's books covering their project. The owner also approves the contractor's accounting system for the project. On lump sum work, however, an owner gives up the right to audit in return for the fixed price.

Payment terms

Payment terms are important to both parties, since they define how the money flows from the owner to the contractor. The owner wants to pay the contractor on the basis of project progress so there must be enough work in place to justify the payment. In the United States we do not usually pay contractors in advance. In some foreign countries that is not the case and advance payments may be necessary to get the project started.

Contractors, on the other hand, do not want to finance projects for owners, so their interest in regular payments is very keen. Contractors usually invoice monthly for payment within 15 to 30 days. In the project design phase, most contractors will finance their own services for 30 days. In the procurement and construction phases, however, the value of the invoices requires another method to avoid the high cost of financing the work.

For example, on an $80,000,000 reimbursable project, the monthly billing for a peak month might be 6 percent of the project cost or $4,800,000. The monthly interest earned on that amount at 8 percent amounts to $32,000 per month, which is a lot of money for anybody to be losing. Over a 10 month period in the life of the contract that amounts to $320,000.

On CPFF contracts, contractors request using a zero-balance bank account for processing project invoices. That way the owner bears the project-financing costs. The contractor forecasts the project expenses for the labor, material, and out-of-pocket expenses for the coming month. As we will learn later, that number is readily available from the project cash flow curve. The owner then transfers the projected amount of money into the zero-balance bank account. The contractor then writes checks against that account for the outstanding project invoices. The account balance usually runs down to zero each month. If the account does not run down to zero in a given month, any balance gets deducted from the following month's cash transfer request.

On lump sum work, the contractor bills the owner monthly or biweekly for costs of work accomplished within that period. Usually the owner, engineer, and contractor will agree to the percent completion of the work before the billing date. That way there will be no surprises to delay the payment.

A key point in setting up the payment terms is to develop a simple and reliable system ensuring a smooth flow of money. Delays can upset the teamwork of the owner and contractor necessary to execute a successful project. It is up to the respective project managers to make the payment system work smoothly. Each party to the contract has a vested interest in setting up a simple and effective payment system at the outset.

Payment terms for international contracts introduce some additional requirements because of problems inherent in collecting money across borders. Also, many international projects have marginal financing. I have also known instances when governments prohibited exporting the cash for making progress payments.

On international projects, contractors must insist on having the owner set up an irrevocable letter of credit at a bank in the contractor's country. Then the contractors are sure that the funds to pay their costs are already in the country. They submit their invoices against the letter of credit as work proceeds. Since the letter of credit is irrevocable, the owner cannot withdraw any funds from the account for any other purpose without the approval of the contractor.

Although the respective accounting departments handle the mechanics of preparing and paying the project invoices, the PMs are responsible for reviewing and approving them for submission and payment. These reviews assure that the invoices conform to the terms of the contract. Then the payment approval should require only a routine arithmetical check by the owner's accountant.

On many lump sum contracts, the owner will withhold a certain percentage of each invoice to build up a retention fund against the contractor's payments. The purpose of the retention fund is to build up a sort of *performance bond* to maintain the contractor's interest through to the end of the project. The amount withheld can range from 5 to 15 percent with the average being 10 percent.

Contractor's consider retentions to be financially onerous because they lose the interest on the funds held by the owner. In the state of Ohio, for example, the contractors have pushed through a law which stops withholding retention funds at 50 percent completion of the work. The owner holds the retained funds in an escrow account with the interest accruing to the contractor. The owner releases the funds upon acceptance of the work. That law applies only to government work within the state and does not affect private sector work.

Another contractor ploy to avoid the cost of retainage is substituting a performance bond for the retainage near the end of the job. The performance bond protects the owner's interest in having the contractor complete the job. However, it costs the contractor only about one-tenth the interest lost on the retained funds. That satisfies the needs of both parties until acceptance of the work by the owner.

Limits of liability

Liability clauses place a limit on any liabilities assumed by either party to the contract. Either party may request a limit-of-liability clause to cover any costs not covered by insurance. The main one usually comes from the party offering the guarantees and warranties on the project.

Agreeing on liability limits can be a sticking point. Contractors prefer to set the limit at something less than their profit or fee on the project. That will allow them to earn something for the part of the job that was done right. Owners, however, want to set the liability limit high enough to keep the contractor's attention should problems arise on the project. This area is more critical to the legal people, who usually work out a settlement without a lot of input from the project manager.

Project change notices

The one thing certain on any project is the changes occurring along the way—sometimes before signing the contract! Every contract must have a specific procedure covering the process for handling changes to the contract. The most frequent type of change met in capital projects is a change in the original scope of work either upward or downward.

Changes in scope, and therefore price, are more critical on lump sum contracts. The contractor has proposed to do a certain amount of work for a fixed amount of money. If the scope changes, the fee also must change. On cost-plus contracts, scope changes automatically go into the owner's account. Therefore, contractors do not absorb the added cost. However, both parties must still record the changes on CPFF jobs to control the original project budget. Also, the scope may change so drastically over the life of the contract requiring adjustment of the fixed fee. We will be discussing the control of change orders in relation to cost control in later chapters.

The contractual procedures for handling project changes should be as simple and direct as possible. It is best to require only the signatures of both project managers without higher approval. Added levels of approval could delay processing the changes, and thus delay the project. Individual company policies usually place a dollar limitation on the project manager's authority to approve changes.

A clear formula for pricing changes is essential to reduce the bickering that normally results when change orders arise. Allowable percentages for overhead and profit, reimbursable costs, out-of-pocket expenses, and so on, should be part of the contract. That will protect the interests of both parties and ease the processing of change orders.

Project changes come in all shapes and sizes in addition to changes in scope. Extra costs can arise from unknown construction conditions, delays due to force majeure, suspension of the work, or some other extra cost. Claims of that type are hard to define and much harder to resolve because they do not add value to the facility.

Delay claims also are difficult to calculate, since they involve such vague subjects as loss of efficiency, cost escalation, and extension of overhead and supervision charges. Reaching agreement between owner and contractor on the validity and value of that type of claim is very involved. Resolution often comes with a settlement just before going to court or arbitration. As the PM representing your company, you must carefully review the contract language covering these matters. You must document the project to prepare or defend such claims when the situation requires it.

Cost-escalator clauses

Cost escalator clauses are not common in contracts for capital projects in the United States. That is because most owners want contractors to include their estimated escalation in the original proposal. Escalation was a serious factor, even in the United States during periods of high inflation following the world energy crisis of 1973–1974. The moderate inflation rates of recent years have calmed things down again. However, escalation clauses are still likely to be found where high inflation is a way of life.

Getting a good, simple, mutually agreeable formula for equal protection of both parties against the effects of inflation is not easy. A simple price adjustment formula involving design, craft labor, and material costs is reprinted from *Applied Cost Engineering,* 2d ed., p. 186, by courtesy of Marcel Dekker, Inc..[7] It offers a possible model for adaptation to your particular project needs for a simple escalation formula.

$$E = P_B\left(\frac{0.1E'}{E_B} + \frac{0.4M'}{M_B} + \frac{0.5L'}{L_B}\right)$$

where E = total escalation to be invoiced
P_B = bid base price
E_B = value of engineering cost index at base bid
E' = value of engineering cost index at centroid of engineering cost
M_B = value of materials cost index at base bid
M' = value of materials cost index at time of procurement
L_B = value of labor cost index at base bid
L' = value of labor cost index at centroid of labor cost

The constants 0.1, 0.4, and 0.5 represent the proportional contribution to the total bid base price. The given values are for illustrative purposes only and actual values must be negotiated in each situation. Of course, the total amount to be paid equals the bid base price P_B plus escalation E. In many situations engineering takes place immediately after placement of purchase order; therefore, escalation on engineering is negligible and can be dropped from the formula. I would prefer to keep engineering in the formula because it can also be subject to delays.

The various cost indices must be those applicable to the area of the project. They should be agreed on and stated in the escalation terms of the contract, along with the formula. A key factor in escalator clauses is that they sometimes result in broad swings in dollar value when there are large swings in the indices. I recommend trying a few sample calculations with the formula. That will test its validity over a range of probable inflation scenarios over the life of the project. In that way you can tune out any really wide swings in the cost of the project.

Suspension and termination clauses

Suspension and termination clauses do not come into play very often on well-conceived and well-financed projects. Their inclusion in the contract is critical because projects are suspended or canceled for a variety of reasons. We always discuss these two clauses together because a termination is really a permanent suspension.

Both the owner and the contractor need to protect their interests when suspension or termination of the work becomes necessary. Largely, the problem areas requiring protection revolve around payments during the suspension, disposition of owner and contractor personnel, ownership of the finished work, and extra costs.

Contractors want contractual rights to suspend the work for nonpayment, or other default, on the part of the owner. In extreme cases, contractors will also want to reserve their rights to terminate the contract if the owner continues to default on payments.

Likewise, owners reserve their rights to suspend or terminate work with or without default on the part of the contractor. Those rights include a clause to cover possible bankruptcy of the contractor. Owners must have the right to complete the work in a timely manner with other contractors.

An important point to consider in that area is the proof necessary when claiming the other party is in contractual default. Usually an official notice is required, along with reasonable time allowed to correct the alleged default before invoking suspension or termination of the contract.

Suspension and termination notices must include reasonable time limits to take effect, since they can have a devastating effect on the project organization. Both parties must plan the temporary disposition of their project forces during and after the suspension. Owners do not like to lose the advantage of keeping their own and the contractor's trained people on the project during the suspension. On the other hand, they often cannot afford to keep nonproductive people on the project.

If the contractor must move personnel to other projects, there is no guarantee that the personnel will be available to return. Assigning new personnel to the project, with the associated learning curve, will reduce efficiency and cost money. On a lump sum project, the contractor will certainly submit a delay claim to recover these added costs.

Suspension or termination notices must provide for timely notice for the event to take place. The minimum length of notice is directly proportional to the size of the project and the human and physical resources involved. On smaller jobs, the notice period can be relatively short, say, one to two weeks. On larger jobs, 30 days is more reasonable. If the termination is more organized owners may grant contractors a longer period to package up the work in progress properly. The organized closeout procedure serves to protect their investment and reduce losses.

Suspension and termination clauses must also address the final payment of the contractor's costs to date and payment of fees and incentives. Resolution of the payment terms permits the owner to gain full title to all of the work in progress according to the contract.

Force majeure

A force majeure clause is an old standard in all contracts; it gives contractors certain rights of excused delays due to matters *beyond their control*. The conditions which are beyond their control are the so-called acts of God. They consist of such things as fires, earthquakes, tornadoes, hurricanes, riots, and civil commotion.

Some other conditions, such as strikes, work slowdowns, and labor shortages, which I call acts of man, are often included in this clause. There can be no argument about acts of God, since they are beyond the control of the contracting parties. The acts of man, however, are open to discussion. The owner may want to place some limits on them. That will put more pressure on the contractor to overcome any adverse effects on project progress that the delays are likely to have.

Force majeure excuses the contractor from performing to the project schedule if the delay is caused by any of the conditions listed in the clause. Causes can vary from minor delays due to labor problems to virtual demolition of the project from a major natural disaster. Fortunately for all of us, the natural disasters do not occur very often, but the contract must consider that chance.

Insurance and risk management

The insurance clauses protect the owner against all potential losses that can occur on a project, mainly in the construction phase. If the contract is for design services only, the contractor's required insurance coverage is only for workers' compensation and motor vehicle liability. The cost of that insurance is borne by the contractors as part of their overhead to meet state legal requirements.

When construction activity is involved in the contract, the owner must set up comprehensive insurance coverage. That includes setting limits for all contractors and subcontractors as protection from claims made by the contractors and damaged third parties. In addition, the owner will need to have a *construction all risks* insurance policy covering the replacement cost of the construction works and materials on the project. Sometimes the owner will request that the contractor take out the policy and will then reimburse the contractor for the premiums. Present-day practice includes taking out a supplementary *umbrella policy* covering all risks not covered by any of the other insurance in effect. So far that has been the ultimate in owner and contractor protection.[2]

In larger companies, a risk management department handles insurance. Thus, the project manager acts as the enforcer to keep the proper insurance in force and make claims when losses occur. The project manager also advises the risk management people of any unusual risk situations in which coverage is not enough or the insured value has changed.

Applicable law

The lawyers insert the applicable law clause in the contract to set up the legal jurisdiction governing this particular project. The owner usually calls for the applicable law to be of the state or the country of their incorporation. In the United States that does not make much difference, since most states are fairly uniform on contract law. On foreign projects, however, it can become a problem if something goes wrong with the project and legal action becomes necessary.

The thought of entering legal action in a foreign court in some third-world country does not set well with the legal people. To get around the problem, international contracts often call for settlement of contract disputes by binding international arbitration. That avoids lawsuits in unfriendly courts. There is really nothing project managers can do about that clause except to be aware of its presence. They may want to take particular pains to document the project to suit the applicable legal jurisdiction.

Arbitration

Settlement of contract disputes through arbitration has been getting more common since the courts have become so expensive and overcrowded. The arbitration clause merely states that both parties agree to submit contractual disputes to arbitration. That can be done without giving up any legal rights to sue if they feel the occasion warrants it.

The arbitration machinery is already in place. The procedure starts by contacting an arbitration service like the American Arbitration Association (AAA). A neutral group, such as the AAA, sets up a tribunal of three members from existing panels of experienced arbitrators. The panel hears the dispute and renders a judgment much like a court without the expense. Each side selects an arbitrator from the panel, and the arbitration board appoints the third member. The proceedings are much less formal and time-consuming than going to court. You can, in fact, present your own case before the tribunal without legal counsel if you so desire.

The operative word in any arbitration clause is "binding." When two parties agree to *binding arbitration,* both agree in advance to accept the findings of the tribunal. If the basis is *nonbinding,* one party is still free to sue in a court of law. Information on U.S. arbitration procedures is available from the American Arbitration Association, 140 West 51st Street, New York, NY 10020. For information on international arbitration services you can contact one of the international tribunals in Paris or Geneva.

Subcontracts

Most contracts for capital projects have a clause stating the owner's policy on subcontracting part of the work. Normally, construction contracts permit the subcontracting of various specialty trades with the owner reserving the right to approve the subcontractors.

Design-only contracts do not usually include the right to subcontract. Occasionally, plans call for outside specialist help through subcontracting. That is a decision the owner must make when drawing up the prime contract.

Subcontracts also fall within the purview of the project manager, since execution for part of the project passes over to others. Subcontracting design services usually risks dilution of project control, so approach it with care.

The purchasing department usually handles the commercial aspects of subcontracting. The most critical factor in subcontracting is making certain that all key prime contract requirements extend to the subcontractors. That will ensure that the prime contractor does not get stuck with paying for missing items.

PMs are more likely to find themselves reviewing the subcontracting procedure rather than developing the details of the subcontracts. That role is quite different from the deep involvement in the prime contracting procedure. The reviewing duties are especially critical when subcontracting in the design area!

Contract Administration

We have spent a lot of time talking about arriving at a mutually satisfactory contract document, but that is only half the battle. PMs also have prime responsibility for overseeing the contract for the life of the project. A very large project may rate a contract administrator to ensure the enforcement of all contract provisions. Most of us, however, will never experience the luxury of having a contract administrator. Therefore, we must work at becoming more contract-oriented without becoming lawyers in the process.

A good way to remind yourself of key contractual requirements is to set up a *contract tickle file.* It automatically reminds you of key notice dates and performance required by the contract. PMs must document the project work according to the terms and conditions of the contract. Keep in mind that the contract sets up only the minimum documentation required. A professional PM should document the project thoroughly enough to defend any adverse claims that are likely to arise.

Treat the contract documents as company confidential material. Limit the distribution to the members of the project team who have a need to know. Make a copy available in the project file for ready access by the PM and the top project staff. Other members of the project team should get their information from the project procedure manual, which summarizes the contractual information in everyday language.

Letters of intent

The one remaining contractual matter for us to cover is the letter of intent. There are times when owners want to start work immediately but it will take several months to work out the contractual details. Issuing the contractor a letter of intent authorizing the work to proceed on a limited basis, pending the contract signing, can solve the problem. Such a letter is only a page or two long, and it spells out the limited basis for starting the work. Referring to the RFP and the proposal to clarify the desired payment terms, scope, budgetary limits, and so on, for the temporary authorization is normal. Figure 2.4 shows a typical letter of intent.

It is not advisable for either party to continue working under a letter of intent for longer than 90 days. Never use a letter of intent as an excuse for not completing the contract negotiations and getting the project on a fully committed footing.

In this chapter we have touched on some of the major points in the proposal and contracting environment from the project manager's viewpoint. There is not much formal literature on these subjects, and most of what does exist is written by sales or legal people. These people do not consider the subject from the project manager's viewpoint. If you do study contracts in more detail, I hope that you will not get too involved in the strictly legal aspects. That could lead you to become so entangled with the legal risks that you might never get the work done. Let alone completing it on schedule and within budget!

ACE DEVELOPMENT COMPANY
200 Casino Avenue
Las Vegas, Nevada

March 30,1989

Mr. Sam Smith, President
XYZ Capital Projects, Inc.
1330 Winner Street
New York, New York

Subject: Super Project USA

Dear Mr. Smith:

This letter will confirm our intent to negotiate a contract with your firm to furnish design, procurement and construction management services in connection with the subject project within the next 60 days. The scope of work and the contract format will conform to that described in our RFQ dated February 1, 1989 and your proposal dated March 15,1989.

Since our scheduled completion date is very critical to us, we are authorizing you to proceed with the conceptual design immediately. We are herewith authorizing you to spend $30,000 for home office services prior to the execution of our contract.

Our project manager will expect your project team to meet in our offices to initiate the project on April 15, 1989. Meanwhile, we will be sending you our proposed contract draft within the next several weeks so that we may resolve the outstanding differences as soon as possible.

We are looking forward to working with your people on this important project and hope that by working together as a team, we can successfully complete the work on time and within budget.

Sincerely yours,

Mr. James Jones, Vice President
Ace Development Company

Figure 2.4 Sample letter of intent.

References

1. "Project Management Starts before Contract Award," *Proceedings of the 10th Annual Seminar/Symposium,* Project Management Institute, Anaheim, Calif., October 1978, pp. II-L.1 to II-1.10.
2. The Business Round Table, *Construction Industry Cost Effectiveness Project Reports,* 200 Park Avenue, New York, NY 10166.
3. John T. Gallagher, "A Fresh Look at Engineering Construction Contracts," *Chemical Engineering,* Sept. 11, 1967, p. 218.
4. J. W. Hackney, "Pros and Cons of Construction Contracts," *Chemical Engineering,* June 21, 1965, p. 160.
5. S. J. Bresler and M. J. Hertz, "Negotiating with Engineering Contractors," *Chemical Engineering,* Oct. 11, 1965, p. 209.

6. Sidney M. Levy, *Project Management in Construction,* McGraw-Hill, New York, 1987.
7. Forest D. Clark and A. B. Lorenzoni, *Applied Cost Engineering,* Marcel Dekker, New York, 1985.

Case Study Instructions

1. As the owner's project manager, prepare a proposed contracting plan for the execution of your selected project.
2. Assume that your management has decided to have your project handled by an outside firm. Prepare your plan for the request for proposals (RFP) and the selection of the prime contractor for your project.
3. As a contractor's project manager, prepare a plan for the preparation of the proposal requested in (2). Give special attention to any factors enhancing your firm's position to win the award even though your backlog of work is quite full at the moment.
4. Draft the main points of the pro forma contract which you will want to include in the above request for proposal. Base this draft on the contracting plan developed under (1).
5. Now put on your contractor's hat. List the points in the pro forma contract prepared in (4) to which you want to take exception to or negotiate.

Chapter

3

Project Planning

We are ready to explore a *total project management* approach to executing capital projects. Planning is the first step of our total project management philosophy for planning, organizing, and controlling the execution of capital projects.

First, we will discuss the methods used to prepare a project execution plan (or master plan) followed by scheduling detailed execution of the work in Chapter 4. Later we will cover the financial and project resources plans for the project.

I specifically separated project *planning* from project *scheduling* to stress the point that these are two separate and distinct functions. The project manager and the key staff members prepare the master plan, and the scheduling people put the plan on the time schedule.

Planning Definitions

There are several good definitions for planning which I have used over the years. I am going to list all three of them because each one delivers a slightly different message.

- Planning is a bridge between the experience of the past and the proposed action that produces a favorable result in the future.
- Planning is a precaution by which we can reduce undesirable effects or unexpected happenings to eliminate confusion, waste, and loss of efficiency.
- Planning is the prior determining and specifying of the factors, forces, effects, and relationships necessary to reach the desired goals.

The first definition reminds us to make use of our prior experience, often gained from past mistakes, to avoid repeating them in our present endeavor. It also says that we should not reinvent the wheel on each new project. The second definition cites the advantages of increased productivity by planning the unexpected and undesirable happenings out of existence before starting to work. The third one stresses making a *conscious effort* to find and control the variables in a capital project. We must do that *before* starting work if we are to meet our project goals. It also indicates the need for an organizational phase to execute the plan.

All of that points to the obvious conclusion that the first move on any project assignment is to do the necessary planning. Furthermore, that applies to every one of your activities throughout the project!

Planning Philosophy

Planning should be done *logically, thoroughly, and honestly* to have a chance to succeed. Previous experience has honed the basic planning logic for capital projects to a fine art over many years. There is no point in trying to develop a new planning logic for each new project. Also, the owner has already set up the basic project execution format in the contracting plan.

After selecting the normal planning logic for your project, you should examine the work for exceptional features affecting the normal logic of your plan. Look for any special problem areas which may be different this time around, such as unusual client requirements, an out-of-the way location, or potential delaying factors which are likely to affect the normal logic of your plan.

Work these potential problem areas over in detail to reduce their negative effects on the master plan and later the schedule. You may leave the details of the schedule to your planning specialists, but you must give their scheduling logic an intensive review.

Earlier I spent a lot of time telling you to opt for simplicity in project management. Now I want to stress being *thorough* in the area of planning. Each aspect of your plan needs individual scrutiny. Discuss it with your staff to get all of the expert input available. Small details passed over in the early stages can rise up to smite you mightily when least expected later in the job. Enough things will go wrong by accident, so there is no need to increase the chances by overlooking problems in the planning stage.

Honesty in planning is also very important. Remember, *you* are the one responsible for carrying out the project plan. *You* will be the one finally held accountable for its success or failure. You may blame a late project on your planning people, but your management will not.

Maintain a delicate balance between optimism and pessimism during this early planning exercise. A project plan without some float or contingency for unforeseen events is a trap. It will spring shut on you eventually unless you have an inordinate amount of good luck on your side.

On the other hand, a fat plan is wasteful of time and money. Parkinson's law tells us that "the work always expands to fill the time allowed." I have always felt that the highest project efficiency results from being slightly understaffed. Working on a little tighter than average schedule coupled to a suitable incentive program also helps.

Good project managers must learn to work under pressure; if they do not, they will not survive. It turns out that we are lucky in this regard because there are always plenty of people around to sharpen the schedule.

Types of Planning

Several types of planning are involved in any capital project. Let us define the types and see how the project manager fits into the overall planning process. The three major types of planning are:

- Strategic planning involves the high-level selection of the project objectives.
- Operational planning involves the detailed planning required to meet the strategic objectives.
- Planning and scheduling puts the operational plan on a time scale set by the strategic objectives.

Does the project team do the strategic planning? No, that is done by the owner's corporate planners. They decide what project to build and what the completion date has to be to meet the owner's project goals. The development phase involves a great deal of strategic planning. It requires the input of market analysis, financing planning, project feasibility, and so on. Those areas need thorough study before the project can get the green light. Sometimes, the strategic completion date is fixed before issuing the decision to proceed without allowing more time on the end date. That uses up the schedule float on the front end and results in a tight schedule for the remaining work.

The project team formulates the master project execution plan within the guidelines called for in the strategic plan. This chapter will explain how to make an operational plan for a typical project. The job of putting the plan onto a time schedule falls to the project schedulers. That points up the difference between the terms "planning" and "scheduling," which are so often mistakenly thought to be interchangeable.

Operational Planning Questions

Operational planning usually raises some interesting questions for resolution in the project master planning phase.

- Will the operational plan meet the strategic planning target date?
- Are sufficient project resources available to meet the objectives?
- What is the impact of the new project on the existing workload?
- Where will we get the resources to handle any overload?
- What company policies may prevent the plan from meeting the target date?
- Are unusually long delivery equipment or materials involved?
- Are the project concepts and design basis firmly established and ready to start the design?
- Is the original contracting plan still valid?
- Will it be more economical to use a fast-track scheduling approach?

We must answer those and any other pertinent questions in preparing the project master plan before the detailed scheduling can start. Preparation of the project schedule before making a logical project execution plan is a waste of time and money.

The Project Master Plan

The master plan must address how we will *plan, organize, and control* the major work activities to meet our goals of finishing the work *on time, within budget, and as specified.*

A major consideration is the contracting plan. Are we able to do this project ourselves, or will we contract for outside help? Do we need more resources for engineering or only for construction? What are company policies in this regard? What government and social restraints come into play in the execution of this type of project?

We must answer those questions during the development of the project execution plan. Some of them are answerable immediately; others must wait for information which will develop as the project progresses.

Project execution plans are subject to review and evaluation as the work progresses. Minor variations are common, but consider major changes with extreme caution. Changes to the plan often bring on a great deal of trauma and should not be made lightly. Since all parts of the plan are so closely interdependent, a change in one major part can affect the interaction of all of the other parts.

After completing the master plan and getting it approved, we can start thinking about more detailed plans. They are:

- The time plan (schedule)
- The money plan (budget)
- A project resources plan (people, materials and services)

The time plan, which results in the project schedule, will be the subject of the next chapter. There we will address the activities which we will schedule, the scheduling methods available to us, and how to select the best system to suit conditions.

The project manager's role

The project manager is the prime mover of the project master plan. The PM accomplishes the master planning through the judicious use of the complete project team's many talents. The PM gets input from the key people in the design, procurement, construction, and project services groups. Lead people from those departments have the responsibility for planning the portions of the work normally performed by them. It's the PM's duty to see that the key groups have a sound basis for their part of the plan to satisfy the strategic and contracting plans.

The PM organizes and chairs a series of planning meetings early in the project. The various contributors to the plan can interact with one another to resolve any operational differences which may arise. The differences not readily resolved by the meeting participants must be resolved by the project manager.

The master planning effort should include early establishment and dissemination of the specific project goals. That allows the master plan contributors to formulate their individual plans to meet the overall project goals. The PM is also responsible for getting all missing information and decisions required to complete the master planning effort on time.

After receiving all parts of the plan, the PM integrates them into a written master plan for presentation to top client and company managements for approval. The PM makes any changes resulting from the final approval stage, and issues the plan as part of the project procedure manual.

I want to remind readers that discussions in this book assume a full-scope, turnkey project as being the most complex case to manage. If your projects have a lesser scope, you need use only the portions of the discussion that apply to your scope of work.

Project Execution Format for Capital Projects

Unfortunately, there is no single project execution format that fits all the various types of projects shown in Figure 1.1. The only way to clearly define the various project formats is to discuss a process-oriented project and then follow it with a typical nonprocess project. I hope all readers will have some interest in how the different groups of the capital projects industry execute their projects. One never knows when crossing the line into another type of expertise may become necessary. Also, there is a good possibility that you might find a new and valuable technique to use.

The involvement of manufacturing processes creates the difference between process and nonprocess projects. A chemical or mechanical process weaves its way through the entire design, procurement, construction, and project activities necessary to complete a process facility.

Another way to state the difference is that manufacturing facilities are basically process design driven, whereas nonmanufacturing projects are architectural or civil design driven. Architectural and civil works projects involve satisfying such human and infrastructural needs as schools, office buildings, hospitals, highways, dams, and public transportation systems. In process projects, the civil and architectural input serves to house and support the needs of the processing operations.

The basic difference between the two types of projects also extends to the field of project management. Although PMs use the same management techniques in different technical environments, they rarely seem to cross over the line to handle projects outside their fields of specialization.

A typical process project format

The development of any project follows the major phases shown on the project life cycle curve shown in Figure 1.3. Our discussion model has the owner using the services of a captive central engineering department or an outside contractor to design, procure, and construct a major process facility.

The scope of the major work activities required to execute a typical process project is shown graphically in Figure 3.1. Each of those major activities breaks down into several individual work activities such that a large project can have as many as 3000 to 4000 activities to schedule. If the project team is to meet its goals, it must successfully plan, organize, and control those work activities so they are performed in proper sequence and on time.

A brief description of the major project operations should give you a

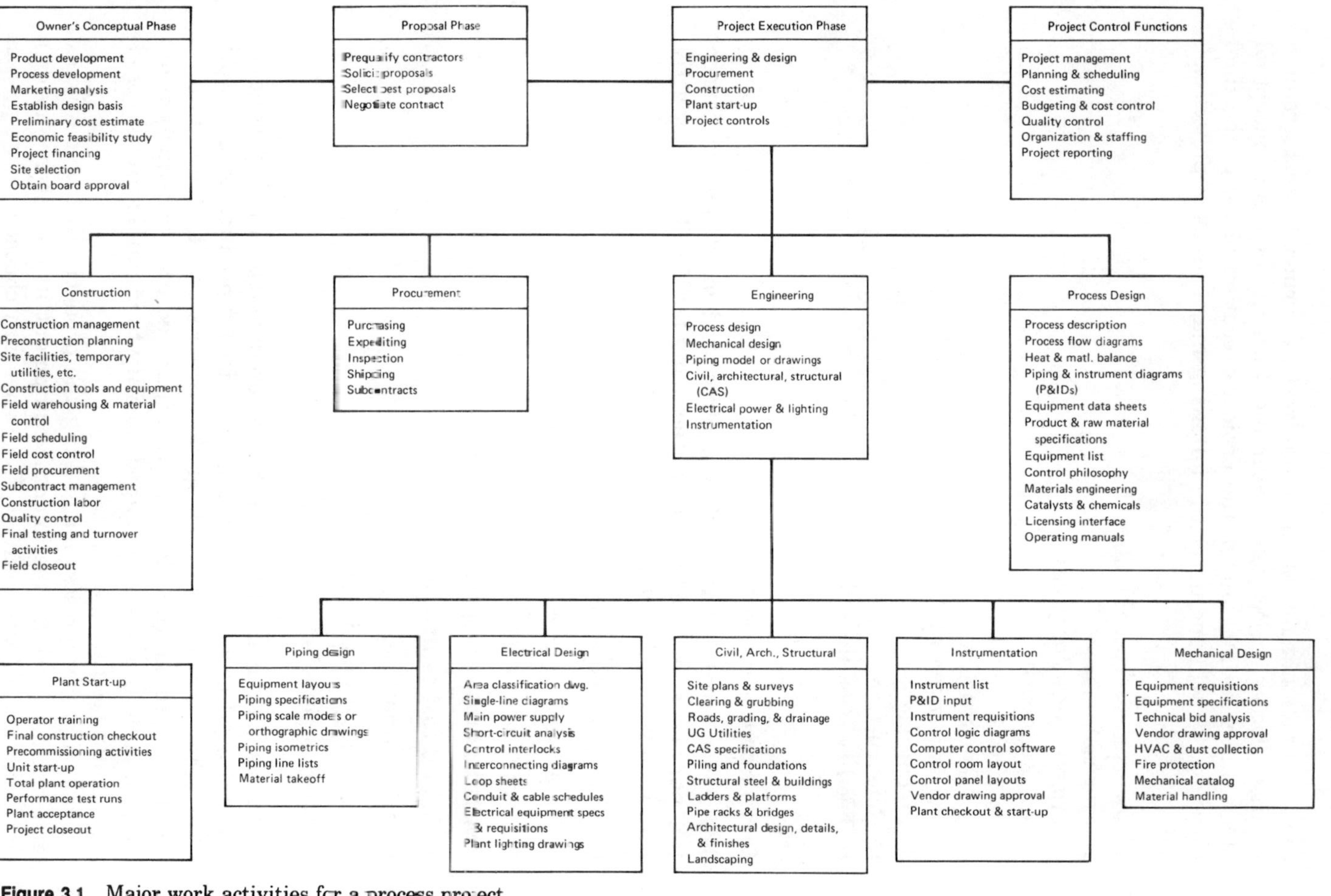

Figure 3.1 Major work activities for a process project.

common understanding of the actual work performed by each group in a large process project team. Starting with the owner's conceptual phase and working through the project execution phase to the final turnover, we will encounter the following major areas of project execution:

1. Owner's conceptual phase
2. Proposal phase
3. Project execution phase
   a. Engineering
   b. Procurement
   c. Construction
   d. Project control functions
4. Procurement
5. Construction
6. Facility start-up and turnover

Owner's conceptual phase

The owner usually assigns a PM to direct and coordinate the conceptual phase of a proposed new manufacturing facility. The major activities involved in this phase are:

- Product development
- Process development or process license
- Marketing surveys
- Setting project scope and design basis
- Capital cost estimating
- Project financing
- Economic feasibility studies
- Board approval of the project

The owner's marketing or R&D group develops a new product or an improved process for making an existing product. If that work offers the chance for increasing the owner's profit, there is an incentive to pursue the venture. An organized and well-thought-out plan to build a new (or revamped) facility to manufacture the product should result.

First, a proven process for making the product must be developed and tested by using experimental equipment in the laboratory or a computer model. Alternatively, the owner may choose to buy a process license from an outside firm specializing in this process. The following R&D steps are not required if the owner buys a license.

A newly developed process may then be expanded into a pilot plant large enough to make test market sample quantities of the product. Assuming all goes well with the pilot plant work and the test marketing, the pilot plant data are used to scale up to a full-size plant. At this point we have preliminary flow diagrams and some major equipment specifications to make a preliminary plant cost estimate in the ± 40 percent accuracy range.

While the R&D work is progressing, the sales department makes the marketing survey to determine the projected market volume, probable selling price, market entry costs, and other expenses involved in bringing the product to market.

The plans to finance the project are going on concurrently with the above technical work to make available the data necessary to perform a feasibility study. The feasibility study is the culmination of all the work done during the initial part of the conceptual phase. The conceptual group presents the study to the board of directors for preliminary approval to proceed with further development of the project.

The initial *strategic planning* results from this first phase of the work. The strategic project completion date is established along with the basic project financial plan in which the project budget plays a major role. Good planning dictates some contingencies in the financial and time plans to cover the usual errors and omissions which will show up as the plan progresses.

Since we are still very low on the overall project life cycle curve, the financial risks at this point are still controllable. The project team makes further evaluation of the project as the detailed development work evolves along the life cycle curve. The final "go-no-go" decision occurs when the detailed design is about 30 percent complete and commitments for equipment purchases are ready to proceed.

The elapsed time for this part of the project can run from several months for a simple project to 3 years or more for a complex one. Since the amount of work in that early stage may not warrant a full-time PM, he or she may handle several conceptual projects simultaneously. The conceptual PM rarely carries the project through the execution phase to completion. The specialized skills developed in handling the diverse conceptual phase activities are so different from those required to execute the detailed project execution phase that not many people can make the transition.

When the project baton passes between these two phases, it is extremely important that the project definition and the feasibility of making and marketing the product at a profit are well proven and documented. Many a project has started on the road to disaster by incomplete or sloppy workmanship during that critical phase. Some of the worst disasters that I have experienced have occurred because the owner tried to skip the conceptual phase altogether![3]

Proposal phase

With the board's approval to proceed with the project, the owner is ready to enter the proposal phase and select a contractor. This phase involves the following activities:

- Prepare a contracting plan
- Prequalify contractor slate
- Prepare a request for proposal (RFP)
- Receive and analyze the proposals
- Select the best proposal
- Negotiate a contract

These activities were described in detail in Chapter 2, so it is not necessary to repeat them here. Suffice it to say that the owner has still not committed a great deal of money to reach this point.

Project execution phase

The slope of the project life cycle curve begins to increase sharply with the initiation of the project execution phase. That is because of significant commitments of project financial resources to the following work activities:

- Engineering design phase
- Procurement activities
- Construction
- Project control functions

The engineering design phase covers those activities required to generate the plans and specifications for the procurement of the equipment and materials and the construction of the facility (Figure 3.1). The various technical disciplines involved are:

- Process design
- Mechanical design
- Civil, architectural and structural design
- Piping design
- Electrical design
- Instrumentation design

All the design disciplines also make a general input to the master planning effort as well as the detailed engineering schedules.

Process design

The design basis for the plant's process and utility systems appears in a process design package developed by the process group. This includes developing the product and raw material slates and specifications for purity. These requirements are set by the owner and are based on the market survey.

The contractor's process group starts with the owner's original design basis (or the process license design package), which was established during the conceptual phase. They further develop the design package into working documents for use by the other design disciplines. PMs should remember that the process group rarely, if ever, issues any construction drawings used in constructing the plant. However, the process design work is urgently needed by downstream design disciplines and procurement before they can start their work. That makes the process work very schedule-sensitive since it can hold up the whole detailed design effort. PMs must give the process design work their undivided attention during the early phases of the project.

The process flow diagrams (PFDs) delineate the major process streams and include a heat and material balance of all the process operations. The PFDs are the basis for the piping and instrumentation diagrams (P&IDs), which show all pipelines, valves, equipment, and instrumentation. The P&IDs are seminal design documents used by the other disciplines to prepare equipment layouts, plant models (or drawings), and the electrical and instrumentation designs.

Another key activity of the process group is preparing equipment data sheets for the procurement of the tagged process equipment. They summarize the process equipment on an equipment list which gives the equipment number, name, key operating conditions, and materials of construction.

The process group also develops the plant utility balances, effluent treatment systems, and catalyst and chemical requirements. At the end of the process design phase they write the plant operating instructions.

Peak activity of the process work is fairly short in duration and occurs at the front end of the engineering work. It plays a key role in getting the project started on a good footing. Since the other design disciplines build on the output of the process group, the PM must give the process schedule top priority during *project planning and execution.*

Detailed design

The work done in detailed design involves preparing the construction drawings and specifications for purchasing the plant materials and equipment. This group assures that the correct amounts of the proper materials and equipment are ordered and delivered to the field for installation according to the construction drawings.

The following design disciplines are involved in process plant projects:

- Mechanical
- Civil, architectural, and structural (CAS)
- Piping
- Electrical
- Instrumentation

Figure 3.1 shows a breakdown of each discipline's major work activities in the designated block. Brief descriptions of those activities will ensure that all readers understand what this work covers.

Mechanical design

The mechanical group usually specifies the vessels, heat exchangers, rotating equipment, and packaged equipment to requisition that equipment for purchase. To produce the requisitions the mechanical group adds the mechanical requirements to the equipment data sheet produced in the process group. Some design organizations handle this work right in the process group.

After receiving the bids, the mechanical (or process) group does a technical bid evaluation on the vendor's quotes to select the best offer. As a follow-up activity, this group also reviews the vendor's shop drawings and handles any necessary design revisions which tend to occur in this process.

The mechanical group is also responsible for designing any heating, ventilating, air-conditioning, dust collection, materials handling, fire protection, and utility systems required for the plant.

Civil, architectural, and structural (CAS)

The CAS groups handle all design for the site, structures, and buildings involved in the facility. Figure 3.1 lists the individual work activities in the CAS block.

The civil team designs the clearing and grubbing, roads and drainage, topographic and boundary surveys, fencing, and underground utilities required for the project. The architectural group designs such buildings as laboratories, change houses, control buildings, plant offices and factory structures. The structural group of the CAS team specializes in the design of the piling, foundations, and structures necessary to support the equipment and frame any necessary buildings.

Most process plant buildings are very utilitarian in nature and often require only minimal aesthetic design. Many manufacturing plants, however, do try to make a statement of the owner's standing in the industrial community through aesthetically pleasing buildings.

Piping design

The specialized nature of the process plant requires a large group of people devoted exclusively to the design of the complex piping systems needed for a process plant. The piping group is responsible for the design, specification, takeoff, and requisitioning of all piping systems and materials for the project. This group usually has the largest design personnel budget for a liquid-processing plant. The piping work is almost always on the critical path of a process plant schedule.

The piping group, in conjunction with the process group, generates the equipment layouts and plot plans for the project. These drawings set out the space requirements for the plant and show the sizes of all buildings and structures needed to support or house the plant equipment. The layouts are also seminal documents used throughout the design discipline groups.

Using the equipment layouts and the P&IDs, the piping group then builds a scale model showing the piping systems needed to support the process. If scale models are not used, the piping group prepares orthographic drawings of the piping systems to show the systems in plan, elevation, and sectional views.

After reviewing and approving the piping models or drawings, the piping group makes piping isometric drawings that show each pipeline in detail. Today, piping isometrics are drawn by computer, so a piping material list is available automatically. The material list records each foot of pipe and each fitting and forms the basis for the piping material requisition. Early attention to this phase of the work ensures there is enough piping material when it is needed for this critical activity. The isos also permits shop prefabrication of the piping subassemblies for later erection on the job site.

Electrical design

The electrical group handles all phases of the plant's electrical design from the main power source to the final point of use. It is responsible for the electrical load studies, motor lists, single-line diagrams, and area classification drawings as well as all interlock and relay ladder diagrams, interconnecting wiring diagrams, and conduit and cable schedules. This group also designs all facility lighting systems. In addition to the drawings, the electrical group specifies and requisitions all of the electrical materials and equipment as well as the installation standards.

Instrumentation

Most process plants have sophisticated control systems to safely and automatically handle process operations. The instrumentation group does not usually exist on a nonprocess project. It handles all the plant automation and control based on the control philosophy set up by the process group. It is responsible for adding the measuring elements and associated control loops and control valves to the P&IDs.

Instrumentation specifies and requisitions any computerized control systems needed for the plant and also develops any software necessary to operate the systems. Present-day PMs should be aware that software development adds a heavy element to the instrumentation budget. Just another reason why design labor costs haven't declined over the years!

In addition to the design of the control systems, the instrumentation group specifies and requisitions all of the process instrumentation equipment and materials required for the project. It also lays out the control panels and consoles as well as the necessary control rooms. Field checkout and calibration of the control systems before plant start-up is a critical instrumentation activity.

Procurement activities

The activities of the procurement group are closely interwoven with those of the engineering and construction groups. Procurement is basically responsible for getting the materials and equipment from the design stage to the construction site as specified and on time to meet the construction schedule. A key early output from this group is the project procurement plan, which plays a key role in making the project master plan. The procurement plan is also a major part of the overall project materials management plan.

PMs sometimes overlook the procurement activities because they get too wrapped up in the engineering and construction activities. No matter how well the engineering and construction teams perform, you cannot build the plant on schedule without the materials and equipment. I will raise this point about procurement several times in this volume to emphasize the importance of procurement in effective project management.

Construction phase

The construction activities involve the largest single commitment of resources on the project, about 40 to 50 percent of the total project budget. Before starting any field work, the construction team performs preconstruction activities and supplies construction input to the master planning effort. This preliminary work usually conforms to the contracting philosophy and the field labor survey made during the proposal phase.

At about 40 percent completion of the engineering, field operations can start. This involves installing the temporary construction facilities and utilities, as well as planning the field site layout and contractor laydown areas. The field warehousing facilities are set up to receive, store, and control the construction materials and equipment delivered to the site. This includes all construction tools and equipment as well as materials bought in the field. The warehousing function is a major part of the project materials management program.

The field engineering group handles all construction document control, setting the lines and grades, and the construction quality control program. The field engineer also originates and controls the field change order log.

The construction group's main mission is the construction of the plant to a point of mechanical completion ready for start-up. Field schedules and cost control procedures are set up to ensure that the project budget and schedule are met in accordance with the project execution plan.

The construction group supports the start-up team with maintenance services until the owner's plant forces can take over. After plant acceptance, the construction team closes out the field operations.

Project control functions

A team of control specialists performs the necessary project control functions to assure the project goals relating to budget, schedule, and quality are effectively met. Cost engineers will be monitoring all project commitments and expenditures to see that they conform to the budget and cash flow projections. A monthly project cost report presents the data to the project team. The scheduling group monitors the project schedule on a regular basis and reports any drift off target at least monthly. The project team should reevaluate the schedule periodically and rework it whenever 25 percent or more of the target dates have fallen behind schedule.

The various groups responsible for each type of material closely monitor and report the status for their particular areas on a regular basis. Material quality is checked regularly by the responsible people in engineering, procurement, and construction.

The PM coordinates all those control functions and issues monthly progress reports to the managements of the owner and the contractor. Any off-target items are highlighted and discussed along with recommended solutions for any problems.

The onset of the project execution phase triggers a significant commitment of major human, physical, and financial resources to the project as we start to climb the life cycle curve. Engineering costs, for example, can run from 8 to 12 percent of the total plant cost. The placement of purchase orders for process equipment and materials involves an even greater commitment of funds. This is a good time for the prudent owner to look at the project financial plan to see if the return on investment is still valid. Using the expanded design data base to make a capital cost estimate with an expected accuracy of ±15 percent would be a good investment. That would be a big improvement over the ±40 percent estimate used in the original feasibility study.

A Typical Nonprocess Capital Project Format

The type of project that we will discuss here is similar to the projects listed in columns 5 through 9 in the matrix of capital projects shown in Figure 1.1. These facilities usually have little or no process content and involve mainly architectural and civil design. The absence of process design and its associated piping, instrumentation, and procurement activities reduces the number of work activities in the overall nonprocess project format.

Owners of these facilities operate in the areas of light industry, government, and commercial development. Owners in these areas usually prefer using separate design-and-build contracts to design and construct with the procurement performed by the constructor. The owner often retains the architect or engineer to help in letting the contract and inspecting the construction work. These split contract arrangements naturally serve to break the overall project format into smaller packages. That reduces the number of the work activities under the control of the respective design and construction project managers.

Figure 3.2 shows a schematic representation of a nonprocess project

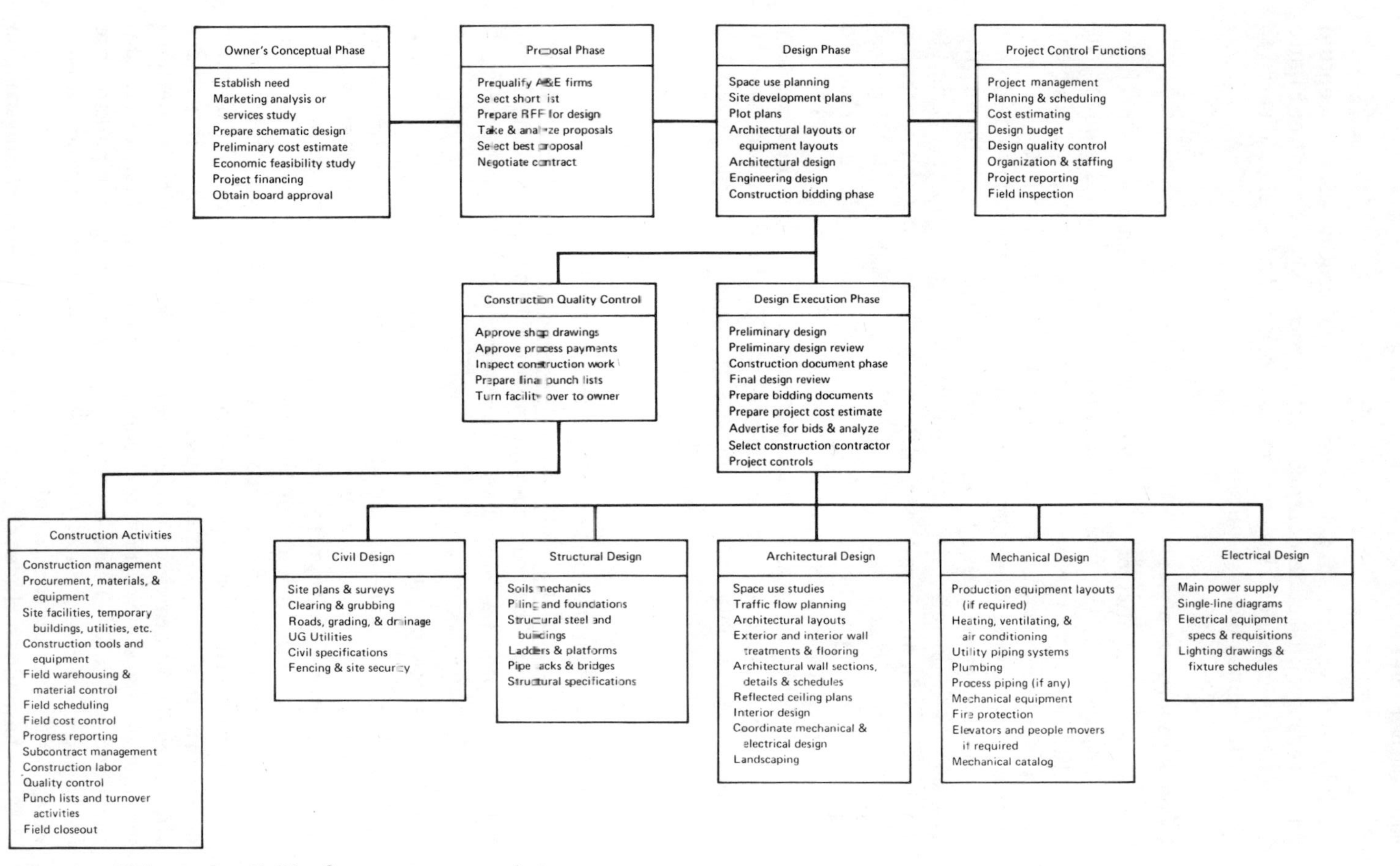

Figure 3.2 Major work activities for a nonprocess project.

work breakdown structure. It covers a typical project from conceptual phase through project acceptance in the process-type format of Figure 3.1. A brief verbal description of these work activities should serve to clarify the actual work required.

Owner's conceptual phase

The owner of a facility of this type must first determine the need for and the purpose of the facility. A market survey should support the need study to prove the project's feasibility. That is true whether the project is a commercial venture or a public service facility.

After quantifying the need and market, the owner should have a schematic design prepared to outline the main features of the facility. The schematic design forms the basis for a preliminary cost estimate which is used in an overall feasibility study to prove out the economics of the facility. In addition to the facility cost estimate, all other cost factors such as property, zoning approval, applicable construction codes, and operating costs must be considered.

The feasibility study is the main document used to get the necessary public or private financing for the project. The final step in the conceptual phase is obtaining the necessary approval of the governing bodies responsible for financing the venture.

Proposal phase

The proposal phase for this class of projects usually involves the selection of the design firm, since the contracting plan calls for separate construction bids. The proposal and contracting procedure discussed in detail in Chapter 2 also applies to making this selection.

The design execution phase

The work to be done in the design phase appears in the scope-of-services section of the design services contract. The first step is usually a preliminary design to expand the schematic design into more detail. The detailed review of the preliminary design sets the basic project design parameters before starting the construction drawings.

The construction drawings phase consists of producing the detailed design drawings and specifications for building the facility. It includes site development drawings, architectural layouts and details, mechanical and electrical drawings, and the structural design for all foundations and buildings. These documents form the basis for making the construction cost estimate (often referred to as the engineer's estimate) by the design contractor.

This phase also includes the preparation of the detailed construction specifications, general conditions, and the bidding documents for the solicitation of construction bids. The owner usually retains the design contractor to solicit and evaluate the construction bids. The A&E also prepares a slate of successful bidders.

Architectural design

In an architecturally driven project, the architectural group leads and coordinates the total design effort. Its work should begin with an information gathering or *programming session* to clarify the goals of the schematic design.[2] Its work covers the basic building design based on space use and traffic flow studies. The architects make layouts of all areas including equipment layouts if they are required. The architectural floor plans serve as the backgrounds for the structural, mechanical, and electrical drawings described below.

The architectural group also sites the building on the property based on the topographic and boundary surveys developed in the civil group. The work includes details of interior and exterior wall treatments, floor covering, wall sections, reflected ceiling plans, and door and window schedules.

The group also handles both the interior design of the occupied spaces and landscaping the grounds. It coordinates the preparation and printing of the general project specifications and the construction bidding documents to include those of the other design disciplines.

Civil design

If the facility under design is a civil works project such as a dam, highway project, bridge, or subway system, the civil design group leads the design effort. In that case, it prepares the layouts, writes the specifications, and coordinates the other design disciplines to get the complete design package.

Regardless of who is leading the project design effort, the civil group handles the design of the site work such as clearing and grubbing, site development plans, access roads, grading and drainage, fencing, and site plans. Its work usually includes any underground utilities as well as the civil specifications.

Structural design

The structural group handles all work related to the building frames, structures, and foundations. The work starts with the site borings and the soils report needed to design any subsurface structures, pilings, caisons, and foundations. It then proceeds to the building frames, whether they are of wood, steel, or reinforced concrete. Any required pipe or personnel bridges, ladders and platforms also fall into this group's work. All structural specifications and approval of shop fabrication drawings are done here.

Mechanical design

A mechanical engineering firm might lead a project involving light or heavy manufacturing if most of the work is mechanical in nature. In that case, the lead design firm would require some architectural input for the building shell. The basic design documents then consist of equipment layout plans, elevations, and sections along with any associated piping or materials-handling equipment.

Regardless of the type of project, the mechanical group also handles any heating, ventilating, air-conditioning and refrigeration, minor process and utility piping systems, plumbing, elevators, production equipment, and fire protection systems involved in the project. Any specifications for the above systems would also be prepared by this group.

Electrical design

Some capital projects can be primarily electrical in nature such as a major electrical substation, a rail line electrification system, or a high-tension electrical transmission line. In these specialized cases, the main design effort falls to an electrical group. More often, however, the electrical group supports the other disciplines in performing the electrical design portion of the project. Its work includes the main power supply, single-line diagrams, power distribution, grounding, and lighting systems fixture schedules. It also provides the electrical equipment and installation specifications.

Project control functions

As in the process-type project, the project control function plays an important part in having a successful project. The project management function on nonprocess projects is normally headed up by a project director, project architect, project manager, or project engineer, depending on the size and nature of the project.

The project controls group handles the project planning and scheduling, cost estimating, cost control (change orders), and project reporting. It is also responsible for preparing and monitoring the project design and construction budgets.

Project organization and staffing are a primary concern of the project management group. Another concern is for the quality of all work performed on the project. The project control group also sets up and controls the project review meetings in both the design and the construction phases if the design firm is inspecting the construction work. This includes the final acceptance of the work and the release of retention funds to the constructor.

For nonprocess-type projects, the moment of truth for the owner arrives with the opening of the construction bids. If the bids are much higher than the engineer's estimate, the owner may have to scale down the project to better match the financial plan. The project manager had better have a good explanation of how this condition arose to defend the project design effort and the cost estimate.

The construction phase

The construction phase is a very important part of any project, especially when it is done under the split contract format. Up to 80 percent of the total investment cost is spent in this phase, so cost and schedule control become even more critical than in the design phase. With the split contract mode, a new project manager takes over to carry out the construction of the facility. The new PM assumes the responsibility for planning, organizing, and controlling all activities listed in the Construction Activities block in Figure 3.2.

Having a complete design package simplifies the planning for this part of the work. It eliminates the schedule interface between the construction activities and the design activities. All the design documents are ready to go on day 1. The interface between procurement and construction still exists unless the owner or the design group has prepurchased the major equipment for the project. If prepurchasing has not been done, the construction PM must make purchasing the equipment and materials priority number 1, starting with the long-delivery items.

The invitation to bid stated the strategic end date for the construction portion of the work. During the bidding phase, the constructors should have made a preliminary plan for executing the work within the given date. If they feel the time allowed is too short, they should take exception to the bidding documents.

After the construction contract has been signed, the field activities for the process and nonprocess plants are quite similar. The field team has to organize the temporary facilities and services to support the field operations. The field personnel and subcontractors must be mobilized as called for in the schedule.

Cost, quality, and schedule control systems are required to control field operations. The field project manager (FPM) must issue a project procedure manual covering the working rules for field operations.

The nonprocess project usually does not have a performance test, so the project closeout and turnover are a little simpler. The main problems closing out a nonprocess project revolve around the clearing of the final punch list. That is a list of unfinished items prepared by the contractor, the design inspection firm, and the owner. All the punch list items must be cleared before the owner will accept the project and release the retention. When the project is accepted, the contractor's PM closes the field operations and leaves the site. The warranty period for the facility starts at this point.

Summary

In this chapter we have discussed the project manager's role in the overall project planning effort. Sound planning is the cornerstone of effective project management. Planning the broad range of activities involved with any type of capital project is essential to meet your project goals. No one can be *lucky* enough to have a successful project without a well-conceived project plan!

References

1. Arthur E. Kerridge and Charles H. Vervalen, *Engineering and Construction Project Management,* Gulf Publishing Company, Houston, Tex., 1986.
2. William Pena, *Problem Seeking—An Architectural Programming Primer,* 3d ed., AIA Press, Washington, D.C., 1987.
3. Peter Hall, *Great Planning Disasters,* Weidenfeld and Nicholson, London and New York, 1980.

Case Study Instructions

1. As the owner's or contractor's project manager, prepare a detailed outline of the proposed project execution plan based on using a turnkey prime contractor to perform the design procurement and construction for your selected project. Prepare an alternative plan using a third-party constructor with design and procurement done by the design firm.
2. As the owner's or design contractor's project manager, prepare a detailed outline of the project execution plan based on a typical split contract with a public bid-opening procedure. The construction contractor will also perform the procurement. Now put on your construction project manager's hat and do the same thing for the construction phase.
3. Make an evaluation of the overall time you feel it will take to execute your selected project by using the three different execution plans. Make a comparison chart showing the key milestone dates

for the three cases. Were you able to maintain the strategic planning completion date in all three cases?

4. Using a copy of The Project Services Checklist, Appendix B, fill in the blanks for the services required on your selected project.

Chapter

4

Project Scheduling

In the preceding chapter we discussed the critical project master planning effort, which lasts a relatively short time at the beginning of the project. In this chapter we will discuss the project scheduling effort, which will be with us throughout the project life cycle.

The scheduling people are responsible for the detailed project scheduling effort, which is based on the project execution plan. The project schedule puts all the work activities onto a time line beginning at the project start date and ending at the completion date. The project scheduler should participate in the master planning sessions, so he or she is conversant with the master plan and the project goals.

Start the scheduling effort with a kickoff meeting chaired by the PM and attended by the design, procurement, construction, and project control group leaders. A review of the master plan should head the meeting agenda to assure that everyone has the latest thinking on the project execution.

Then turn the meeting over to the scheduling people to gather the detailed information that they need to prepare the schedule. Now the PM's role is to see that definite assignments and dates are set for the information flow. The project manager acts as the enforcer on this occasion.

Because of the inherent differences in content and approach for process- and nonprocess-type projects, it will again be necessary to discuss scheduling in two different contexts as we did in Chapter 3. The basic underlying scheduling techniques are the same for all types of projects. However, there are some nuances in their applications which make separating the discussion more acceptable to PMs in the various specialized capital project arenas. This time we will start with a discussion of scheduling a nonprocess-type project.

Scheduling a Nonprocess Project

For clarity, let us define a nonprocess project as one which falls in columns 5 through 9 in Figure 1.1. These projects are normally done by an A&E firm in which the principal design discipline may be architecture or one of the engineering disciplines. The contracting basis for these projects sometimes include design, procurement, and construction operations, but, more than likely, the work will be split into separate design and construction contracts. The discussion here covers a project with a total scope of design, procurement, and construction. Obviously, when dealing with a project of lesser scope, the reader will use only the portions of the discussion which apply to that type of project.

Major activities to be scheduled

- 1.0 Schematic design phase
  - 1.1 Programming session
  - 1.2 Schematic design preparation
  - 1.3 Approval of the schematic design
- 2.0 Design development phase
  - 2.1 Principal design discipline
  - 2.2 Other key design disciplines
  - 2.3 Code review and approval
  - 2.4 Approval of design development phase
- 3.0 Construction document production phase
  - 3.1 Construction drawings
  - 3.2 General specifications
  - 3.3 General conditions and bidding documents
  - 3.4 Construction cost estimate
- 4.0 Procurement
  - 4.1 Historic delivery dates
  - 4.2 Vendor promised delivery dates
  - 4.3 Schedule long-delivery equipment and materials
- 5.0 Construction bidding and project follow-up
  - 5.1 Advertise for bids
  - 5.2 Receive and analyze bids
  - 5.3 Approval of recommended bidder
  - 5.4 Negotiate contract
  - 5.5 Issue notice to proceed
  - 5.6 Approve shop drawings
  - 5.7 Inspection of construction
  - 5.8 Approval of progress payments
- 6.0 Construction phase

- 6.1 A&E, owner, and contractor(s) kickoff meeting
- 6.2 Contractor mobilization
- 6.3 Prepare detailed construction schedule
- 6.4 Buy equipment and materials
- 6.5 Open construction site
- 6.6 Execute construction
- 6.7 Punch list and completion phase

7.0 Project closeout

- 7.1 As-built drawing changes
- 7.2 Project acceptance
- 7.3 Release of final payments
- 7.4 Close project files

This list covers the major activities that must be scheduled for any project. Each one of these activities is a macro activity that breaks down into the other submacro work activities listed in Figure 3.2. These submacro activities in turn break down into individual tasks which may result in a list of several hundred to several thousand scheduled activities.

Since we are only interested in how the scheduling activities affect the project manager, our discussion will deal only with the macro-subjects in the list of major activities to be scheduled. Readers interested in more detailed scheduling procedures can find more information in the reference material listed at the end of this chapter.

Schematic design phase

The schedule for the schematic design phase should allow time for a brainstorming session to make sure that the design team really understands the owner's needs and goals. One approach to handling this type of session is well described in a book called *Problem Seeking, an Architectural Programming Primer.*[1] These programming sessions can last for 1 day or as long as a week on larger projects. In any case, some form of information exchange session is necessary at project initiation to eliminate false starts and major rework on this critical phase.

The schedule input for the schematic phase should include all of the disciplines contributing to the design effort. The scope of the deliverables planned for this phase must be clearly defined before we can accurately schedule them. The scope of the schematic design phase should only cover that work needed to set the basic parameters of the project. If one of the parameters is cost, adequate definition for a suitable cost estimate will be necessary.

Allow some time for approval and minor rehash of the schematic design phase by the owner. If possible, get the owner to agree to a time limit for the approval process. Progress on the design will be essentially stopped during the approval phase unless the owner gives some form of limited go-ahead to the design team. The approval process is on the critical path, so any delay over the scheduled time will extend the completion date accordingly.

Design development phase

In this phase the project is developed in further detail by the principal design discipline along with the input of the supporting design disciplines. Major decisions affecting project cost occur during this phase, so a preliminary project cost estimate may also be in order. All disciplines making a contribution to the work must be consulted in setting up the elapsed times for this part of the schedule.

The review of the development phase by the governing zoning, architectural review boards, and building departments can be difficult and time-consuming. Therefore, start these activities as early as possible to prevent schedule slippage due to approval delays and any mandated redesign. If the schematic design work meets the approval requirements, start the approval procedure using the schematic material.

Final approval by the owner can be fairly routine at this phase if there are no surprises in the cost estimate. If major redesign does occur at this point, the schedule must be revised accordingly. As with most approvals, this activity is on the critical path, so any delay will ripple throughout the design schedule.

Construction document production

This is the area where we expend most of the design effort because all the design disciplines are involved. Design subcontractors working outside the offices of the principal design firm are also spending money. Give special attention to integrating their activities and milestone dates into the overall schedule.

The bulk of the work in this section consists of making the detailed drawings and specifications for the construction trades involved. List each document and schedule it separately while allotting time for preparing, checking, approval, and updating. Schedule basic design drawings such as floor plans, layouts, and plot plans for early issue to permit the later design work to proceed on schedule.

Near the end of the design work, begin preparing the construction bidding documents and the construction cost estimate. Often, this work is done concurrently with the owner's final approval of the construction documents to save time on the overall schedule. Meeting the scheduled bid advertising date (a key milestone on the critical path) is vital to avoid delaying project completion.

Procurement

If the scope of work includes procurement of equipment and material items, we must consider the procurement and delivery times in the overall project schedule. Base the first-pass schedule on historical delivery dates for the purchased items, and change to vendors' promised delivery times when they become available.

As with any design, procure, and construct contracting basis, the procurement must be sandwiched between the design and construction phases. The field-required date is the critical date for procurement. By working backwards from that date through the procurement phase, we can set the late start date for the design work on that item. Special attention is always necessary for long-delivery items which we know will appear on the critical path.

Procurement of the physical resources for any capital project plays a key role in the preparation of the project schedule because it affects the completion date. The PM must consider the effect of procurement on the completion date whether the procurement falls in the design phase or the construction phase. My experience has been that many project managers do not give enough attention to the scheduling and execution of the procurement program. Since it is impossible to construct the facility without the physical resources, late deliveries will result in a missed schedule.

Construction bidding and project follow-up

This phase signals completion of the design work and the start of the construction phase. By entering the bidding action, the construction contractors start the construction phase. In responding to the bid, contractors must become familiar with the project and prepare a preliminary construction schedule. This will be the first confirmation (or denial) of the preliminary construction schedule made during the design phase. Since the completion date appears in the contract, any disagreement between the strategic completion date and the successful construction contractor's scheduled completion date must be resolved before letting the contract.

As mentioned earlier, progress on the project is minimal during the bidding phase. Any unexpected delays during the bidding stage will result in a direct extension of the project completion date. Delays can come in various shapes and sizes such as issuing a series of design change addendums, requests for extensions by the bidders, and failure to meet the bidding schedule.

Perhaps the most serious delay encountered during this phase is having to throw out the construction bids for some reason. The most common reason is the bids coming in over estimate and thus upsetting the project financial plan and necessitating major redesign to meet the budget. In public work, failure to meet the voluminous governmental regulations can cause rejection of the bids on purely legal grounds. This sort of major glitch in the bidding proceedings can cause a nearly catastrophic upheaval of your otherwise well-planned schedule.

After qualifying the bids and selecting the successful bidder, the owner must approve the selection and proceed with executing the contract. Use of an AIA (or equivalent) short-form construction contract should allow these activities to proceed without delay. But remember, nothing concrete is going forward on your project during this period until the contractor receives a notice to proceed or a letter of intent. When that happens, the clock starts running on the construction schedule and the project is moving forward again!

The scheduling action now shifts to the construction work, and the design activity goes into a coordinating mode. There are, however, several schedule-sensitive activities performed by the design team which affect construction progress, so be sure to follow up on them. They consist of such items as approval of shop drawings, resolution of design interferences, approval of construction progress payments, timely inspection of construction work, and any other construction schedule-sensitive activities.

Virtually all the activities in this part of the work affect activities on the critical path. Therefore, they take on even greater significance than their seemingly innocuous role in the overall project. Also, they occur during a major transition period on the project, so they often fall between the cracks during schedule preparation. As PM, it is your duty to see that this does not happen on your project!

Construction phase

On a split design-and-build project, the main thrust of the project management will shift from the design project manager to a construction project manager. The design project manager shifts into the role of owner's representative responsible for monitoring construction progress and quality but with no day-to-day responsibility for construction operations. The responsibility for construction operations is in the hands of the contractor's project manager.

On integrated design-and-build projects, the project manager is customarily responsible for the construction operations from the standpoint of cost, schedule, and quality. The construction PM is assigned to the project by the construction department, but reports *functionally* to the project manager. We will cover this subject in more detail in the chapter on project organization.

The construction department or contractor phase involves a greater commitment of human, physical, and financial resources than any other part of the project. That puts even greater pressure on performing a good job in scheduling this part of the project. Since the owner pays for this major commitment of resources, it must give a lot of attention to developing an effective construction schedule.

The construction contractor is responsible for generating the detailed construction schedule for the construction activities from start to finish. The construction scheduling picture is more complicated when the owner lets individual contracts for the general construction, mechanical, and electrical trades. In that case, it is wise to pay the general contractor a fee for coordinating the work of the other trades to smooth the construction process. Coordination of the other trades should include supervising the overall construction schedule to ensure integration of the construction activities. If owners do not take this action, they must assume the overall construction schedule supervision themselves. If schedule supervision is lacking, all sorts of loopholes for the contractors missing their schedule dates will open up.

If procurement of the project equipment and materials is part of the construction contract, the procurement activities have to start immediately. Again, the longest-delivery items must receive special attention for early order placement to avoid construction delays. Procurement activities shall, of course, proceed concurrently with the start of construction activities. The contractor is generally most willing to start procurement promptly to protect the prices used in preparing the lump sum bid.

Mobilization of the construction facilities, equipment, and labor takes time. The contractor must evaluate them in relation to the accessibility of the site and utilities when scheduling this activity. Depending on project size, this activity can range from driving a pickup truck onto the jobsite to building extensive temporary construction facilities. Adverse weather conditions at the site are also a factor to consider in scheduling the mobilization.

Scheduling major construction operations must be done in enough detail and in coordination with all trades to develop a workable document for controlling all phases of the work. The construction team uses the overall schedule as a guide to prepare the weekly detailed field schedules to control the day-to-day site activities.

The completion date for the construction occurs when the owner signs off on the acceptance of the project, but interim dates of *beneficial occupancy* or *substantial completion* also are important milestones on the schedule. These terms mean that the owner can move into the facility and start to use it before completion of the final cosmetic touches. Early occupancy allows a little more time to carry out the project closeout activities listed under 7.0 of the major activities to be scheduled.

Scheduling a Process-Type Project

Process-type projects are usually handled by engineering-construction or A&E firms with a strong process design department. Process projects are driven by process, piping, and instrumentation engineering disciplines with the electrical, civil, structural, and architectural groups playing significant supporting roles.

The contracting basis for process projects usually includes design, procurement and construction in a single contract to reap the benefits of a fast-track project schedule. In recent years owners have opted to alter the total project concept with the introduction of third-party constructors to gain the cost benefits of open-shop construction. In that case the procurement is usually split between the design firm and the construction firm, with the former buying the process equipment and the latter the bulk materials. This allows retaining the fast-track schedule while offering the owner a potential cost saving.

Use of a third-party constructor introduces some additional intercompany coordination to coordinate scheduling and scope activities. A detailed procedure manual covering the interface between the two major contractors is a must to prevent problems in this area. We will discuss that procedure later under project controls.

Let us start the discussion of scheduling a process-type project examining the major activities the PM is responsible for on this type of project.

Major Activities to Be Scheduled

The major areas to be scheduled, who makes the input, and the overall effect on the success of the schedule are worth looking into. For the detailed work breakdown structure for each of the activities listed below, refer to Figure 3.1. We discussed a nonprocess-type project earlier in this chapter.

- 1.0 Engineering phase
  - 1.1 Process engineering
  - 1.2 Detailed engineering
  - 1.3 Design documents schedule
- 2.0 Procurement activities
  - 2.1 Historical delivery dates
  - 2.2 Vendor-promised delivery dates
- 3.0 Construction activities

3.1 Overall construction schedule
3.2 Detailed field schedules
4.0 Overall project schedule
4.1 Target completion date
4.2 Plant start-up sequence
4.3 Project execution philosophy
4.4 Scheduling philosophy
5.0 Proposal schedules

We recognize the three major areas of project effort in the engineering, procurement, and construction (EPC) activities at the beginning of the list. The overall scheduling requirements appear in paragraph 4.0. Each of the major EPC sections needs its own working schedule for monitoring and controlling the activities performed in that section. The overall project schedule results from the integration of all the subsections. The integration process provides an opportunity to handle any interfacial coordination among groups.

Let us examine each of the listed areas in more detail so we can see how the various major sections interact. We can also point out some of the more sticky interfaces where potential scheduling problems are most likely to arise.

The Design Phase

After the contract award, the scheduling activities of the design phase are of prime importance because they require immediate action. The details of the facility will not be well enough defined to make a meaningful final construction schedule at this stage. Naturally, if your project is for design only, this is the only part of the project which you will be scheduling in detail.

Engineering Schedule

The major subdivisions of the engineering design for a process plant which require scheduling are process engineering, detailed engineering (construction drawings), and the design document schedule. Figure 3.1 lists the work activities in each engineering discipline box. Each of the disciplines makes a detailed schedule for controlling the work in its respective area. Let us examine how the various pieces of the engineering schedule fit together and interact.

Process engineering

The process group supplies vital basic design data to all the general engineering disciplines. Therefore, we must closely schedule the process activities within the limits of the human resources available. Figure 4.1 shows graphically the dependence of the start of the detailed design disciplines on the output of the process group. Very little detailed design can begin before process issues the process flow diagrams and equipment data sheets.

The key item to look for in scheduling process work is the date the owner or licensor will issue the process design package. The process design package contains all of the basic information developed on the project so far. It can be in the form of a design book developed by the owner's R&D group or a basic design package purchased from a process licensor. Nothing much can move on the process design work until this information is available. If we are fast-tracking the project by

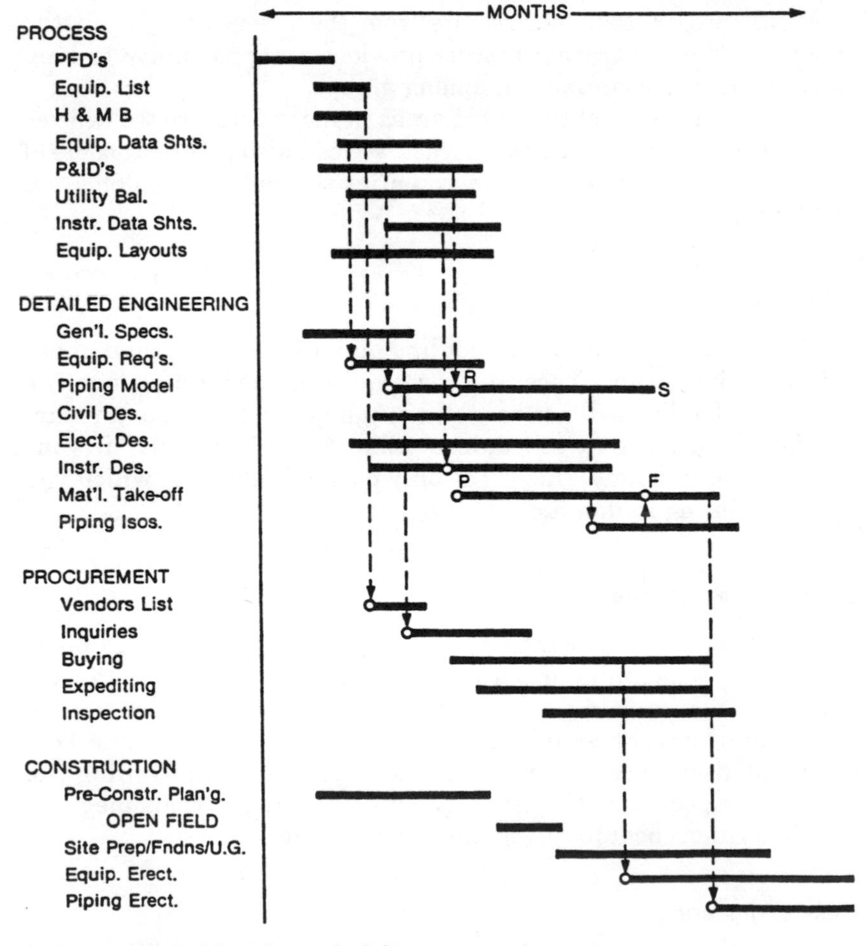

Figure 4.1 Typical bar chart schedule.

releasing the data piecemeal, the flow must be continuous until all of it is available or the schedule will suffer. If the process design package is developed in-house, add the development time to the front end of the process design schedule.

The process schedule is a key area where many engineering schedules get into trouble. Process groups are notorious for failing to keep their schedules, which in turn makes the succeeding design activities late. The problems with keeping the process schedule usually fall into two areas: (1) shortage of personnel and (2) lack of management skills among process group leaders. Over the years (until recently) there has been a chronic shortage of chemical and mechanical engineers, who form the backbone of the process department. Often two or three projects are breaking at the same time, so scheduling the limited talent becomes difficult. Make sure you have the necessary process staff to meet the proposed schedule!

Many technically qualified process engineers get promoted to process group leaders despite their lack of management training. That is a prime cause for missed schedules and budgets. The only recognized cure for this is better management training for the group leaders and closer supervision from the project management team.

In defense of the process group, however, let me also say that they are always under heavy schedule pressure to get the follow-on work going. Also, the process design area is subject to the greatest number of changes in the early stages of project development. The process area has the largest potential of any design area for saving or wasting money. We will discuss this fact in later chapters devoted to cost control. The input by the owner's project team plays a very significant part in the success of meeting the process design schedule, so don't overlook the owner's process obligations.

Another point to remember when scheduling the process work is that the front end work must reflect the needs of the later activities, such as delivery of long-delivery items and the start-up sequence of the plant. Schedule the long-delivery items out of process design first to get them on order as early as possible. Often this equipment must be ordered by using a preliminary specification which is updated later as more exact definition becomes available.

Sometimes the owner may start purchase of that type of equipment before letting the design contract to get it into the vendor's production schedule early. After letting a contract, the owner usually turns the order over to the contractor for completion of the delivery. Some examples of the equipment are paper machines, complex chemical reactors, or complex automated production lines.

To keep the process work on a fast track, it is not absolutely necessary to deliver the complete process design package as a unit. As sections of flow diagrams, equipment lists, and equipment data sheets become available, make preliminary releases so other disciplines can start their work. However, be sure that the process work is finished before issuing any detailed design documents for construction.

Make the process schedule by listing all the work activities in their order of occurrence. Then assign the estimated labor-hours required to accomplish each one. Only those people who can be productively employed on each operation are assigned to the task force. Knowing the number of personnel and the estimated time required to complete each activity, we can estimate the elapsed time to complete each operation. The schedulers enter the elapsed times into the CPM schedule.

Again, let me stress the critical importance of this part of the schedule and the ripple effect that any late issues will cause throughout the project. As we will learn later in this chapter, schedule float is a very valuable commodity. Do not squander it early in the project!

Detailed engineering

Detailed engineering covers all the other design disciplines on the project, such as mechanical, piping, CAS, electrical, and instrumentation. These disciplines prepare the construction drawings and specifications for building the project. Each design discipline on the task force must make its own detailed schedule for controlling its performance against the schedule. Each discipline schedule is then integrated into an overall engineering schedule. The integration process should consider any areas where the work of the various groups is interdependent.

Although starting the main work in the design disciplines depends on the output of the process group, there are a few independent activities that should start immediately. Some of these areas of early work are the property and topographic surveys, subsurface exploration and soils report, overall site planning, design code investigations, and the general specifications. The general specifications set the basis for the quality of the facility and thus affect its overall cost.

Coordination with any outside agencies, such as railroads, local government officials, utility sources, and environmental permits, should start in the early stages of the work. Quite often an early site preparation contract is desirable to beat impending seasonal bad-weather situations.

Any work on the long-delivery equipment and material items should be given priority scheduling as soon as data become available from process. Follow these items closely through order placement to avoid losing time saved in earlier stages.

The design document schedule

A few years ago we used to call the design document schedule the drafting room schedule, but with the switch to computer-aided design (CAD), the drafting room has virtually disappeared. It has been replaced by CAD terminals located in dimly lit cubicles, each with an operator hunched over an input terminal in front of a CRT screen and operating a mouse.

Despite the new drawing method, it is still necessary to list each major document required to define the work. Each entry must have an estimated time for completion, a scheduled start date, about four columns for showing percent complete, and a scheduled completion date. Each of the listed entries should have double spaces in the column to enter the actual start and completion dates, along with the scheduled dates.

The Estimate to Complete column is important for the design supervisors to enter the time estimated to complete the document. The computer compares the new total hours to complete to the original budgeted hours. Any difference appears in a Variance column, which shows whether there is a projected over- or underrun. In addition to using this document to control the budget, we can also use it to calculate the overall percent completion of work in progress. We will go into the use of the Variance column later in project control.

Personal computers using standard spreadsheet software are excellent tools for creating and updating the document control schedules. The main source of problems in using this method is the failure of the design supervisors to accurately forecast the estimated hours to complete. Consistently underestimating the time to complete can lead to serious and disconcerting overruns of the budgets and schedules for that phase of the work. Project managers must give this area continual surveillance to avoid nasty surprises at the end of the design work.

The detailed engineering and the design document schedules require close coordination with the procurement activities. Prompt action by procurement to supply the vendors' data to the design groups is especially critical. One of Murphy's laws is that "Any design effort will be vendor-data starved."

As the detailed design effort feeds on the process data at the front end of the schedule, so it also feeds on the vendor data in the middle and at the end! PMs are hereby warned of another place that design schedules usually slip. I cannot overemphasize the criticality of expediting the vendors' data.

Procurement Activities

The procurement schedule shows expected times for vendor inquiry, bidding, bid analysis, ordering, and delivery of the equipment and material. The procurement schedule conforms to the procurement plan developed as part of the project master plan.

The first pass on the procurement schedule uses historical delivery dates of various types of equipment and materials, considering existing market conditions and previous experience. After order placement, the schedule is updated with vendor-promised delivery dates. These dates are in turn reviewed and revised periodically through expediting reports as manufacture of the purchased items progresses.

The procurement schedule starts with the *field-required date* necessary to support the construction schedule. By working backwards from that date, we can determine the latest order date for the material. It is wise to include some safety factor for possible slippage of a supplier's promised delivery date. Items with average delivery times are the ones which will show a normal float in the CPM critical item sort.

Items with the least float must receive top priority and attention when preparing this phase of the overall schedule. The long-delivery items will always fall on the critical path. Therefore, they must have top priority all the way back through detailed engineering and process design.

An important point for PMs to remember is that equipment fabrication does not start until placement of the order and approval of the shop drawings. After that, a good expediting program with progress reports (particularly on long-delivery items) will assure meeting your goal of getting the physical resources to the field on time.

Again, I want to stress the importance of the joint procurement and engineering effort needed to assure the availability of vendor data during the design effort. Specific actions are needed to thwart Murphy's law that the design effort will be vendor-data-starved. Expediting vendor data must start automatically after order placement.

Construction Schedule

It is imperative to base the logic of the construction schedule on the execution philosophy developed in the project master plan. The construction schedule should also reflect the start-up sequence of the various units of the total facility. Naturally, units that start up first must finish first.

In the early stages of the design phase, the construction schedule must be broadbrush, since most of the construction details are still undefined. At this point, a macro approach is necessary just to test the feasibility of the strategic end date for the project.

At the 30 to 40 percent completion stage of the design phase, a more detailed construction schedule is possible. The new schedule uses information gathered during the constructibility analyses and from the preliminary plans and specifications. This is the time to open the field operations on a fast-track schedule for a turnkey contract.

It is best to base the construction schedule on the input of the people who are going to manage the construction operations. They will calculate the elapsed times for the various construction operations on the basis of standard historical labor productivity rates. They should use a productivity factor applicable to the exact site location to fine-tune the labor rates. Although this master construction activity schedule is part of the overall project schedule, it is not suitable for controlling the day-to-day operations in the field. For that reason I have shown Detailed Field Schedules as item 3.2 in the above list of major activities to be scheduled. The detailed field schedules break the overall construction schedule activities down into itemized bar charts. The bar charts are used to control the field work as laid out in the weekly construction planning meetings.

Overall Schedule

The first pass of the overall project schedule results from integrating the individual schedules into one schedule. To check the feasibility of the overall schedule, start at the target completion date and work the major activities back toward the front. Use the unit start-up sequence to set the unit priorities throughout the schedule. Units that come on line first must receive top scheduling priority.

Since the long-delivery items are on or near the critical path, test them against the target completion date to see if they fit. If the long-delivery items do not fit on the first pass, check alternatives to improve delivery. If suitable delivery is not possible, you may have to discuss revising the target date with top management.

Shortening the critical path to suit the strategic completion date usually costs more money. You should evaluate these extra costs and present them as part of the proposed solution to the scheduling problem. Occasionally a schedule might show that the target date can be pulled forward. Before making such a recommendation, be sure to retain a reasonable float in the schedule.

After resolving the major problem areas, circulate the overall schedule through all the project groups for review and comment. After incorporating all the review comments into it, circulate the schedule to the project disciplines, the owner, and the project managers for signature. This signifies that everyone understands the *project time plan* and agrees that they can meet the schedule requirements at the times stated. The signature stands as evidence that the participants cannot later plead ignorance of the time plan!

As the project progresses, job conditions may force schedule revisions. We may have to change some key milestone dates to meet the new conditions. An example of such a change could occur when the final construction schedule is issued later in the project. However, do not make significant changes to the original schedule without agreement of the original signers. Keep a copy of the original master schedule intact as a record copy to compare with the current schedule.

Project execution and scheduling philosophy

The project execution and scheduling philosophy are two items listed under the overall schedule category in our list of major activities to be scheduled. The project execution philosophy is actually the project execution plan, which we discussed in Chapter 3. This plan lays down the basic ground rules for project execution. It also answers such questions as: Who will supply the process design? Will engineering be subcontracted? Who does procurement? Will we subcontract construction? Each answer affects the schedule format.

The term "scheduling philosophy" refers to the selection of the scheduling system. For example: Will we use bar charts or CPM? How often will the schedule be cycled? What is the project scope? How many activities are there?

Before starting the project schedule, we must set the project execution and scheduling philosophies. This allows all the normal questions, such as listed above, to be decided during the first scheduling meeting. It is the PM's responsibility to establish these two philosophies and get top-management approval. Changing either of these philosophies later can be very expensive and disruptive to the project work.

Proposal schedules

I have included the subject of proposal schedules in the list of scheduling subjects because they often figure in the development of the project schedule. Proposal schedules, as the name implies, first appear during the inquiry and proposal stage. These schedules are almost always broadbrush efforts which show a sketchy interrelationship of the major project parts and their start and end dates. Schedules submitted in proposals are often overly optimistic because the contractors are anxious to sell their services.

Proposal schedules sometimes get caught up in the contract negotiations and take on more credence than they deserve. As PM, you should be careful that these preliminary schedules do not gain universal acceptance as the real project schedule. Proposal schedules may, however, sometimes be useful as guides (with qualifiers) before the first-pass project schedule becomes available.

Scheduling Systems

Now we are ready to discuss the various methods for scheduling capital projects which are available to us. This discussion will not be in enough detail to turn you into a trained scheduler, because we are looking at scheduling only from the viewpoint of the project manager. We just want to make sure that you get a good understanding of the basic theory involved to use schedules effectively in planning and executing your projects.

Many of the examples in this section are taken from *CPM in Construction Management*.[3] This volume, or one of those listed in the references at the end of this chapter, would be a valuable addition to your project management reference library. Attending a good seminar on project scheduling is another excellent way to increase your basic knowledge of project scheduling should you wish to learn more about it.

The two basic methods we will be discussing are bar charts and logic-diagram-based schedules. Both methods are used extensively and sometimes interchangeably in project work, and each has advantages and disadvantages. Knowing when to select the correct method is half the battle in successfully making and controlling your project schedule.

Bar Charts

In analyzing the history of project scheduling, it appears that preplanned written schedules came into use on capital projects in the early twenties. The forerunner to the bar chart was developed by an industrial engineer, Henry L. Gantt, for scheduling production operations during World War I. The name "Gantt chart" is still in use today to designate certain types of bar charts. It was sometime after World War I that bar charting was adapted to the scheduling of construction projects.

There appears to have been little formal scheduling done for the large capital projects developed over the preceding centuries. Since no schedules have been found among the documents in the Egyptian tombs, we must assume they were planned in the minds of the builders. The supervisors of the construction, however, had certain methods for improving schedule which we do not have available today. When things got behind in those days, the king just sent out his army to round up more slaves to get the work back on schedule. Present-day project managers cannot fall back on that practice to get their jobs back on schedule, as we will discuss later under the subject of labor relations.

Bar charts are the simplest form of scheduling and have been in use the longest of any systems we have available. They offer the advantage of being cheap and simple to prepare; they are easy to read and update; and they are readily understood by anyone with a basic knowledge of the capital projects business.

The main disadvantage of the bar chart is its inability to show enough detail to cover all the activities on larger, complex projects. On large projects, the number of pages required to bar-chart the project becomes cumbersome, and interrelation of work activities becomes difficult to follow from page to page.

Figure 4.2 illustrates those difficulties, as does Figure 4.1. By combining these bar charts for a process-type project, we can condense a list of macroactivities for the project on one page. Each of the macroactivity bars, however, really needs to be broken down into a series of individual tasks to control the work. That means each bar would in turn take another series of bars to show, for example, each of the 50 or so flow diagrams normally found on a larger process project. Showing the project in that sort of detail could easily require 15 to 20 pages of bar chart schedules to list all of

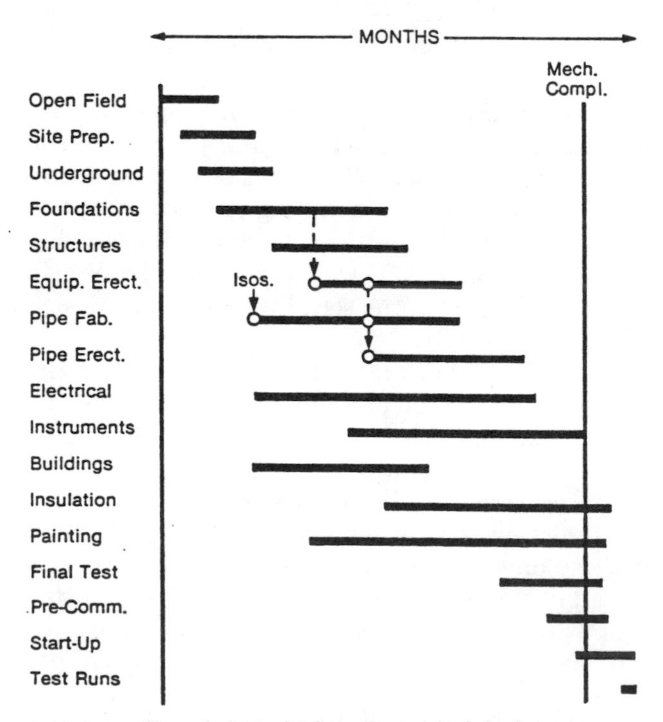

Figure 4.2 Second part of a bar chart schedule for a process type project.

the individual tasks. The problem could be solved by using a PC spreadsheet to record progress on each flow diagram and reporting only *overall progress* on the bar chart.

When we try to schedule a larger project in that sort of detail with bar charts, we quickly lose most of the advantages that we listed earlier. The schedule becomes unwieldy and difficult to interpret, and we run the risk of losing control of the project time plan.

I have tried to illustrate, in Figures 4.1 and 4.2, the difficulty of showing the need for completing certain activities before subsequent activities can start. These precedence requirements are shown by vertical dashed lines. For example, the preliminary issue of the P&IDs must be completed before the piping model can begin. This sort of notation makes the bar chart more cluttered and difficult to read.

I recall working in the late forties and early fifties as a project engineer on a megaproject which cost about $90 million, and I cannot remember having seen an integrated schedule for the project. Since no one ever mentioned a strategic completion date, I assume that we did the scheduling by a seat-of-the-pants approach similar to that used on the Pyramids! I am also sure that we would have finished that megaproject earlier if we had used a more sophisticated form of scheduling.

As the size and complexity of projects grew in the late fifties and sixties, finishing projects late became the rule rather than the exception. Late finishes, along with their associated cost overruns, caused increased pressure on owners and contractors to develop improved scheduling techniques.

Logic-based schedules

Fortunately, about that time the network schedule and the computer came on the capital projects scene. We now had a tool available to make the many repetitive calculations and a place to store and sort the data needed to control the large number of work activities.

As I remember it, the U.S. Navy and the DuPont Company independently developed two different logic-diagram-based scheduling systems at about the same time. The Navy's system was called PERT, which stands for program evaluation and review technique. Its first successful application was on the Polaris missile program. Scheduling that highly complex weapons system program with bar charts alone would have been nearly impossible.

At about the same time, DuPont first used its critical path method (CPM) of logic diagram scheduling on refinery turnarounds and rehabilitation projects, which required a minimum of downtime to be economical. The method proved to be so successful on the short-turnaround jobs that DuPont quickly used it on new construction projects as well.

Other owners and contractors lost no time in adapting these new scheduling methods to their projects to improve on their timely completion. The CPM system is simpler than the PERT method, so it soon became the system favored for use on capital projects. The KISS principle triumphed again!

Shortly after the introduction of the PERT/CPM systems in the early sixties, the pendulum swung to the side of overly detailed, computerized schedules. That did not work out as well as the early successes with the systems seemed to indicate. If a little bit of CPM was good, more had to be better! Everyone promptly started to schedule in too much detail on each activity. The result was reams and reams of computer output that virtually inundated many untrained users. I knew several old-time construction managers who happily tossed piles of computer printouts into the trash can after looking over the first page for a few minutes. Their only recourse was to run the job by the seat of their pants as before, with the expected mixed results.

This was an outstanding example of installing a new and complex system without first properly training the users. Most of the construction managers of that period were people who worked themselves up through the ranks. In many cases, they were untrainable in the new technology of computerized scheduling.

Fortunately, some of the users of the newly developed techniques remembered the KISS principle and developed some easy-to-use systems before the paper flood turned everybody off. Several good mainframe programs came on the market: MCAUTO's MSCS system, Metier's Artemis system, and IBM's PCS system, to name a few of the 100 or so programs offered. As computer capacity and new software bloomed, the programs developed and improved rapidly over the next twenty years.

In the eighties, personal computer hardware developed so rapidly that PCs have now virtually taken over the CPM scheduling market serving capital projects users. Mainframe computers are now required only on the very large projects.

Basic Network Diagraming

To understand how a logic network works, one must understand some basic network diagraming techniques. I will briefly cover some of the basic practices here, but I strongly recommend that you pursue the subject more deeply in the references listed at the end of this chapter.

The two systems which I will address here are the arrow diagraming method (ADM) and the precedence diagraming method (PDM), both of which are applicable to the CPM system. Most of today's CPM schedules use the PDM method because of its greater flexibility in using lead and lag factors, which are discussed later.

Logic diagram scheduling has its own terminology which has developed over the years. I have included a brief glossary of some of the more common logic diagraming terminology in Figure 4.3.

Term	Definition
PERT method	Program evaluation research technique
CPM schedule	Schedule using the critical path method
PDM schedule	Schedule using the precedence diagramming method
Arrow diagram	CPM diagramming method using arrows
Logic diagram	Arrow diagram of complete project or section of a project
Time-scaled chart	Logic diagram with a time scale
Activity	Any significant item of work on a project
Activity list	List of work items for a project; also work breakdown structure
Activity duration	Elapsed time to perform an activity
Optimistic time	Earliest completion: shortest time
Pessimistic time	Latest completion: longest time
Realistic time	Normal completion: average time
Activity number	Number assigned to each activity
Early start date	Earliest date activity could start
Late start date	Latest date activity could start
Float	Measure of spare time on activity
Free float	Time by which activity can be changed without affecting next activity
Total float	Total free time on any activity or project
Negative float	Time a critical activity is late

Figure 4.3 Glossary of CPM terminology.

Arrow diagraming

In arrow diagraming the basic unit is a work activity which occurs between two events or nodes. The events or nodes are numbered sequentially, and the activity is identified by the beginning and ending event numbers. These numbers are also designated as the i-j number as shown in Figure 4.4. The sample activity is designated by its i-j number 1-2.

Each activity has an elapsed time necessary to accomplish the work involved. The estimate of elapsed time for the activity must consider the scope of the activity and any historical data available from previous similar activities. If there are no historical data on the activity, the scheduler must use the best estimate (guess) of the value based on input from the people most experienced in performing that activity. The value estimated for the sample arrow in Figure 4.4 is 3.5. The unit of this value can be hours, days, weeks, or months, depending on the scale selected for the schedule. Most capital project time values are done in weeks, but short-turnaround schedules are often done in hours.

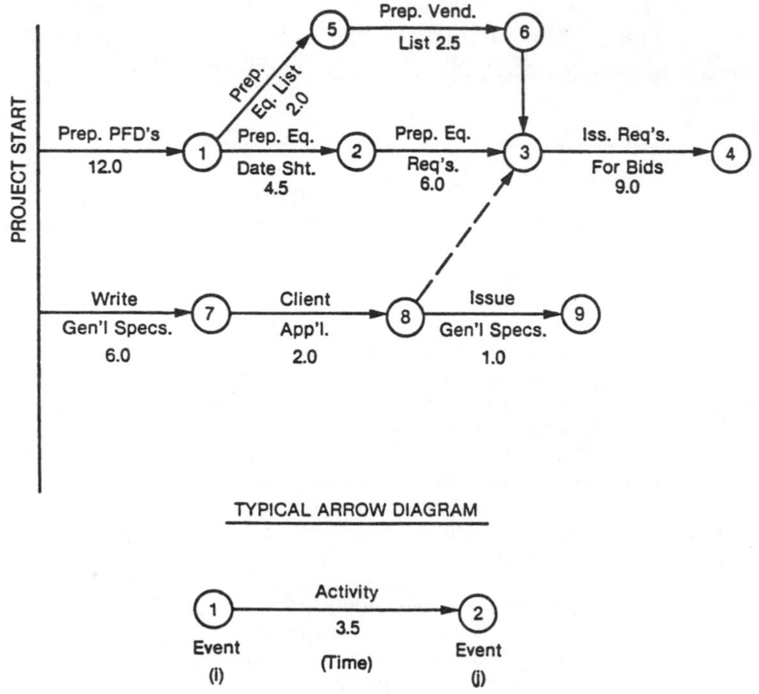

Figure 4.4 Basic arrow diagraming terminology.

In the PERT system the elapsed time is calculated by assessing an optimistic time and a pessimistic time and then calculating an average time. The computer program automatically calculates the average time. Most CPM programs do not bother with this averaging exercise but merely select the most likely time for the elapsed time input.

Obviously, the proper selection of the elapsed time for the work activities is an important part in preparing the CPM schedule, since it directly affects the critical path and all of the float times. That points up to the project manager the importance of getting the most educated guesses from the most experienced specialists available on his or her staff. What makes this method work is the averaging out of offsetting pluses and minuses over many work activities. Hopefully, we won't get all-pessimistic or all-optimistic inputs for our elapsed times. That could skew the schedule too far one way or the other.

After running the first pass on the computer (or manually), it is a good idea to recheck the elapsed times of the critical and near-critical activities to ensure that the estimates are reasonable. Also, consider any abnormal factors, such as adverse seasonal conditions, conflicting workloads, critical facility usage, shortages of material and human resources, and transportation times which could adversely affect an otherwise normal elapsed time estimate.

The construction of any network must start from a work breakdown structure (or detailed activity list) showing each activity to be scheduled. Sometimes this looks like an activity tree as one major item is subdivided into its major parts. For example, we might have an activity called equipment procurement and delivery which could be further subdivided into tanks, columns, and pumps. The columns could be further subdivided into small columns (short delivery), large columns (medium delivery), and field-erected columns (longest delivery).

By breaking out the longest delivery items in that way, we can see which part of an overall activity (column delivery) falls on the critical path and which parts have float. If we had several field erected columns with varying completion dates, we might even break that critical item into individual columns to track the latest one.

A comparison of the amount of detail covered in a CPM work breakdown structure and a comparable bar chart will clearly show the difference in the amount of detail covered by the two systems.

Returning to the arrow diagram example shown in Figure 4.4, let us start to construct an arrow diagram for some of the interacting work activities shown in the bar chart Figure 4.1. Beginning at the Project Start line, we add activity arrows proceeding from left to right. The goal of this partial diagram is event 4, issuing equipment requisitions for bids.

The work activities which precede that activity are the preparation of process flow diagrams (PFDs), equipment lists, vendor lists, equipment data sheets, requisitions, and general specifications. Since this is not a time-scaled diagram, the length of the line can vary to meet the needs of labeling and joining to the following event. The number selection is optional, the only restriction being not to use the same number twice.

Activities which must be done in series are drawn end-to-end. Parallel operations can start simultaneously from either the project start line or branch from a single-activity node as shown in activities 1-2 and 1-5 originating at event 1. In this case, we can start writing the general specifications while starting the PFDs, which are parallel operations. We then add the activity description and elapsed times in weeks to the activity arrows.

Sometimes an activity on one path restrains completion of an activity on another path, as in activity 8-3. This restraint is known as a *dummy activity,* and it shows as a dashed line with zero elapsed time. It shows us that the activity "Issue Requisitions for Bids" cannot proceed until the general specification documents are ready for use in the bidding packages.

To analyze the diagram for the critical path, we must add the elapsed times for each possible path from the project start to the last event as follows:

(1) Path 0-1-2-3-4	12+4 + 6 + 9	=31 days
(2) Path 1-5-6-3-4	12+2 + 3 + 3 + 9	=29 days
(3) Path 0-7- 8-3-4	6+2 + 9	= 17 days

Path 1 equals 31 days, which is the longest. That means the critical path for getting the equipment out for bids passes through activities 1, 2, 3, and 4. If we are to improve the schedule for issuing requisitions for bid, we must shorten the activities on path 1. If we were able to shorten path 1 by 3 days, we would have two critical paths of 29 days each. If we were able to shorten path 1 by 4 or more days, the critical path would shift to path 2.

Another way to describe the difference between paths 1, 2, and 3 is to say path 2 has 3 days of float and path 3 has 14 days of float. That also says activity 1-5 can be started 3 days later than activity 1-2 and still finish without extending the critical path of 31 days. Likewise, we could delay finishing activity 0-7 by 14 days. Also, the 3- and 14-day total floats could be spread over all of the activities on the noncritical paths.

Another form of logic diagraming notation is the *precedence diagraming method* (PDM), which grew out of the arrow diagraming method to overcome some of the ADM's faults. Most present-day CPM scheduling programs are PDM based systems.

Figure 4.5 shows the basic terminology for PDM diagraming. Three different notations are available to show a variety of start-to-finish relationships to fit any type of activity. This gives more flexibility to notation when building the logic diagram.

PDM shows the activity number in the block and not on the line as shown in the sample precedence diagram in Figure 4.6. Placing the activity in the block leaves more room to describe the activity in greater detail. The activity number also appears in the block as a simple code number instead of a multidigit i-j number. The elapsed time appears below the activity block so the activity connecting line is short. This makes the PDM diagram more compact.

Actually the PDM diagram in Figure 4.6 covers the same activities as the arrow diagram in Figure 4.4. I think you will agree that the PDM diagram is cleaner and easier to read than the ADM diagram. The strongest advantage of the PDM system is its ability to show lead and lag factors when an activity can start before the preceding one is fully completed. Figure 4.5 describes the lead and lag notations which makes PDM a more powerful scheduling tool.

Activity 120 in the PDM diagram in Figure 4.6 illustrates the use of a lead factor. In that case we show that the equipment requisitions (activity 300) can start before completing all the equipment data sheets in preceding activity 120. To show that potential for shortening

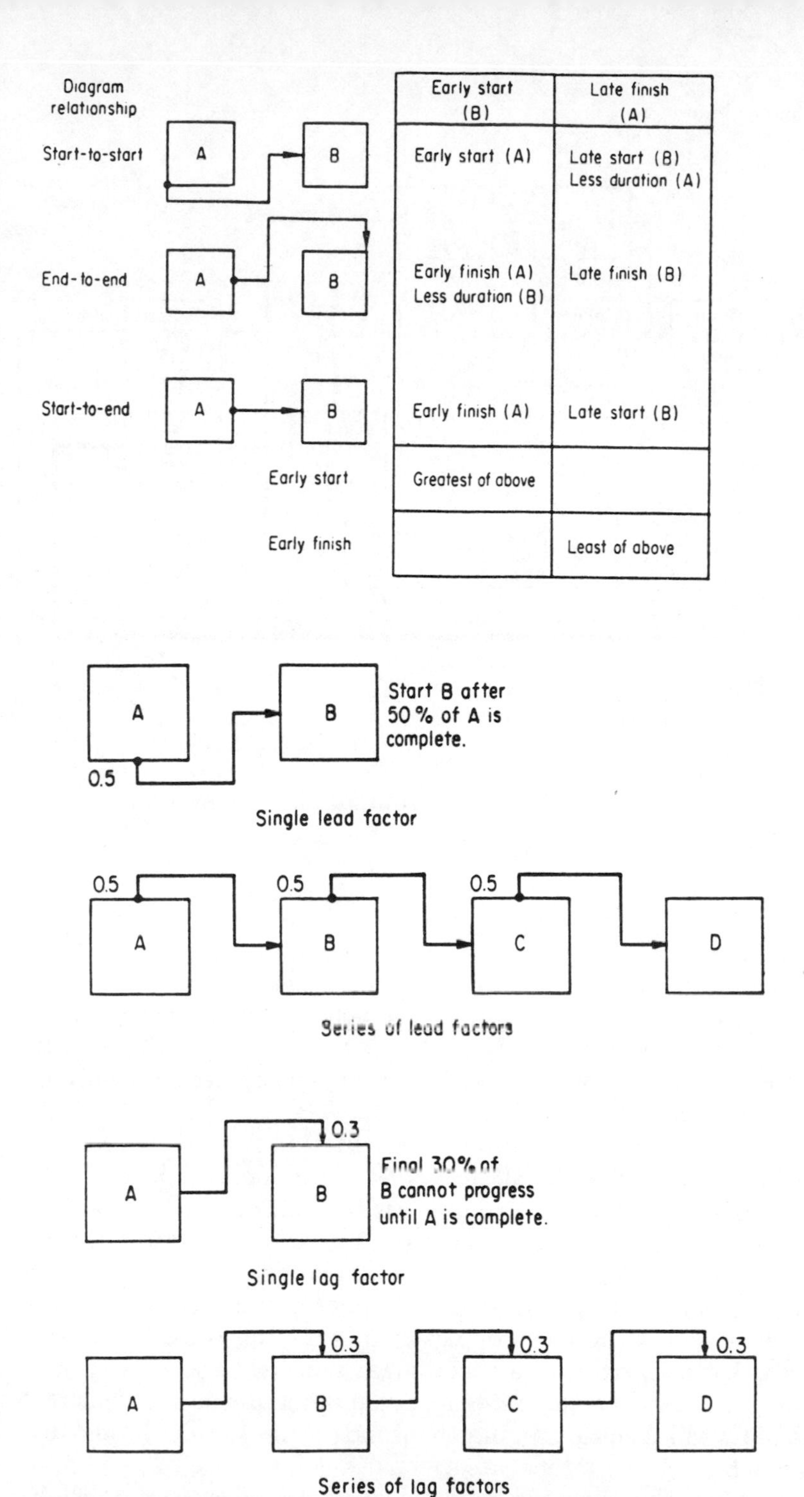

Figure 4.5 Precedence diagraming terminology. (*Reproduced with permission from* CPM in Construction Management, *McGraw-Hill, New York, 1984.)*

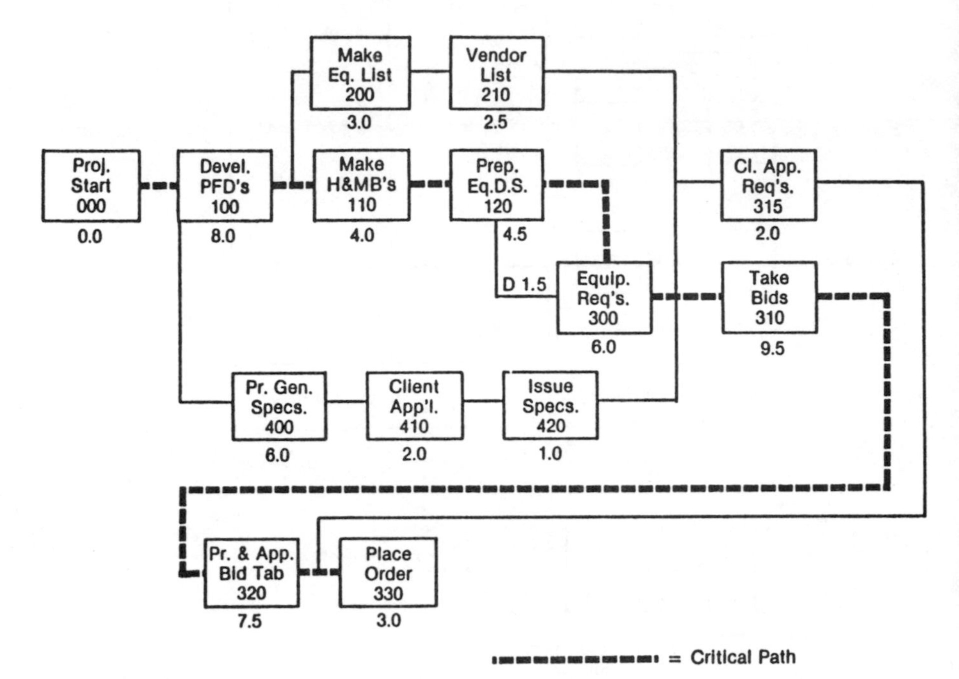

Figure 4.6 Sample precedence logic diagram.

the critical path on the ADM diagram, we would have had to break activity 1-2 into two arrows with an intermediate node.

A much more detailed account of precedence networking appears in Chapter 5 of *CPM in Construction Management*.[3] When referring to the book, don't waste time studying about the use of punch cards to input your data. That system has long since been superseded by direct input on a terminal.

Whichever the diagraming method, calculating float time and the early and late start dates for all activities in the logic diagram is the same. The calculation is quite simple but too voluminous to cover in this discussion. Mr. O'Brien covers the principles very well in Chapter 2 of the *Scheduling Handbook*.[2] I recommend that you study that chapter to learn how to calculate a critical path manually if you do not already know how. It becomes apparent in studying these early and late start calculations that they are indeed manifold and repetitious. That makes the use of a computer program mandatory in all but the simplest of CPM schedules. Doing thousands of these calculations by hand is much too labor-intensive.

I have reduced these sample logic diagrams and discussion to their simplest forms to give any nonscheduling readers an introduction to the workings of logic diagraming. My simple example is multiplied by up to several hundred times in a large project of several thousand activities. On larger projects the CPM logic diagrams can become large enough to cover the walls of a typical project management war room. The scheduling people like to have the complete diagram displayed so they can easily check the logic of the diagram in detail. The project managers may have need to refer to it occasionally to resolve particularly knotty problem areas. Usually, however, PMs work from the early and late start date and float-sort printouts for day-to-day control of the project.

Normally, less than 20 percent of the project activities will fall on the critical path, so CPM is a management-by-exception technique. When more than 20 percent of the activities fall on the critical path, it is a sign that the schedule is in trouble and needs to be recycled. Having activities with negative float also is a sign that the schedule is no longer achievable.

Recycling the schedule should not be confused with the monthly progress evaluation. Recycling might be necessary over a three or four month period when your short term goals are not being met. The recycling procedure addresses how to get the work back on track without extending the completion date.

Advantages and Disadvantages of CPM and Bar Chart Methods

An evaluation of the advantages and disadvantages of bar charts vs. CPM will allow us to select the most effective system for a given project. That will lead us into some simple rules that are applicable to the selection.

Advantages of CPM

The number 1 advantage of the CPM system is its ability to handle many work activities on complex projects with ease. Let me introduce a word of caution on this point: *Do not fall into the trap of using more activities than necessary just because it is easy to do so.* Your schedule can get bogged down in too much detail, which makes it harder to use and costs more money to operate. This is the same trap that almost killed the CPM system in its early days! One way to avoid this problem is to break out some of the less-complicated scheduling areas and use bar charts for them. They could be offsite areas such as small office buildings, warehouses, tank farms, and roads. A blend of the two systems can result in a simpler and more effective project schedule.

Another outstanding advantage of CPM is the intangible benefit of forcing the project team to dissect the project into all its working parts. That forces the early analysis of each work activity. The CPM logic diagram is actually a dry run for all phases of the project. For that reason alone, it behooves the PM to play a major role in planning the CPM schedule and checking the resulting logic diagram.

The actual scheduling phase, such as calculating the early and late start dates and the associated float, is best left to the scheduling technicians and the computer. It is usually necessary to run the first pass of the schedule several times to test and debug the logic diagram before the final version is ready for review and approval.

The large menu of output sorts is another big advantage of a computerized CPM schedule. It allows the various interested members of the project team to order the output sort best suited for their work. Most CPM programs will yield a sort menu as follows:

1. Total float per activity
2. Limited look-ahead sorts
3. Critical path sort
4. Critical equipment sort
5. Project milestone sort
6. Bar chart printout
7. Human resource leveling

Most PMs, for example, will find the sort of activities by total float most valuable for their needs. This sort starts with the low-float (critical) items listed first for immediate attention. The less critical high-float items show up later on the list. By using the look-ahead sorts, one can also home in on specific time periods. A 30-, 60-, or 90-day look-ahead sort will list only those critical items which will occur in the next 30, 60, or 90 days.

Other members of the project team will need other types of data sorts to make their work more productive. For example, the expediters will find a critical item sort more useful in finding their required delivery dates than extracting those dates from a milestone chart. As the revised delivery data and actual progress are fed into the computer, revised printouts quickly reflect the changes and their effect on the schedule.

Construction people usually find that the key milestone date sort better suits their needs. That section of the CPM schedule is the basic document that the field scheduler uses to make the detailed weekly work schedules in the field.

Most CPM scheduling software even delivers a bar chart printout, which is most convenient for upper management's purposes in reviewing project progress. Management people do not have the time nor inclination to delve into the details of the logic diagram. Simplified bar charts are usually included in the progress reports to give a graphic view of actual progress against the schedule. On smaller projects, a simple time-scaled CPM diagram might be used in the progress report.

The rapid turnaround of data by the computer also allows the project team to perform what-if exercises with the logic diagram. When scheduling problems arise, the project team can try alternative solutions by changing elapsed times for problem activities. That generates new early and late start dates and a new critical path. The computer calculates a new critical path in a matter of seconds, with immediate access to the new output data right on the computer screen.

A CPM/computer system also simplifies recycling the schedule. Recycling becomes necessary whenever deviations grow to a point at which some of the intermediate goals are in jeopardy. Recycling involves revising any target dates that may have slipped beyond repair or when a significant change in scope occurs. Exercising some what-if options should allow you to obtain the most effective recycling option.

Human resources leveling

Another important advantage of the CPM/computer system is the ability to level peak personnel requirements, which occur during the project's design and construction phases. By taking advantage of the available float and rescheduling the start of noncritical activities, it is possible to shave personnel peaks. Leveling the personnel requirements leads to more effective use of the project's human resources.

Disadvantages of CPM

There are only a few disadvantages to using the CPM method for project scheduling, and even these can be avoided with proper attention from the PM. However, overlooking any of the disadvantages can scuttle your efforts to control the project schedule!

It is extremely important to have your key project people trained in CPM techniques. That includes all levels from the project management team through the design group and into the field. Remember the story about the construction managers disposition of the CPM/computer output!

I don't recommend controlling a large project with only a newly trained crew using a new software system without running your old scheduling system in parallel, at least until the new system has been proved to work. If the new system breaks down for any reason, you will be without any means of controlling the end date on the project.

The cost of running a CPM schedule is likely to be higher than that of using bar charts, particularly on smaller projects. That is especially true of running the CPM schedule on a mainframe computer. In recent years, the relatively modest cost of PC hardware, software, and training has enabled us to expense off the cost of computerized CPM scheduling on a medium to large project. It might take several small sized projects to cover writing off a PC scheduling system.

The minimum hardware requirements to accomplish the CPM scheduling should consist of an IBM compatible PC of 640K memory, a 40-megabyte hard disk, and a dot matrix or laser printer. Fancy graphics require a plotter, but it is not essential to do a decent job of schedule control.

Any one of the 100 or so scheduling software programs which suits your type of project will do the job. The cost of the training could be the sleeper in the scheduling cost budget, depending on the experience and computer literacy of your project people. Regardless of the cost, the training is the linch pin of the whole system, so don't pass over it.

The real savings in using a PC computerized schedule is that it will generate a good deal more data than is possible with bar charts. That means the unit cost of the data is low. However, if the data are not being used, *or worse, are being improperly used,* you will not be getting your money's worth. It is up to the PM to see that the computerized schedule output is used in a cost-effective manner.

The cost-effectiveness of using CPM

It is difficult to accurately quantify the cost-effectiveness of using CPM scheduling systems on capital projects. First, there is no absolute measure of the time saved by using CPM vs. bar charts. Second, the value of the time saved must be balanced against the value to the owner in having access to the facility earlier. Any comparisons of this nature have proved to be highly speculative and difficult to verify.

Most owners and contractors accept any additional cost of using CPM scheduling systems as a way to improve the odds of completing their projects on time. Those who do not believe that CPM saves money and ensures a project's earlier completion date can continue to use manual bar charting with reasonable hopes for success.

O'Brien's book, *CPM in Construction Management,*[3] contains a chapter on costs and some expected savings from using CPM. On average, the cost of applying a CPM system to a project is about 0.5 percent of the total facility cost. The major cost areas for using the system are schedulers' time, software cost, and computer time.

PMs must be aware of the type of scheduling system that is being proposed for their projects, so they can budget funds to cover the cost. Small projects can be done with a part-time scheduler; medium-size projects need at least one person full-time; and larger projects will require two or more schedulers to handle the workload. Include all computer costs (including training) in the project budget, especially if a mainframe computer is used.

Advantages of bar chart schedules

As I said earlier, bar charts are inexpensive to produce and are easily understood by people with a minimum of scheduling training. I heartily recommend them for small, less complex projects, as being entirely suitable and cost-effective. The more comprehensive CPM system is often too complicated and represents unnecessary overkill when used on small projects. Some good PC-based CPM programs are available for scheduling a series of small projects which draw from the same resource pool.

The only thing that threatens the economic advantages of using bar charts on small projects has been the advent of PCs, along with less complicated scheduling software. It is easy to be tempted into the use of a PC with the project manager or engineer acting as the project scheduler. That can be all right if the project leader does the scheduling work in his or her free time. If, however, the project leader gets so involved as to let the rest of the project direction go its own way, the project is doomed to failure.

Disadvantages of bar chart schedules

As we showed in Figure 4.1, bar charts have limited ability to show many detailed work activities and their associated interaction. They become bulky and unwieldy on larger projects with as few as 100 activities.

Bar charts cannot show clearly the interaction between early start and late finish dates of activities and the resulting float of noncritical activities. There is no clear identification of the critical path through the project that appears with the CPM system. Also, it is impossible to develop the wealth of scheduling detail with a bar chart that can be developed and manipulated with the CPM system.

Computer vs. manual scheduling methods

The major factor in selecting computer over manual scheduling methods are project size and complexity. Small projects are best done manually, since good time control is possible at low cost. However, a complex plant turnaround project with a relatively low budget but working three shifts on a tight schedule, definitely warrants a computerized approach.

On larger projects using CPM, computer operation is a must to perform the many repetitive critical path calculations in a short time. Manipulating and sorting the expanded data base of project information is well worth the additional expense if the system is properly applied. The arrival of PCs and minicomputers, with their associated scheduling software, has brought the cost well within acceptable limits. The simpler operations of the PC-based systems has also reduced the cost of the necessary CPM training.

Scheduling System Selection

Our discussion of the advantages and disadvantages of the available scheduling systems should allow us to develop guide lines for selecting an effective scheduling system. The selection involves a variety of such factors as:

- Size of project
- Complexity of project
- Scope of services required
- Sophistication of user organizations (i.e., client, field organization, subcontractors, and so on)
- Available scheduling systems
- Scheduling budget
- Client preference
- Mixing schedule and cost

Size of project

We have already discussed this point earlier. The rule of thumb is *bar charts and manual systems for small projects and computerized CPM for medium-size and large projects.* The level of sophistication of the system tends to become greater as the projects become larger.

Complexity of project

Even small complex projects can well use computerized CPM schedules if the smaller number of activities take place in a very short time span. An example is a plant turnaround worth $1 million or less with only 2 weeks to do it. On the other hand, a $1 million project with a 12-month schedule probably would not need a computerized CPM schedule.

Scope of services

Full-scope design, procurement, and construction projects lend themselves to more complex scheduling methods because of the extra interfaces among the many design, procurement and construction activities. A project involving just one of these macroactivities could have effective control with a less sophisticated and less costly system.

Sophistication of user organizations

The sophistication of user organizations is probably the most overlooked factor in selecting a scheduling system. Often the need to produce a full-blown CPM schedule exists. However, one key project group may not be experienced in CPM to properly interpret their part in it. Assure yourself that the failure of this group to perform properly will not defeat the proposed scheduling method.

One such example could involve working with a client in a developing country. If the client's people lack experience in the proposed scheduling system, they might not feel comfortable using it to track job progress. Also, they may not keep their contributions to the project on schedule. Another example might be an inexperienced construction force not being able to use the output of the CPM scheduling system, such as we discussed earlier.

The worst possible case would occur if the PM were not versed in the selected scheduling system! That underscores the need for present-day PMs to stay current on the latest CPM scheduling methods available in their companies and the marketplace. I definitely recommend you success-oriented PMs do further in-depth study of logic-based scheduling than I have presented here.

If there is a shortage of CPM know-how in your organization, it is possible to hire a CPM consultant to handle your project scheduling. At least one member of your team, however, should have enough knowledge of the system and the project to monitor the consultant's work. That is the best way to ensure that the resulting schedule is effective for your project.

Existing company systems available

The availability of company systems is important because we want to use a system which has been in use within the organization and has been thoroughly tested on prior similar projects. Introducing a new system on a project can often cause more problems than it solves. As an owner's project manager, it is a good idea to assure yourself that the contractor is proficient in the system before allowing its use on your project. It is also not a good idea to force the use of your corporate standard system on the contractor just because your organization is familiar with it. A much wiser course is to train your people in the use of the contractor's scheduling system.

Scheduling budget

If the home office services budget does not allow for the expense of a sophisticated scheduling method, you are going to come up short of money. Most computerized CPM scheduling costs have a tendency to grow and overrun their budgets. A common problem is job stretch-out which increases the schedule cycles, which in turn runs up the scheduling personnel hours and computer time. A factual estimate of the total cost of the proposed scheduling system is needed to select an effective system for the project.

Client preferences

Owners who want computerized CPM schedules, and are willing to pay for them, are entitled to have them. If the owner does not specify a preference for a scheduling system, some common ground for limiting the cost of a suitable system will have to be found.

In recent years most federal government contracts have required adherence to a strict contractual standard calling for use of CPM schedule control and reporting. Careful attention must be given to investigating the latest scheduling requirements for any federal work on which you may be proposing.

Mixing cost and schedule control

Lately, many software companies have been promoting programs that claim to control cost along with schedule. This may work on very simple projects where we can tie project costs closely into the schedule. On larger projects this has never really been very successful. The budget numbers are broken down into different categories than the scheduled activities. Most cost-reporting systems seem to need the accuracy level required by good accounting practice. Combined schedule and cost reporting systems have not yet been able to generate this type of accuracy. I would be wary of using a system tying these two areas of project control together and running the risk of losing control of two very critical areas of project management.

References

1. William Pena, *Problem Seeking: An Architectural Primer,* 3d ed., AIA Press, Washington, D.C., 1987.

2. James J. O'Brien, *Scheduling Handbook*, McGraw-Hill, New York, 1969.
3. James J. O'Brien, *CPM in Construction Management*, McGraw-Hill, New York, 1984.
4. J. J. Moder, C. R. Phillips, and E. M. Davis, *Project Management with CPM, Pert, and Precedence Diagraming*, Van Nostrand Reinhold, New York, 1983.
5. Ira H. Krakow, *Managing Your Projects Using the IBM PC*, Brady Communications Co., Bowie, Md., 1984.

Case Study Instructions

1. The basic activities, with their elapsed times (in working days), required to construct a habitable house are listed below:

Clear site	2	Interior plumbing	9
Excavate basement	3	Interior painting	6
Rough walls	9	Wallboard	8
Exterior siding	7	Exterior fixtures	2
Pour basement	6	Exterior paint	6
Roof	4	Flooring	5
Exterior plumbing	4	Roofing	4
Electrical	8	Landscaping	6

   Arrange the work activities according to precedence requirements and draw the arrow diagram.
   a. Calculate the early and late start dates.
   b. What is the earliest completion time?
   c. What is the critical path?
2. Assuming that you have access to a personal computer and an applicable project management software program, input the data in instruction 1 above to check your manual work on that problem. Even a demonstration disk program should be sufficient to program the above simplified schedule for the house.

   Practice the various sorts and options available in the program, including some what-if exercises to try to shorten the house completion date.
3. Evaluate your selected project in regard to the type of scheduling method you would recommend to your client or management. List the reasons for each of your recommendations.
4. Referring to your project master plan developed in your case study at the end of Chapter 3, assume some best-guess elapsed times for the major activities and long-delivery items for your project. Based on your assumptions and your knowledge of activity precedence, sketch out a logic diagram to indicate the anticipated critical path through the project.

   Make a list of recommendations that would help to shorten the completion date, along with an explanation for each recommendation.

Chapter

5

The Project Money Plan

The *project money plan* is the financial forecast for the project, and it sets the basis for the control of project costs and cash flow. Developing the money plan involves the functions of cost estimating, budgeting, cash flow, and cost control systems. Taken together, these functions make up the field of cost engineering.

This chapter considers cost engineering from the project manager's viewpoint. We don't expect PMs to become full-fledged cost engineers, but they had better know how to deal with cost engineers when they are functioning on their projects. Those of you who made your own estimates on smaller projects will have a definite advantage when you become a full-fledged project manager. I strongly recommend adding one or more of the books listed in the reference section of this chapter to your library to use when you are making your project money plans. Of course, reading the books before consigning them to the library is also an excellent idea!

Preparing the project financial plan cuts across virtually every line of endeavor on the project, as well as the total organization. Therefore, the project manager must be very careful in handling the human relations aspect of developing the project money plan. A sound project financial plan also holds the key to reaching a major goal of the project manager's creed of *finishing the project within budget.*

Project Cost Estimating

The foundation for the project money plan is a sound project cost estimate, which is a reasonable anticipated cost of executing the work. It should be neither optimistic nor pessimistic, and to be truly effective, it must also be produced at a reasonable cost.

Types of cost estimates

Cost estimates fall into four basic classes based on the purpose the estimates serve during the project's life cycle. The names of the estimates vary widely across and within the various disciplines dealing with capital project execution. Those that I have chosen for our discussion are given in Table 5.1.

In addition to the differences in the names of the four classes of estimates, one can expect a good deal of variation in the expected accuracy figures. That is why I have shown a range of expected accuracies for each type of estimate except the definitive estimate. The exact percent used depends on company policy and the degree of project definition available at the time of the estimate. For example, feasibility estimate accuracies could even run as high as ±50 percent if only imprecise design data were available at the time.

Percentage of accuracy

The use of percent accuracy is perhaps the most misunderstood concept in the field of cost estimating. The figure is not a percent of contingency for the estimate. Rather, it is an indication of the probability of overrunning or underrunning the estimated cost figure. The estimate should already include a contingency allowance for any unexpected errors of omission and commission in making the estimate.

Since these estimates are made as the percent completion of the design work increases, the percent of accuracy becomes progressively more refined. The increase in accuracy is due to the improvement in project definition and improved pricing data which become available to the estimators. The lower percent accuracy numbers show that the probability of overrunning the estimate is becoming lower.

The diagram in Figure 5.1 is a graphical representation of the classes of estimates and where the estimates occur in the life cycle of the project. Please remember that the nomenclature, the timing of the estimates, and the duration of the project are highly variable in different project environments. You can tailor the diagram to your individual project.

TABLE 5.1 Names of the Types of Estimates

Type of estimate	Purpose	Accuracy, %
Feasibility	Determine project feasibility	± 25 to 30
Appropriation	Obtain project funding	± 15 to 25
Capital cost or budget	Project control budget	± 10 to 15
Definitive	Final cost prediction	± 5

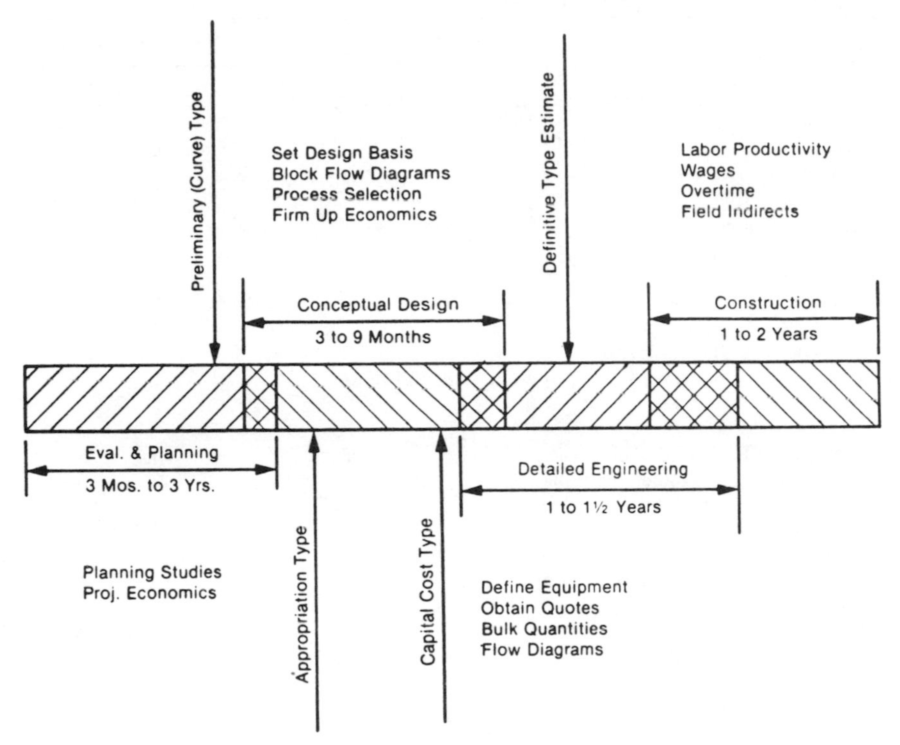

Figure 5.1 Classes of estimates (time line). (*Reproduced by permission from* Chemical Engineering, *July 7, 1975, p. 263.*)

Project definition

The amount of project definition is the single most important factor controlling the accuracy of any estimate. Table 5.2 lists the various items of design data that are needed for the four classes.

The class 1, or feasibility estimate requires only the minimum design data such as the products (or services), a brief process description, the unit capacity of the facility, and a general location. Because of the minimal design data available at this stage, the time spent to estimate the cost is low and the percentage of possible error is high.

For a class 2 estimate, the available design data becomes greater with the addition of block flow diagrams, general layouts, preliminary equipment lists, and a specific site location. The design becomes somewhat more refined with a preliminary plot plan, basic mechanical and electrical designs, and an estimate of the design costs. The additional project definition increases the expected accuracy of the estimate. Even some very early construction cost factors emerge at this time.

The major risk factor in the estimate's accuracy at this point is that the available data are still subject to revision during the schematic de-

TABLE 5.2 Estimating Data Checklist

	Types of estimates			
	1	2	3	4
General				
Products	X	X	X	X
Process description	X	X	X	X
Plant capacity	X	X	X	X
Location—general	X			
Location—specific		X	X	X
Basic design criteria		X	X	X
General design specifications		X	X	X
Process				
Block process flow diagram	X			
Process flow diagram (with equipment size and material)		X		
Mechanical P&Is			X	X
Equipment list		Xp	X	X
Catalyst/chemical specifications		X	X	X
Site				
Soil conditions		X	X	X
Site clearance			X	X
Geological and meteorological data			X	X
Roads, paving, and landscaping			X	X
Property protection			X	X
Access to site			X	X
Shipping and receiving conditions			X	X
Major cost is factored	X			
Major equipment				
Preliminary sizes and materials		Xp	X	
Finalized sizes, materials, appurtenances, and drivers				X
Bulk material quantities				
Preliminary design quantity takeoff		Xp	Xp	
Finalized design quantity takeoff				X
Engineering				
Plot plan and elevations		X	X	X
Routing diagrams			X	X
Piping line index				X
Electrical single-line and area class.		X	X	X
Fire protection systems			X	X
Underground systems			X	X
H.O. services—factored estimate		X		
H.O. services—detailed estimate			X	X
Catalyst/chemicals quantities		X	X	X
Construction				
Labor wage and fringes		X	X	X
Labor productivity and area practices				X
Detailed construction execution plan				X
Field indirects—factored estimate		X	X	
Field indirects—detailed estimate				X

TABLE 5.2 Estimating Data Checklist *(Continued)*

Schedule			
Overall timing of execution	X		
Detailed schedule of execution		X	X
Estimating preparation schedule		X	X
Miscellaneous			
Freight cost			X
Start-up		X	X
Insurance and taxes		X	X
Royalties	X	X	X
Import/export duty		X	X
Financing data	X	X	X
Escalation			
Escalation analysis		X	X
Contingency			
Accuracy analysis	X	X	X

Xp = partial data for major elements. Types of estimates: 1 = feasibility; 2 = appropriations; 3 = capital cost; 4 = definitive.

sign phase. Watch estimates made during this early phase closely for design changes that adversely affect project costs. Set up a monitoring system to flag significant changes immediately and determine their effect on the cost estimate.

By the time of the class 3 estimate, most of the design data are so well defined that the accuracy improves rapidly. Try to freeze the design at this point and permit only minor changes with a minimal effect on the estimate. In nonprocess projects, the estimate corresponds to the so-called engineer's estimate that is made at the lump sum construction bidding stage.

The class 4 estimate should be made when the design is almost finished. By that time, most of the materials have been ordered and only minor unknowns in the construction cost area are still outstanding. The major difference between the class 3 and class 4 design data is the improved accuracy of the material takeoff and price information. That permits accurate pricing of the equipment and materials budget.

Table 5.2 is a good checklist for information that your estimating team must have available to produce the desired estimating accuracy. If most of the data are missing, you are not likely to produce the desired accuracy regardless of how much time you spend on the estimate.

How to Make the Four Classes of Estimates

The approach to the different classes of estimates varies considerably, and it is worthy of some general discussion. Each estimate serves a different purpose and depends on the available project definition.

Feasibility estimates

During the conceptual phase of most projects, management must know how much the expected capital investment is going to be to test the commercial viability of the project. These early estimates are also known as screening estimates because owners often use them to compare the cost of comparative capital projects or improvements.

Estimators are usually very short of information about the project at this point. They may have only the proposed capacity of the plant, a block flow diagram of the processes, and an approximate location. For an A&E type of project, the estimator may have only some general information such as total square footage of buildings, number of rooms or beds for an office or hospital, KW capacity of a power plant, or cubic yards of fill in a dam project.

One approach to use is to plot the capacity vs. the actual cost data of previously built projects with similar characteristics. If enough units of different capacities and their cost figures are plotted, a good practical cost per unit capacity curve will result. Plotting the data on a logarithmic scale will yield a straight line, which is easier to read. To improve the accuracy of the method, it is necessary to factor out the effects of inflation and project scope differences.

The total installed cost of the units must reflect the same scope of work, or the curve will be distorted. One cannot compare a grass-roots project with an addition to an existing facility with a developed site and infrastructure already in place. The site and infrastructure costs must be estimated and factored in or out, as required, to get comparable costs.

In many industries there are several good sources of unit costs per stream day of capacity for various types of processes. However, the most reliable ones are those that can be generated within your own company and are based on the type of facility built to your firm's standards. These numbers are generated by dividing the plant cost by the tons, barrels, pounds, and so on, of daily throughput. The dollar value will increase over the years as inflation affects overall plant construction costs. Be sure your estimator is using current figures.

The high percentage of inaccuracy (± 25 to 30 percent) expected with screening estimates is a direct reflection of the approximate nature of the pricing methods. It is unfortunate that owners must use the most inaccurate cost estimate for the go-no-go decision at this early stage. The project's commercial viability or its cost/benefit ratio can only be based on this relatively inaccurate estimate of the expected project cost. On the other hand, the flexibility of this cost estimate allows those evaluating the project some leeway in calculating the project's return on investment.

PMs must use good judgment at this stage. An overly pessimistic estimate can lead to passing up economically sound projects. Likewise, overly optimistic estimates can lead to wasted investment in uneconomical projects. This early in the game, I lean toward the optimistic school of thought. Several other cost checkpoints along the way provide more accurate estimates before we reach the point of no return on project funding.

The good news about feasibility estimates is that they are very inexpensive to make while we are still low on the project cash flow curve. One cannot spend too much money getting some basic curve or unit cost figures together to arrive at a feasibility cost estimate.

Appropriation estimates

As we move to the right along the time scale of the project life bar diagram shown in Figure 5.1, more detailed project definition becomes available. That permits some refinement of our estimating procedures. Since we are going for board approval with this number, it had better be reasonably accurate.

Looking under the heading of class 2 estimates in Table 5.2, we see that some very important decisions which further define project costs have been made. A specific site has been selected, and preliminary plot plans have been developed. Flow diagrams, major equipment specifications, and overall layouts are available. Process flow diagrams with preliminary heat and material balances have been developed, and they permit the preliminary sizing and specification of the major process equipment items.

Over the years, many cost factors have been developed for various processes, and they make it possible to factor the total plant cost from the process equipment cost. We will discuss this estimating approach in more detail later in this chapter.

The additional design data available can open the door to at least two different estimating approaches at this stage. It may be wise to run an approximation by two different methods and use the resulting figures to check the earlier feasibility estimate. If the numbers jibe pretty well, confidence in the estimate will be increased. If the numbers do not reinforce one another, it's time to find out why.

For A&E types of projects which are not equipment-oriented, appropriation estimates can be made from rough quantity takeoffs of bulk materials by applying historical unit cost figures to them. Other project costs such as site development, HVAC, special finishes, electrical, and plumbing can be factored into the estimate to obtain the total facility cost.

This class of estimate still carries a fairly high percentage of inaccuracy to cover errors of omission and commission because of the preliminary nature of the design documentation. Some of the doubt over the accuracy of the estimate can be removed by carrying a higher percentage of contingency to offset the potential errors. The amount of contingency carried is often a function of the owner's policies on such matters; it could run as high as 20 percent or as low as 10 percent.

The appropriation estimate is considered a budget-type estimate because it does contain a breakdown of several major project cost elements. For example, the estimate may have been made from separate cost estimates for design, site development, foundations, structural, electrical, and mechanical systems, etc., to arrive at the total figure. It is usually the first estimate with enough detail to use it as a project budget. The design team uses this budget to control costs for subsequent phases of the project.

Capital cost estimates

The next estimate along the time line in Figure 5.1 is the capital cost estimate, which is sometimes a further refinement of the earlier appropriation estimate. On process projects, we now have fully developed P&IDs, firm prices for major equipment, and preliminary piping models or drawings. The civil, architectural, and structural (CAS) design is well enough along to support some bulk material takeoffs and pricing of the work in that area. Pricing of preliminary piping takeoffs or factored estimates from comparable projects give us the first refined look at the important piping cost figures. The large amount of piping in a chemical plant represents a large cost item and is often a budgetary problem area.

At this point, we can approach an accuracy of about plus or minus 10 to 15 percent. With further refinement, we are also able to evaluate the accuracy of each major subdivision of the estimate to get a better handle on the overall contingency allowance. This estimate is refined enough to be used as a detailed budget for controlling project costs. We will discuss how that happens later in this chapter under budgeting.

Definitive estimates

The definitive estimate is the final estimate in the series for a process job. It also corresponds to the engineer's estimate made at the end of the design phase on A&E types of projects. All of the major equipment has been ordered, and firm bulk material prices have been established and applied to actual takeoff quantities. The only remaining major cost unknowns are the rate of field labor productivity, field indirect costs, and any possible construction change notices. The contingencies for these outstanding items can be readily evaluated, however, which brings the percent accuracy for the definitive estimate down to ±5 percent of total project cost. On A&E types of projects, in which the construction is being bid lump sum, the percent accuracy of the final cost estimate will even drop below the 5 percent figure after the bids have been received and the successful bidder is accepted.

The figures developed in the definitive estimate are used to fine-tune the existing cost control budget for the whole project. Naturally, the contractor's lump sum bid price will become the project construction budget for controlling the construction costs.

Cost trending

Cost trending is a procedure which has been developed over the years to predict how the project costs are moving during the time period between estimates. On larger projects, the project team is likely to find itself in "blind spots" when predicting project costs during the design development stage. The project scope may have undergone a series of growth changes during the period between the appropriations estimate and the capital cost estimate. That can lead to financial shock when a later estimate turns out to have grown beyond the feasibility cost for the project.

That rude financial shock can be prevented by using a cost-trending procedure. A group of project and cost engineers is set up to monitor the cost of any design changes which may occur during the design development phase. Details of the establishment of a cost-trending group are given by Clark and Lorenzoni[1] starting on page 151.

The trending group issues a biweekly or monthly cost trend report which in effect updates the latest estimated project cost to the current design basis. This relatively inexpensive system gives the project manager early warning of any adverse cost trends while there is time to act on them. It also permits the project managers to be more confident about the project cost forecast when responding to their management's inquiries between estimates.

The series of estimates and the cost-trending procedure give owners a set of control points by which they can recheck the financial feasibility of the project while still low on the project resources commitment curve. With design costs running from 6 to 12 percent of total project cost, the owner can cancel or downscale the project scope before passing the financial point of no return. That can be done before placing equipment orders or starting construction. When the project is canceled after the commitment of equipment and construction funds, however, substantial cancellation costs are likely to be incurred.

Phased release of funding based on these key estimating milestones is fairly common in capital project execution. The need for this type of controlled project release is often set out in the contract documents and is definitely covered in the *Project Procedure Manual*.

Estimating cost-effectiveness ratio

The cost of performing a project estimate is generally directly proportional to the degree of accuracy desired. As pointed out in our preceding discussions, a project may require several estimates to maintain the desired control of the project costs. Therefore, estimating services can be a significant cost factor in the home office services budget.

Figure 5.2 is a graph of the cost of estimating vs. the accuracy of the result. The time line at the bottom shows the three basic phases of the project, and the vertical axis shows the cost and percent accuracy. In the planning stages, when percent accuracy is high, the estimating cost is very low. Even if we have to develop some basic cost curves, the amount of estimating labor is not significant. As more detailed design data become available, we can spend more labor doing detailed material takeoffs, getting equipment bids, and extending the costs for the various parts of the project. As the estimating cost rises, the percent of estimating accuracy can be expected to improve. Please note the phrase "can be expected to improve." The expected improvement of the accuracy will happen only if we do the estimate right!

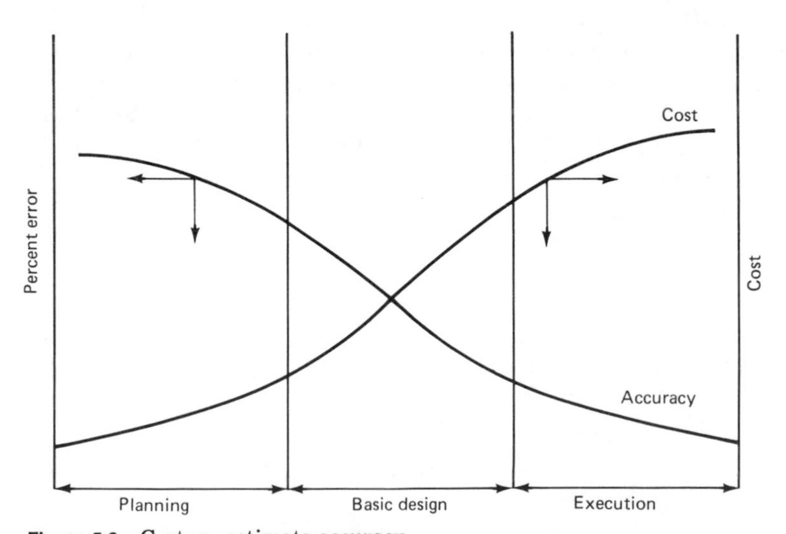

Figure 5.2 Cost vs. estimate accuracy.

To put the cost of doing an estimate into more concrete terms, I have included Table 5.3, which rates the cost of estimating for a range of project sizes related to the percentage of accuracy. The expected cost is given in terms of a percentage of total project cost, so the chart is independent of inflationary effects. Please observe the usual caveat that the numbers are representative and may have to be adjusted for your particular type of project.

Estimating Methods

The tools and methods available to the estimator are tailored to the type of estimate to be performed. They are also functions of the estimating capabilities and the data available to the project team. Some firms have complete in-house estimating departments with extensive files of historical data relating to the specific types of projects done by that firm. Others may have only a few reference books containing average unit costs across the country.

Most large engineering and construction contractors have extensive estimating capabilities which can be drawn upon by an owner who has such a firm under contract. Accordingly, most contractors regard their information as highly proprietary. It is virtually impossible for an owner to have access to those files except as they pertain to the specific contract being estimated. Developing and maintaining historical cost files is expensive, and it is one of the key advantages that a contractor can offer a client.

An alternative to building a costly in-house historical estimating file is to use one of the outside estimating services. Some such services

TABLE 5.3 Estimating Cost Chart

Level of accuracy, %	Costs of estimates as percents of these total project costs:				
	$500,000	$1,000,000	$5,000,000	$15,000,000	$20,000,000
5	4.00	3.70	1.00	0.70	0.65
10	1.60*	1.50	0.46	0.34	0.30†
15	0.76	0.70	0.21	0.17	0.15
20	0.44	0.37	0.13	0.10	0.08
25	0.28	0.24	0.08	0.07	0.05
30	0.21	0.17	0.06	0.05	0.04
35	0.16	0.3	0.04	0.03	0.03
40	0.12	0.10	0.03	0.02	0.02
45	0.10	0.08	0.03	0.02	0.02
50	0.08	0.06	0.02	0.01	0.01

*An estimate accurate to within ±10 percent of a $500,000 project would be expected to be about 1.6 percent, or $8000.

†On a $20,000,000 project, an estimate accurate to within ±10 percent would be expected to be about 0.3 percent of the total project cost, or $60,000.

can be purchased on an annual subscription basis and will then supply a battery of cost files at a nominal price. An example is Richardson Engineering Services, Inc. of San Marcos, California. Its cost indexes are computerized and updated regularly, and the data are tailored for specific parts of the country.

An example of a complete outside estimating service which will do a specific estimate complete for a fee is the Icarus Corporation of Rockville, Maryland. You give them your design data and estimate basis, and they give you the complete estimate broken down to suit your code of accounts.

On A&E types of projects, the architects and engineers are often responsible for preparing the project cost estimates in the absence of a resident estimating department. In that case, standard cost indexes such as Means[3] and Dodge[4] are used along with specific local knowledge from similar recent projects that may be available within the firm. The accuracy of the unit prices used in that type of estimate is very critical; the estimate will be used to compare the contractors' bids, which are often made in more detail than the engineer's estimate. If the engineer's estimate and the low contractor's bid are not within ±10 percent, the owner will probably throw out the bids and rebid the project on a revised design. As we discussed earlier, that has very embarrassing consequences to the project schedule.

Estimating Tools

Many estimating tools and shortcuts have been developed over the years and are described in the literature listed at the end of this chapter. The comprehensive bibliography of cost engineering literature, which is included in Appendix E of Humphreys and Katell[2], expands the list of available cost engineering literature even further.

Computerized estimating tools

The advent of the computer, and especially personal computers (PCs), has added to the estimator's arsenal of tools. With the use of a mouse and an estimating menu, material takeoff, listing and pricing become easier and more accurate. Combination programs can also develop lists for the materials associated with the basic item being taken off. For example, when making piping takeoffs, insulation and painting costs can be taken off simultaneously by indicating whether the line is insulated or painted.

Another example of computerized estimating tools is the use of computerized design programs for the preliminary sizing of vessels and heat exchangers. When the basic design data and code information are input, the program will print out the area of the heat exchangers and the weight of the vessels. On a chemical plant, use of this information with the proper material cost multipliers gives an excellent handle on the equipment cost. In turn, the equipment cost forms the basis for a more accurate factored type of estimate early in the project. Using this approach results in a more accurate and cost-effective estimate earlier in the project.

The first thing which comes to mind is that use of the new computer tools should reduce the cost of making the estimate. That does not appear to be true, because the trade-off has been to generate more takeoff and cost data with improved accuracy for roughly the same cost.

Perhaps the best tool of all for project managers is to develop a gut feeling about their project estimates. Unfortunately, that can come only with time and experience in dealing with the estimating environment. As a project manager you will never have time to become an accomplished estimator yourself, so do not try to dig into all of the details of any estimate you are reviewing. Look at the various major cost accounts to see how logical the numbers appear to be. Also, compare the interrelations of the various major accounts by standard rule-of-thumb values which apply to your area of practice. Any accounts which appear to be out of line should be investigated in detail to confirm or deny your suspicions.

Estimating Escalation and Contingency

Escalation and contingency are the last two major estimating items that we must deal with before we wrap up our estimating discussion. They both are somewhat esoteric items that seem to defy any fixed rules for accurately estimating them.

Escalation

Price inflation has been with us for such a long time that we now consider it a normal way of life, but the rate of inflation can swing wildly with economic conditions. To improve our cost projection, we must make some sort of educated guess of its value over the expected life of our project.

Some people try to take the easy way out by writing escalation for wage-price inflation out of the estimate. That is done by stating that the estimate covers "the cost in effect at the time of the estimate," not at the time of project completion. That is fine, but I have never found owners who would even let their own project people off the hook that easily—let alone the contractor's people. The owner will almost certainly insist on a best estimate of the anticipated escalation of project costs regardless of the prevailing economic conditions.

Estimators have a number of ways, including a variety of computer programs, to project cost escalation. The advantage of a computer program is its ability to evaluate individual accounts and apply a system of factors to get an overall weighted average escalation for the whole project. In any case, any projection of escalation has to be based on past history coupled with a forecast of economic conditions over the expected life of the project.

Getting past history is easily done by examining indexes of past escalation trends, which are available from a variety of sources. The Bureau of Labor Statistics (BLS) is probably best known for a broadly based national index. Since we are interested in indexes pertaining to capital projects, we will want to look at those which cover the engineering-construction industry. Examples are those published by F. W. Dodge and *Engineering News Record*. For process projects, the economic indicators page in *Chemical Engineering* is a good source of economic data including past escalation information.

Extending the past inflation curve into the future is probably the most simple approach to projecting escalation. The practice can be hazardous, however, if there happens to be a sudden shift in the rate during the course of the project. An outstanding example of that type of situation was the escalation experienced during the oil embargo of the early 1970s. In a few months, the normal escalation of about 5 percent per year suddenly vaulted to 20 percent. In addition to the sudden increase in prices because of the Arab oil embargo, companies were buying up pipe and fittings just to hoard them for future possible needs. That created further shortages of piping materials which drove prices even higher.

I was involved with a lump sum project to deliver engineering and hardware for a petrochemical plant to an eastern bloc country. The materials had to be procured in the United Kingdom because of the local financing arrangements. Project costs were suddenly running $4 million over budget and rising because of the unexpected escalation. Fixed-price bids on the equipment were suddenly out of the question, and on some items we were unable to get bids at all. All prices were quoted with price in effect at time of shipment.

Projects with Iron Curtain countries have longer than average durations because of the red tape involved, so we were caught by the sharp inflation spike brought on by the oil embargo. Since there was no escalation clause in the contract, and the firm did not choose to press a claim for additional costs brought on by the unusual market conditions, there was no recourse but to absorb a substantial loss on the project.

I don't recall hearing many people predicting the oil embargo before it hit, so I doubt that many estimators or PMs could have been expected to allow for such a wide swing in their estimates. Using 20/20 hindsight, however, the contract should at least have had an escalation clause to cover the lump sum cost exposure on a project that could be expected to experience long administrative delays.

In the final analysis, it is virtually impossible to accurately predict the rate of inflation over a long period of time. The best way to approach the matter is to study the recent escalation history and then weigh that figure against an economic evaluation of the market conditions which can be expected to prevail in the project location. Are many projects going to be competing for the available material and labor during the execution of your project? If you are executing a project in a tight sellers market condition, you can expect adverse escalation pressure on your project costs.

One thing on the plus side is to remember that the inflation rate does not have to apply to the total project cost. Monies committed and spent early in the project do not carry the inflation rate for the duration of the job; only the monies spent later in the project are subjected to the full exposure of price increases.

Contingency

The contingency allowance in any estimate is always a point of discussion. We must remember that contingency is a factor which is added to cover the two major unknowns present in all estimates:

- Errors due to inaccurate design data
- Errors of omission and commission in estimating

Regardless of how hard we try for perfection, we are likely to experience some estimating errors that are due to materials being taken off inaccurately or omitted, along with mathematical errors. That is not as bad as it sounds, since these types of plus or minus errors tend to offset each other and have only a minor effect on the final number. The percent allowance for inaccurate design data is a function of how far along with the design we are. At the beginning of the project, the number is higher, and it gradually tapers off as the design phase develops.

Contingency factors do differ for various types of capital projects, but they usually run about 15 percent early in design and can drop as low as 3 or 5 percent when the design is complete. In many cases we do not have a free hand in applying contingency to our estimates. Often, the estimated cost is being compared with other costs to determine the selection of a project, a process, or a contractor. In order for the estimate to be competitive, the contingency must be a factual representation of the anticipated accuracy of the estimate.

Estimates can sometimes be subject to a pyramiding of contingencies, which makes the final estimate far too high. A common time for that to happen is when making labor estimates. They are based on historical *standard labor-hours,* which already contain a nominal allowance for contingency. The discipline group leader or craft supervisor will generally add further contingency in the form of extra labor-hours to the estimate. The next higher manager may crank in more safety factor based on the group's recent poor performance on a project. By the time the final figures go to estimating for pricing, when another contingency factor is added, the total contingency can reach a figure which includes up to 30 to 40 percent fat.

That type of problem can best be detected by applying some industry standard rule-of-thumb comparisons which reflect the interrelations of certain costs to the overall project cost. A good example is the percentage relationship of the total home office costs to the total facility price such as that in Figure 1.1. When the rule-of-thumb test turns up suspicious numbers, check them out in detail for contingency pyramiding or just plain errors.

When escalation and contingency are considered in the aggregate, they can easily amount to 25 to 30 percent of project costs in the early stages. That definitely makes them worthy of a project manager's consideration in successfully meeting the project goals.

Foreign currency fluctuations

The foreign currency contingency arises only on projects which have an international flavor. We may be doing a project in a foreign country or procuring major equipment and materials from overseas, which brings the factor of foreign currency exchange into the estimate.

In the case of buying equipment and materials overseas, we can protect against currency fluctuations by buying "futures" in the currency of payment for the goods. If the value of the seller's currency goes up and the value of ours goes down, we can still pay the original price with the prepurchased funds. If the seller's currency declines, we can sell the currency, pay the original purchase price, and pocket the difference!

If you ever have the opportunity to manage a project with a large budget in foreign exchange, you should get the best advice possible for setting up the financing and payment terms. It is critical to ensure that the terms of all the agreements involved are set up to protect your firm's investment and the project budget.

Estimating Home Office Services

An important estimate that design project managers are often called on to make is the home office services budget. It includes design, procurement, cost engineering, and home office construction management. Although the budget is not a large percent of the total facility cost (6 to 12 percent) for a design/build project, it is highly visible to top management. Most of the design budget is spent early in the project's life cycle and sets the tone for the rest of the project, so the budget definitely deserves careful attention from the project manager.

The best way to describe the preparation of a home office services budget is to follow a typical format for such an estimate as shown in Figure 5.3. The first page (Figure 5.3, page 1) summarizes the total labor input, and Figure 5.3, page 2, covers the out-of-pocket expenses and the total estimated cost. The example used here is a manual format which is ideal for a small project, but it can just as easily be computerized on a PC spreadsheet.

The largest cost item in the estimate is the design labor, which is covered in lines 1 through 14 in the estimate format. The format shown is for a process-oriented project, so you will have to adapt the design categories to the scope of services involved in your particular type of project.

As in any estimate, it is very important to have the scope of services well defined at the time of the estimate; Appendix B is an excellent checklist for this purpose. Getting all involved parties to sign off on the checklist provides a solid basis for the home offices services estimate and the subsequent control budget. It is equally important to make all preliminary project scope and design data available to the estimating people for better project definition.

In contractor organizations, the project manager and the design manager usually are responsible for making the estimate as part of the project proposal effort. The fee structure for the project (i.e., lump sum or cost-reimbursable) has a profound effect on the accuracy required of the estimate. Lump sum costs must be more accurately estimated because the bidding firm will be at financial risk if the estimate is too low. An owner, on the other hand, will normally consider the estimate too high because of a commitment to pay the agreed-upon price no matter how much contingency the contractor has included! Cost-reimbursable fee structures allow a little more leeway in the estimate because neither party is assuming a fixed financial risk.

Page 1 of 2

HOME OFFICE SERVICES COST ESTIMATE

PROJECT__________ JOB NO.______ DATE______
CLIENT__________ ESTIMATOR:______

NO		Work Description	Cost Code	Manhours	Rate	Cost	% of Total
1	ENGINEERING DEPARTMENT	Project Administration					
2		Process Engineering					
3		Mechanical Equip.					
4		Vessels					
5		Piping					
6		Civil, Arch., Structural					
7		Electrical					
8		Instrumentation					
9		Drafting Systems					
10		Engineering Scheduling					
11		Field Problem Resolution					
12		Other					
13							
14		SUBTOTAL ENGINEERING DEPT.					
15	ENG. SUPPORT	Engineering Clerical					
16		Stenographic					
17		Cost Reporting					
18		Project Scheduling					
19		Other Services					
20							
21	PROC. & CONSTR.	SUBTOTAL ENG. SUPPORT					
22		Procurement - Buyers					
23		Inspection-Expediting					
24		Home Office Construction					
25		Project Accounting					
26		Others					
27		SUBTOTAL SUPPORT GROUPS					
28	OVERLAYS	Total All Manhours / Costs					
29		Escalation @ %					
30		Contingency @ %					
31		Total Escalated Manhours / Cost					
32		Fringe Benefits @ %					
33		H.O. Overhead @ %					
34		TOTAL ALL LABOR COSTS					

Figure 5.3 Home office services estimate.

After all of the scope and design data are assembled, the package of information is circulated among the discipline leaders to prepare the labor-hour estimates for each skill level required. The standard labor-hours from historical cost records give the labor-hours per operation for preparation of such items as specifications, drawings, and esti-

HOME OFFICE SERVICES COST ESTIMATE

PROJECT__________ JOB NO.________ DATE________
CLIENT __________ ESTIMATOR:________

NO.	Work Description	Unit	Rate	Cost	% of Total
35	Overtime Premium				
36	Travel & Living Expense				
37	Reproductions / Printing				
38	Telephone & Telex				
39	Supplies & Misc.				
40	Models				
41	Agency Fee Costs				
42	Subcontracts				
43	Computer Time & Software				
44	Freight & Shipping				
45	Postage				
46	Consultants				
47	Testing Services				
48	Automobiles				
49	Other				
50					
51					
52	Total Out Of Pocket Exp.				
53	Escalation @ %				
54	Contingency @ %				
55					
56	Escalated Out of Pocket Exp.				
57					
58	TOTAL ALL LABOR COSTS (Page 1)				
59	GRAND TOTAL H.O. COSTS				

Reviewed By : ________

Approved : ________ Engrg. Mgr.

Approved : ________ V. P. - Operations

Figure 5.3 Home office services estimate. *(Continued.)*

mates. Each discipline supervisor reviews the labor-hours before the estimate is passed on to estimating for pricing labor costs. Each discipline cost is totaled separately and then rolled up into a grand total.

Special attention is needed for line 9, Figure 5.3, which is labeled Drafting Systems. It refers to automated systems such as computer-aided design (CAD) and computer-aided engineering (CAE). Since each system carries a substantial cost for CAD equipment time, the definition for performing these services must be established as part of the basis of the estimate. The total cost for the systems is made up of machine and operator components. The estimate may be based on labor cost including machine costs on page 1 of Figure 5.3, or the machine cost can be shown as an out-of-pocket expense on page 2, line 43. Machine time is a major cost factor, so it must not be overlooked.

The design support labor for clerical and project control personnel must be estimated and priced in the same way as the technical disciplines. Some smaller operating companies may choose to include it in overhead—which is something that I cannot recommend. Even if the personnel are not working on the project full time, their time should be estimated and put into the project budget and be subjected to the project cost control procedures.

If any procurement and construction services are included in your scope of work, labor-hours for them must be estimated and included in the project total if they are to be performed in the home office. Actual cost data from preceding projects can often give estimators reasonably accurate factors to arrive at these numbers if standard labor-hour values are not available to estimate the actual hours.

The bottom section of page 1 of Figure 5.3 contains the overlays (or markups), which include such items as the escalation and contingency percentages we discussed earlier. Escalation on design and home office services is usually quite predictable over the expected design period of normal projects. Design costs are not generally subject to the wide swings encountered in project hardware costs, and they are often factored right into the estimate—even a lump sum fee estimate.

The handling of contingency in design estimates was discussed earlier when we pointed out the possibility of pyramiding contingency factors as the labor estimate passes up through the organization. Assuming that the contingency has not been added to the estimated labor-hours, a moderate contingency for errors of omission and commission can be added at the bottom of the column.

Allowances for fringe benefits and office overhead are fairly standard in each type of work; they can be verified by the accounting departments of the buyer and seller involved in the home office services estimate. The grand total at the bottom of the first page of Figure 5.3 gives the total labor cost for the work without any fee charges. The fee (or profit) is usually added by top management in a separate pricing meeting.

The second page of the home office services estimate covers the so-called out-of-pocket expenses. They are usually passed on to the client without any markup for overhead or fee, but some of these costs may have markup built into them. Examples are reproduction services and computer time. They are often run as separate profit centers which charge the project at a standard competitive *market rate* for services rendered. Usually, all-inclusive rates of that nature are established during contract negotiations.

Sometimes, out-of-pocket costs are estimated as a percent of the design labor-hours. Percents are collected and averaged over a number of past projects for use on current estimates. The costs can also be estimated directly from bids, estimated number of trips, standard rental rates, and the like, if factoring is not feasible.

Escalation and contingency are added to the out-of-pocket expenses on the basis of the method used to estimate the expenses. That gives the total estimated cost of the expenses, which are then combined with the labor figure from page 1, Figure 5.3, to arrive at the grand total for the estimated home office costs for the project.

The section below the estimate on page 2 is available for the respective management personnel to sign off on the estimate before the estimate is presented to the client or put into a proposal. It is not unusual to recirculate that type of estimate several times before all of the questionable areas are resolved and the estimate is approved by management for final pricing.

A few rules of thumb are handy when project managers are checking home office estimates. The best one is to check the percent-of-total column at the right-hand side of the estimate form. The ratio of the hours of each discipline to the total design hours is usually quite standard for a given type of project. For example, Table 5.4 shows a breakdown for an A&E, nonprocess type of project.

Table 5.5 shows a set of numbers for the disciplines that are typical of a petrochemical type of project.

By comparing the two tables, it becomes apparent that the petrochemical plant has a heavy piping input and a very minor architectural commitment when compared to a nonprocess project. In fact, the 15 percent CAS portion in the petrochemical project is almost all civil and structural hours. A pulp-and-paper project might fall between

TABLE 5.4 Discipline Breakdown for a Nonprocess Project

Discipline	Percent of total
Project management	7
Architectural	38
Civil/structural	20
Plumbing	10
Mechanical	15
Electrical	10
	100

TABLE 5.5 Discipline Breakdown for a Petrochemical Project

Discipline	Percent of total
Project management	7
Process engineering	8
Mechanical	10
Piping	40
Control instrumentation	10
Electrical	10
Civil/architectural/structural	15
	100

these two sets of numbers because it has some process work, but it also has large buildings for housing the process machinery.

Another good factor to remember is the one for estimating total procurement costs, which usually run at 5 percent of the cost of materials being purchased. The number breaks down further into 2.5 percent for buying, 1.5 percent for expediting, and 1.0 percent for inspection. These numbers are subject to some variation depending on the scope of services called for in each of the procurement functions. Project managers specializing in each type of project must develop their own set of project specific numbers for quickly spotting problem areas in their estimates. Estimators are often a good source for such numbers in addition to those you can accumulate from your own experiences.

Requirements for a Good Estimate

We can summarize the discussion of project cost estimating with a checklist of basic requirements for a good estimate.

- A sound, fixed design basis
- A realistic project execution plan
- Good estimating methods and accurate cost data
- Neatly documented detail
- A reasonable estimating budget
- A knowledgeable estimator

Figure 5.4 shows a diagram of data flow for the preparation of a detailed capital cost estimate for a typical industrial project. The diagram also serves as a checklist for the types of input and their sources, which you will need to consider on any major capital project cost estimate. If you can satisfy the listed requirements for a good estimate of your project, you should be able to develop a sound estimate to serve

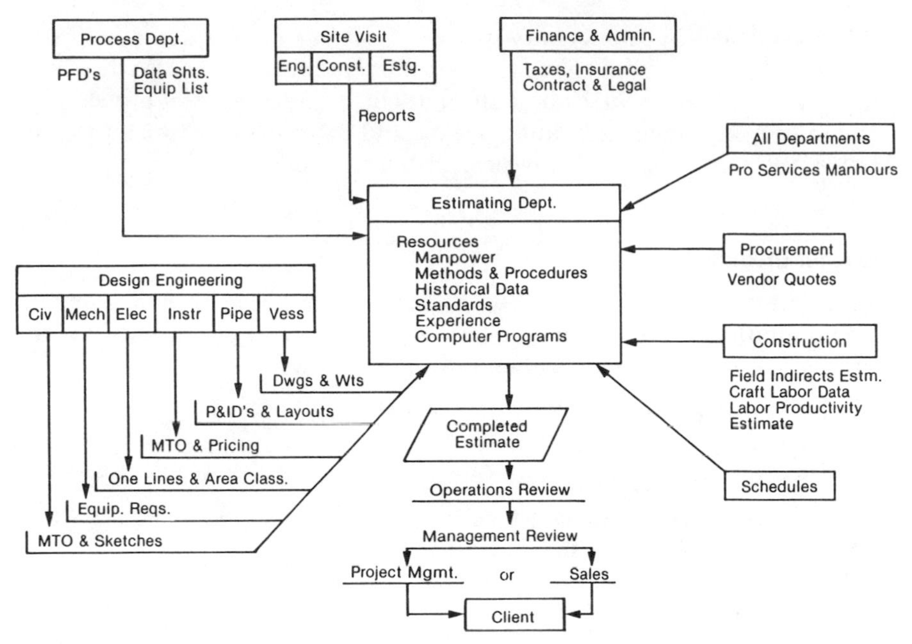

Figure 5.4 Estimating flow diagram.

as a basis for a reliable project budget. Success in those two areas will provide the basis for an effective cost control performance on your next project.

Project Budgeting

The purpose of the project budget is to set a cost, or money target for each item, and the total project. Since all of the financial aspects of the project revolve around it, the project budget must be realistic when compared to the actual expected cost of the project.

We have had a good deal of discussion about how to make an accurate estimate of the anticipated project costs to get a realistic budget. A poor estimate will lead to an unrealistic budget. That in turn will lead to loss of control, cost overruns, and financial problems for all concerned with the project. Loss of cost control will surely prevent our reaching project goal 3: *finishing the project under budget.*

The project budget is the baseline for the cost control program. It is of utmost importance that we set our sights to beat, or at worst, meet each item in the budget. Note that I said "beat" first because we want to be under budget if possible. We know that the odds are against our batting 1000 on a given budget, so we will use some of our winners to offset the losers. The only way we can meet our project goals is to win more budget items than we lose, especially the big ones. With a conservative budgetary attitude and prudent management of our escalation and contingency accounts, we should have every chance to meet our expectations to finish the project *within* budget.

Budget breakdown

To organize the budget for better control, it becomes necessary to group similar items into major accounts. A typical breakdown for an industrial project can be broken down as follows:

- 1.0 Home office services (10 to 15 percent)
  - 1.1 Engineering labor
  - 1.2 Procurement labor
  - 1.3 Out-of-pocket expenses
  - 1.4 Support services labor and expenses
- 2.0 Major equipment accounts (25 to 30 percent)
  - 2.1 Vessels and exchangers
  - 2.2 Rotating equipment
  - 2.3 Packaged equipment
  - 2.4 Materials handling
- 3.0 Bulk material accounts (15 to 20 percent)
  - 3.1 Piping
  - 3.2 Instrumentation
  - 3.3 Electrical equipment
  - 3.4 Structural and reinforcing steel
  - 3.5 Curtain walls
- 4.0 Construction costs (40 to 45 percent)
  - 4.1 Field labor
  - 4.2 Field materials
  - 4.3 Subcontracts
  - 4.4 Field indirect costs
- 5.0 Contingency
- 6.0 Escalation

A typical A&E project budget might break down into major cost accounts as follows:

- 1.0 Home office services (6 to 8 percent)
  - 1.1 Design labor
  - 1.2 Out-of-pocket expenses
  - 1.3 Taking bids
  - 1.4 Construction inspection

2.0 General trades construction (65 to 70 percent)
 2.1 Site development
 2.2 Foundations
 2.3 Structural framing
 2.4 Architectural trades
 2.5 Roofing
3.0 Mechanical trades (10 to 15 percent)
 3.1 Plumbing
 3.2 Heating, ventilating, and air-conditioning
 3.3 Fire protection
4.0 Electrical trades (10 to 15 percent)
 4.1 Power supply
 4.2 Power distribution
 4.3 Lighting
 4.4 Communications, alarms, etc.
5.0 Contingency
6.0 Escalation

The percents in the parentheses are order-of-magnitude numbers given only to indicate the relative portion of the whole project cost that each section might represent. The percents can vary somewhat from those listed depending on the type of facility being built.

In the case of the process type of project, the project manager has responsibility for the budget covering all of the activities. In the case of the A&E type of project, the responsibility may be split between the design and construction phases of the project. In the latter case, the construction project manager will be responsible for the lion's share of the costs.

Code of accounts

In a detailed budget, each of the major divisions given in the above lists is further broken down into much greater detail according to a standard code of accounts. A standard code of accounts is a must for any budget used for controlling project costs. In fact, the code of account format is best introduced during the detailed estimating stage so that the estimate can be readily converted into a project budget without going through a complicated conversion stage.

Codes of account are basically accounting systems in which project managers tend to take little interest. They are important in cost control as we will learn later, so it behooves the project manager to become familiar with their use. Chapter 8 in *Applied Cost Engineering*[1] presents a good description of the reasons for having a workable code of accounts for effective project control.

A word of caution on account codes is necessary because some accounting people tend to go into too much detail. The purpose of the account code is to sort and record costs in separate accounts for use in controlling the budget and establishing a cost history for the project. The history function of the account code provides the best estimating data source for subsequent project estimates. Any cost account breakdown beyond that needed to satisfy the above needs is useless and wasteful of project resources. If your code of accounts is overly detailed, you might want to start a campaign to simplify it for better workability and cost savings.

Budgeting Escalation and Contingency

The handling of the contingency and escalation accounts in the budget is critical to effective project cost control. Remember that we said these two accounts must be prudently managed if we were going to control project costs. The most prudent way to manage these accounts is to keep them separated throughout the project and not distribute them into the budgeted items at the outset. If the escalation and contingency are distributed at the outset, you will be committing funds against inflated budget numbers. That means we are making the target cost higher than we really hoped to pay, which leads us into a feeling of complacency about meeting targeted costs as we travel through the project.

Our first effort in committing funds for project resources should be against the uninflated number. That applies the necessary pressure on the project staff to buy the best value and also to ensure that the staff is operating within the minimum specification suitable for meeting the project's quality requirements.

Sometimes company policy demands that the contingency and escalation be prorated among the accounts at the start of a project. In my opinion, having people committing funds against a fat budget is not the best way to beat a budget. All efforts to change such a flawed policy should be made if you hope to ever have successful projects.

Holding the two blocks of funds in separate accounts allows the project staff to use them to offset inflation and contingency costs as they arise during the project execution. For example, home office salaries usually escalate only once, or at most twice, a year at scheduled times. Thus, any actual escalation of unexpended base salaries and fringes can be factored into the budget as required.

Escalation on hardware purchases depends somewhat on the speed of placing the orders and getting fixed price bids. If the market or project conditions are such that firm prices cannot be obtained, the escalation account must be handled very carefully to avoid financial surprises at the end of the project. When we buy at fixed prices, we are prepaying the escalation, but at least the vendor is assuming some of the escalation risk—especially if the delivery is late.

If needed, contingency is fed into the budget at certain milestone points in the project. The timing for that activity depends on how well the project was estimated in the first place. A less accurately estimated project would have a larger contingency account which might have to be fed into the budget earlier as required. If early use of the contingency fund is not required, it can be adjusted downward as the project progresses.

Conversely, a well-defined estimate and budget might not require the release of any contingency funds until the 50 percent commitment point is reached and a significant overrun condition is perceived. Each area in which contingency is to be applied must be analyzed separately, and the proper amount of funds must be added to each account where the need is justified.

The release of the contingency account is usually considered a very important action, and it is taken only with the concurrence of the upper operating managements of the contractor and the owner. On incentive types of contracts, these two accounts are usually part of the upset price in which the owner and contractor share the savings, so careful administration of these funds is critical to calculating the final fee payment.

The escalation and contingency accounts should be evaluated each month against the remaining work to complete the project. They can be adjusted downward as the project nears completion and the window of exposure to escalation and contingency narrows. Any funds remaining in the accounts at project closeout are added to any budget underruns to arrive at the total budget underrun for the project.

Project managers of lump sum contracts have a special need to properly manage the escalation and contingency accounts because any amounts left over accrue to the project profit column. Since the owner agreed to pay a lump sum price, including any escalation and contingency allowances, the owner can make no claim on any unused balances in these accounts. If the balances come out on the negative side, the contractor can make no claim against the owner for additional costs for escalation and contingency unless there is an escalation clause in the contract.

The key point for project managers to remember when performing any type of contract is not to use the budgeted reserve funds for legitimate project scope changes which may be requested by the client. Project managers must initiate change orders to cover the new costs and not tap their contingency funds, which may be needed later for their intended use.

Computerized vs. Manual Budgets

The decision of whether to use computerized or manual methods for setting up the budget was once based on the size of the project. In those days we did small projects manually to avoid the high cost of the mainframe computer charges. That is no longer necessary because even the smallest offices have access to personal computers. With the use of today's standard spreadsheet programs, it is relatively simple to set up a computerized budget for a small to medium-size project. Only a very large project is forced to use a mainframe computer because of the size of the data bank and the processing speed.

The main problem with using the computerized format is keeping the budget posted as commitments are made and funds are expended. The need for updating is really no different from that when making manual budget entries, but we tend to think that somehow the computer is able to make the ongoing entries by itself. Be sure that you make one member of the project team responsible for making the entries on a routine basis as the paperwork on the project is generated. That ensures that other members of the team will get accurate, current information when they call up the budget on the project PC.

Budget format

Budget formats can vary widely depending on company standards, but a few things are required to make any budget work. Figure 8.5 illustrates a budget format suitable for control of capital project costs. The header contains the necessary identification data and calls out the title of the report, project name, number, location, report run, and period-covered dates, along with the description of the costs covered in that section of the report. Once the original budget is approved, it generally takes on the title of Project Cost Report, since it is a *living* document subject to monthly updates.

The account code and the description of each cost item are listed in the left-hand columns. The items are usually grouped by cost codes in sequential order of the account numbers. By using the same format as in Figure 8.5, each major equipment account is further itemized by equipment tag number. That way, each item purchased is compared against its original estimated cost.

On most projects, the individual cost code groups are "rolled up" or summarized into total costs, which are then shown in an executive summary report as shown in Figure 8.6. For example, each piece of process equipment will be itemized with the budgeted cost in the equipment cost account. Then the total cost of all equipment will be shown on a roll-up summary sheet as the equipment account, along with the cost of all the other materials such as concrete, steel, and piping to give a total equipment and material cost for the project. Using that method makes it possible to show the total cost for the project on a single executive summary sheet for a quick overview by upper management.

The next column, Original Budget, lists the approved estimated cost budget obtained from the original cost estimate. The numbers in the budget should never be changed, because they form the approved baseline for the financial plan against which the final cost report will be compared. Any authorized changes in the original budget can be handled in the next column, called Transfers. This column is used if it becomes necessary to transfer any funds from one account to another for any reason.

The next column to the right is called Approved Extras (or Change Orders); it covers any approved changes in scope or other approved extra costs. The sum of all the columns to the left is then equal to the column entitled Current Budget, which represents the latest budgeted cost for the project. The columns to the right of the Current Budget column represent the recording of project commitments, funds paid out, and the estimated cost to complete. We will discuss the detailed workings of these columns later when we cover the subject of cost control and reporting.

I strongly recommend that all project managers pay close attention to the conversion of the cost estimate to the original project budget. Your acceptance of the original budget means that you agree with the financial baseline for the project costs. These are the numbers against which your performance on meeting the project's financial goals will be judged. If you feel that the numbers are seriously flawed, now is the time to speak up about it. Protesting later in the project that you had a bad budget will not improve your image!

The Project Cash Flow Plan

The final leg of the financial plan is the project cash flow plan. It is the projected rate at which the project's cash resources will be spent. Note that the cash flow plan is based on monies actually *spent,* and not just *committed.*

The project financial people must know when the actual funds are needed to support the project. If the funds for the project are set aside at the start, any funds not immediately needed for the project must earn the highest rate of return prior to their expenditure. The people who administer any construction loans also need a schedule for the expected flow of the funds.

The cash flow plan is developed from the cost estimate and the project schedule. The overall shape of the cash flow curve will follow that of the project life cycle bell curve. Payments start slowly at the front, rise to a peak at the midpoint, and taper off at the end. Plotting money spent vs. time will result in a classic S curve. The projected monthly expenditures also can be shown as a series of vertical bars which will result in a typical bell-shaped curve when the tops are connected. See Figure 8.11.

The first expenditures pay the invoices for or the costs of the design effort. As we noted earlier, these are a relatively small part of the overall project costs. The cash flow curve falls behind the normal percent progress curve because the purchased equipment bills come in after the equipment is fabricated and delivered. On a split-contract A&E type of project, the cash flow for the equipment and material costs does not start until the contractor actually receives the equipment and material on the jobsite or starts the field payroll. That makes the initial slope of the cash flow curve even less steep in the early stages of the project.

An exception to the general trend for the cash flow to lag the progress curve would occur if the equipment vendors needed advance and/or progress payments for their work. That does happen on some foreign projects in which it is customary to have the owner finance the fabrication of the equipment with down payments and progress payments.

The cash flow curve is usually evaluated each month, along with the project budget and schedule. Any significant changes to the schedule and budget are incorporated into the cash flow curve to update the curve each month. Reviewing the cash flow is a fairly routine activity handled by the cost people on the project. The project manager needs to get into the act only if serious cash flow changes, cost overruns, or cash shortages suddenly appear.

Summary

The project manager plays a key role in generating the project financial plan, which establishes the baseline for the control of all project costs. It is the part of the overall project plan that enables us to meet our goal of finishing the project under budget. The project manager must become well informed in the art of cost engineering to ensure that a sound estimate of the project costs is made. The project cost estimate must be converted into an effective project control budget, which serves as a basis for the cost control system. A project cash flow plan is a valuable tool for projecting the cash needs for the project.

References

1. Forrest D. Clark and A. B. Lorenzoni, *Applied Cost Engineering,* 2d ed., Marcel Dekker, New York, 1985.
2. Kenneth H. Humphreys and Sidney Katell, *Basic Cost Engineering,* Marcel Dekker, New York, 1981.
3. Robert Snow Means, *Building Construction Cost Data,* Robert Snow Means Company, Duxbury, Mass., latest edition.
4. *Dodge Construction Systems Costs,* McGraw-Hill Information Systems, New York, latest edition.

Case Study Instructions

1. Make a feasibility (class 1) type of estimate for your selected project. State the basis of the estimate, (i.e., curve, unit tonnage, unit area, etc.), any assumptions made, and the expected accuracy of the resulting estimated cost.
2. The owner of your selected project has informed you that the economic feasibility of the project has not yet been fully proved. The owner has requested you to recommend a series of project checkpoints at which the project economics can be reevaluated prior to committing 15 percent of the total project costs. Prepare a list of your checkpoints and your reasons for selecting them. How would you implement your plan?
3. Prepare a project budget format for your selected project and fill in the executive summary sheet showing the major cost accounts. Prorate the total plant cost into the major accounts in accordance with a typical cost breakdown for the type of project you have selected. Present a proposal to your management (or client) showing how you plan to administer the escalation and contingency accounts in the project budget.
4. Select the contingency and escalation factors you would recommend using for each of the four major estimates that you expect to make for your selected project. Give your reasons for the values which you selected for each estimating stage.

Chapter

6

Project Resources Planning

Up to this point we have prepared a project execution plan, a time plan, and a money plan for the project. Now we are ready to turn to planning the other resources needed to implement the project execution.

There are six key project resource areas that require early evaluation and forward planning on the part of the project manager. Some projects may require only one or two of these areas, whereas larger ones may involve all six.

The six key areas are:

- Human resources
- Equipment and materials
- Services and systems
- Transportation arrangements
- Project facilities
- Project financing

The first two areas are the most important from a financial and planning standpoint, and they have the greatest influence on performance early in the project. Also, they are used on virtually every capital project.

Human Resources Planning

At this stage we are discussing only the types and numbers of people needed and not the way in which we will organize them into a team. Project organization will be handled as a separate subject in a later chapter. Human resources planning breaks down into three major categories on a total design-and-build project as follows:

- Home office personnel
- Construction personnel (supervision and labor)
- Subcontractors (engineering and construction)

Theoretical personnel loading curves

The simplest form of a personnel loading curve is a trapezoid as shown in Figure 6.1. The curve plots personnel needed vs. the scheduled time to accomplish the tasks. Actually, the most efficient way to staff a project would be to immediately staff the average number of people needed on day 1, continue to the end of the work, and then drop to zero. The average number of people is arrived at by dividing the total hours by the unit calendar time, and it is shown graphically by the solid horizontal line in Figure 6.1.

We know that such an ideal loading is not feasible for many reasons such as not having all of the design data ready, not having all the people available, and not having places for the people to work.

Since we have to start from zero and assign people gradually, the next most efficient theoretical curve is the trapezoid. It has a uniform buildup, a level peak, and a uniform build-down. The peak level on the trapezoid must be slightly above the average personnel level to make up for the area under the curve (labor-hours) lost during buildup and build-down periods.

In actual practice, the personnel loading curve takes the shape of a bell curve as shown by the dashed line in Figure 6.1. Since the bell curve tends to fall inside the trapezoid at the start and the finish, its peak must extend above the peak of the theoretical curve to account for the lost hours. Remember, the area under the curve is a constant which is set by the labor-hours estimated to perform the work.

A good rule of thumb to remember is that the bell curve peak usually exceeds the average personnel loading curve (line) by about 33 percent. That rule allows you to make an approximation of the peak

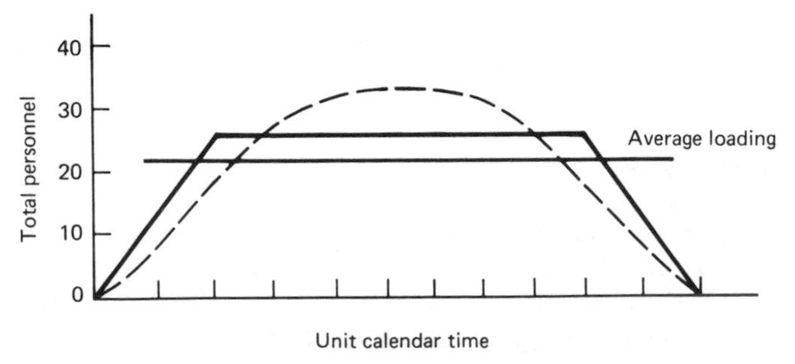

Figure 6.1 Theoretical personnel loading curves.

personnel requirement as soon as the design personnel estimate is completed.

Practical personnel loading curves

So far, we have been discussing theoretical curves, but now let us look at some more likely loading situations which are useful in planning our human resources. In Figure 6.2, I have shown the theoretical bell curve as a dashed line and the forward- and backward-loaded curves as solid lines. The latter two result when the personnel loading occurs earlier or later than normal on a project.

The significance of these conditions becomes apparent when we look at the family of S curves which results from plotting the percent of hours expended vs. the scheduled time as shown in Figure 6.3. The S curve of the normally loaded project has a gradual start and finish, which indicates smooth starting and finishing conditions. The forward-loaded curve shows a rapid project kickoff and an even more gradual than normal phase out at the end. The backward loaded project indicates a more relaxed start and a very steep finish slope on the S curve. The steep finish leads to such problems as inefficient use of personnel and overrunning the budget. The inefficiency results from having too many people working on only a few remaining tasks.

Normally, projects cannot phase personnel off the job so quickly, which means the project will overrun the schedule and finish late. The simple lesson to be learned here is that front loaded projects may slip to a normally loaded mode and still finish on time. There is little or no hope that backward-loaded projects will finish on time. Any slippage during execution further exacerbates the problem and makes the project finish still later than planned.

Project managers must remember, however, that front loaded projects don't happen just by drawing the personnel loading curves that way. All of the necessary start-up requirements of design data,

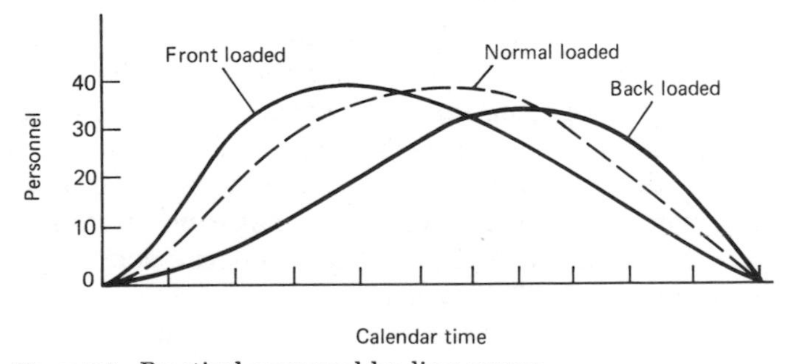

Figure 6.2 Practical personnel loading curves.

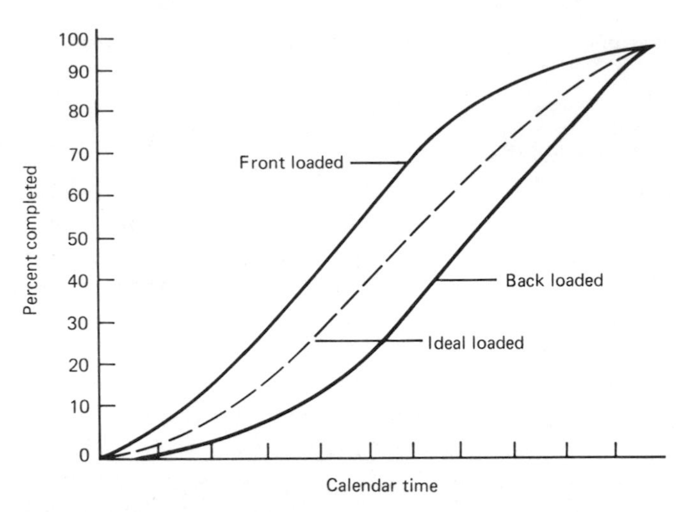

Figure 6.3 S curves from bell curves.

facilities, personnel, and/or materials and equipment must be available to make it happen.

The classic form of bell and S curves introduced here is used throughout the project, as we will see in succeeding chapters. For a more detailed discussion of bell and S curves and their varied applications, see the chapter devoted to that subject in Kerridge and Vervalen's *Engineering and Construction Project Management.*[1]

Home office personnel

To arrive at the total project personnel distribution for the design effort, it is a good idea to plot a curve for each discipline to be used in the design team. A typical example of such a plot is shown in Figure 6.4, which is actually a composite of all disciplines needed for a process type of facility displayed in one graph.

The basis for making the curves is the number of labor-hours for each discipline originally estimated by using the home office estimate format shown in Figure 5.3. Each discipline head has to project the number of people that will be needed to carry out the work scheduled for that week. The number of people for each week is plotted on the graph by using a different line code for each discipline.

The weekly value for each discipline curve is added to plot the composite total project personnel curve. You will note that each discipline tends to peak at a different time in the schedule but the composite curve peaks in about the middle of the project.

The composite curve gives the project manager an indication of the

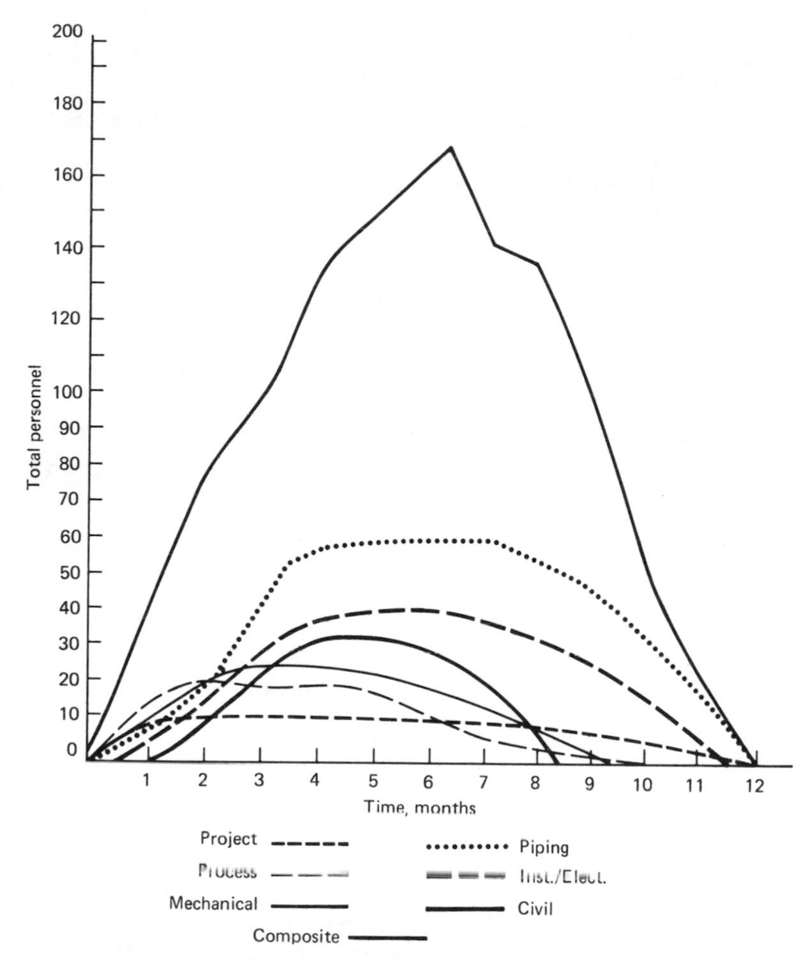

Figure 6.4 Typical design project personnel curve.

type and size of facilities needed for the project as it develops. The S curve, which is developed from the total personnel plot, is used by the project manager to track the overall design team's progress during the control phase of the project.

The early and late start dates for drawing the curves are derived from the project schedule. Thus we see that the project team starts early in the project and builds rapidly to a long, flat peak that holds until the design work tapers off. The process design package will be ready early in the project, which will let the process group build up sharply, peak early, and tail off first. The mechanical group starts at about the same time in order to move the long-delivery equipment requisitions into the procurement department. If the mechanical and process groups are combined, the process line will have a longer duration proportional to the labor-hours involved.

Piping shows an early start with a slow buildup and a later peak because the piping people are only working on the equipment layouts, plot plans, and piping specifications early on; they do not get into the peak piping work until the P&IDs and layouts are available. Civil (CAS) work starts later, since those people depend on the site surveys, soils report, equipment layouts, and equipment weights to start design work. Some of the CAS work must finish early, however, to support the opening of field construction activities.

Instrumentation and electrical are generally late starters because of the development of basic design data, so they wind down later in the design phase. The variation in the early and late start and finish dates among the various disciplines leads to a fairly balanced bell shape in the composite curve.

The most striking thing about the composite curve (Figure 6.4) is its sharp peak. The sharp peak would be a definite cause for concern in a construction labor curve because it could lead to an overcrowded condition on a limited-access jobsite; in a design curve resulting from fairly flat individual discipline curves, it should not present a special problem except for project task force space. I would not spend much time worrying about the peak in the design team. The actual curve will tend to roll to the right because of scheduling problems which normally occur in most design efforts. Except on very large projects, the total elapsed time for the design effort is not great enough to have multiple peaks. The shorter elapsed time in design schedules precludes the "roll-to-the-right syndrome"—one peak rolling over a later peak and thereby creating a super peak. The syndrome could very well occur in the longer total time in the construction schedule. That is the basic reason why personnel peak shaving is more critical in construction personnel loading curves.

Remember, Figure 6.4 is for a process project. A nonprocess project will have another set of discipline curves depending on the type of facility being designed. The curves should reflect your labor-hour estimate (or budget) and project schedule.

Note that Figure 6.4 does not include curves for any procurement or project control people who are a part of the home office services estimate or budget. Since the project control hours are relatively minor, to show a separate curve would unnecessarily complicate the picture. The hours could be included with the project engineering hours, since they follow a similar pattern.

I have found that procurement progress is better monitored through a purchasing commitment curve than a personnel curve, as we will learn later when we consider project controls. A separate control of procurement personnel hours should be maintained by the project purchasing manager.

Each design discipline leader should also keep a separate set of curves for use in planning and monitoring the work in that discipline. A. E. Kerridge's chapter in *Engineering and Construction Management*[1] is an excellent dissertation on the development and use of bell and S curves in project management.

Construction personnel planning

When the project involves construction, either as an integrated project or as a separate construction contract, the field personnel resources must also be carefully planned and controlled. In either case, the detailed field personnel planning is usually a function of the construction department or the contractor. On an integrated design, procure, and construct project (DPC), the project manager monitors the construction personnel planning only as it relates to the overall management of the project, the schedule, and the budget.

The volume of construction labor resources can amount to 7 to 10 times the design hours depending on the type of project. With that number of project resources involved, it is doubly important to properly plan and control the field supervision and labor budgets.

The field personnel break down into two groups of people: field supervision and craft labor. Percentagewise, the field supervision labor-hours are very low compared to the craft labor-hours, so personnel loading curves are not usually needed for the supervisory staff. A simple list of the staff and their proposed duration of assignment is sufficient.

The number and type of field supervisory staff varies greatly depending on the size, contracting plan, and type of project involved. Table 6.1 shows a range of supervisory staff sizes one might expect to have on various types of projects.

TABLE 6.1 Typical Sizes of Field Supervisory Staff

Contracting basis	Number of people		Types of people
	Process	Nonprocess	
Self-perform (direct-hire craft labor)	30–50	15–25	Managers, craft supervisors, foremen, timekeepers, etc.
Construction management (subcontracted)	10–20	5–10	Manager, supervisors, control people, etc.
Third-party constructor	2–10	2–10	Inspectors, engineers, or architects, etc.

The self-perform format means that the contractor is hiring most of the field craft labor directly with a minimum of subcontracting. That leads to the largest field staff because all of the supervisory staff is provided by the prime construction contractor.

A construction management approach requires less supervision because the subcontractors will furnish the craft labor and supervision as parts of their contracts. The construction management firm need only supply the management and administrative personnel to control the field work.

The third-party constructor mode requires only limited coordination and inspection services from the design contractor, so the field organization is the smallest of the three options. This example of the third option does not include the third-party constructor's field supervision people.

Field craft labor planning

Construction craft labor planning usually starts with a preconstruction labor survey in the area of the proposed jobsite. In addition to checking the supply of craft personnel, the study determines the prevailing wage rates, fringe benefits, local working conditions, competing projects, etc.—all factors affecting the availability, cost, and productivity of field labor. All of the information gathered is assembled in a field survey report which is used as the basis for estimating the field labor costs and craft labor planning.

Otherwise, planning for the craft labor is done similarly to planning for the design personnel, as discussed previously, except that the different trades involved are used instead of design disciplines. Since the time schedule for the construction phase is at least double that of the design phase, the individual craft curves tend to have flatter peaks. Since the number of field labor-hours is usually 7 to 10 times as great as the number of design hours, higher peaks can develop in the field. A typical example is given in Figure 6.5.

A major difference in the field labor curves is due to the use of construction subcontractors for a large percentage of the work. For example, steel for the project may be purchased on a fabricate-and-erect basis, which causes the ironworker labor to be reflected as a subcontract. If the job involves a large amount of steel work, it may be desirable to include the ironworkers in the overall field labor planning chart as in Figure 6.5. On the other hand, if the work of a ceramic tile subcontractor is minor, it may not be worth putting in the diagram.

As with the design labor curves, the area under the curve is equal to the estimated number of hours for that part of the work. The start and finish times for the curve are taken from the construction schedule.

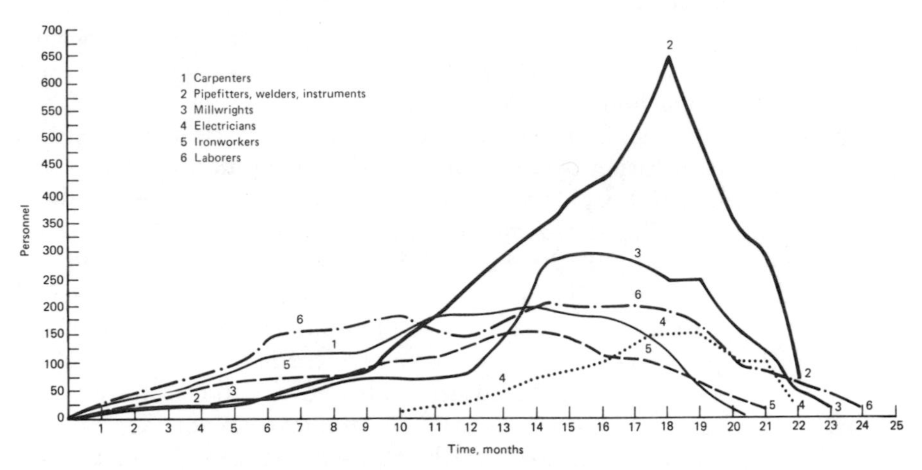

Figure 6.5 Construction labor-planning curve.

Therefore, foundation work comes first and is followed by steel erection, floors, curtain walls, and so on, as the job progresses.

The composite field labor curve is developed by adding the individual craft numbers and plotting the resulting value. As we can see from the sample curve, craft labor peaks do occur from time to time in both the individual craft and the composite curves. If the peaks are found to be undesirable for any reason such as labor shortages or overcrowding of work spaces and field facilities, the schedule can be manipulated to reduce the peak requirement.

Labor peaks can be leveled out by rescheduling the start of some noncritical activities to an off-peak time. Naturally, that can be done only with the activities which have sufficient float and are not on the critical path. If we have the good fortune to be using a computerized CPM scheduling program, the peak shaving can be done automatically with the resource-leveling feature of the CPM program.

It is possible that not all of the peaks will respond to the rescheduling techniques because of the critical nature of your schedule. At least, however, your forward planning of the field labor will give you ample warning of any impending field labor problem areas in time to plan alternatives.

Subcontracting Design and Construction

A convenient way to expand the human resources on a project is through the use of subcontractors or consultants. That means is often used in the areas of design and construction to expand a firm's capability in certain areas of expertise as well as in the area of human resources. The project manager plays a major role in implementing the project subcontracting plan and overseeing the subcontractor's performance. That is another area of project management which sometimes does not get the attention it deserves.

If the project design resources are only moderately overextended, the most ready solution is to expand the staff when needed with some contract personnel. People from almost any discipline can be obtained from agency firms specializing in that sort of business. Fortunately, there are a lot of qualified people who prefer job shopping to regular full-time employment. The services of these people can be added and deleted at very short notice without obligation to pay any benefits or termination pay. The markup on these services is a nominal 25 to 30 percent of base salary to cover all taxes, benefits, and profit. The base pay for shoppers does tend to be slightly higher than that for regular employees, and all rates are usually quoted with premium rates for overtime. On average, however, the overall hourly cost for contract vs. captive personnel is about a wash. One thing that does tend to push overall contract personnel rates higher is the payment of per diem allowances for people brought in from out of town.

Both captive central engineering departments and design contractors have successfully used job shoppers to handle temporary peak staffing problems on many projects. Some companies have tried to establish limits on the number of contract people that can be used without jeopardizing the quality of the design. As long as you have enough qualified permanent employees to direct and supervise the work, I see no problem with going as high as 20 percent. That assumes your people have screened and selected qualified shoppers for the project. Doing the work in-house by using contract people eliminates many of the disadvantages of subcontracting the design, as we will discuss in the following paragraphs.

Design Subcontracting

On commercial A&E projects, design subcontracting is a way of life for smaller firms. Such firms do not have all of the skills needed to design a project in-house, so they must go the subcontractor route on almost every project. Even though it is a way of life, the subcontracting interface is a prime source of technical and commercial coordination problems for the person managing the project. When an integrated design firm subcontracts some of the design on a larger project, the interface problems are likely to be even worse because the ability to work in that mode is not that well developed. The following are some areas requiring close attention by the project manager:

- Increased coordination and supervision time and cost
- Introduction of differing design standards

- Increased quality control problems
- Communication problems
- Dilution of project control

Increased coordination

Chances are good that the subcontractors will be working in their own locations across town or halfway around the world. The number and seriousness of the coordination problems will be directly proportional to the distance between offices. If the subcontract is large enough and remote enough, you may have to consider setting up a group of project coordinators and discipline supervisors in that office to ensure compliance with project schedule and quality standards. The costs for the duplication of staff must be considered in estimating the project design budget.

International projects present by far the most difficult situations for design subcontracting. On most international projects today, the host country insists that all design work possible be carried out in that country. The reasons for the demand are to conserve hard currency and foster the development of local design capability.

Some years ago, I managed a large petrochemical project to be built in a far-eastern developing country. The contract called for the front-end design work to be done in the United States and the detailed design to be done in the host country. It amounted to about a 50-50 split of the design budget. In that case, the owner selected a newly organized captive engineering company with very little petrochemical plant design experience either as a company or as individuals.

We anticipated that it would take a minimum staff of 15 to 20 of our people to monitor the work over a 14-month period. Considering the cost of putting a person and his family on the ground in that country for the required period of time, we estimated the total additional supervisory cost at about $2.8 million. When the negative design productivity factors were applied to the lower pay scales of the host country, there was not much savings available from the local design budget to offset the added supervisory cost.

There has been a rapid growth in the capability of local design firms in some of the developing countries worldwide, but the improvement has not been universal. So, beware when you are forced into that sort of subcontracting situation!

Job standards

Differences in job standards can also be expected when subcontracting design work. Every effort should be made to ensure that the subcontractor conforms to the project design standards. As a minimum, the best goal to set is to make the finished design look as though it had been done by the same company. Most subcontractors are not very enthusiastic about conforming to your standards because it adversely affects their schedules and costs. You will have to include enforceable language to that effect in the subcontract!

Quality control

Ensuring that the subcontractor adheres to the project's quality standards is a serious matter. As prime contractor, the owner will hold you responsible for the subcontractor's work. Therefore, you must ensure that the subcontract has in it the same warranty clauses that appear in your prime contract. Any shortfall by the subcontractor in the areas of quality, cost, and schedule will be held for your account.

Communication problems

Like problems of coordination, those of communication are functions of the distances between offices. The further the offices are apart, the higher the communications budget will be to cover travel and living costs, communication system costs, and the time spent in ironing out misunderstood directions. If things get too bad, resident supervision is the only answer, and often it is not a complete solution.

A lot of improvements in communications that have been developed in recent years must be investigated for application to solving communication problems. Document faxing has probably been the most cost-effective means of solving interoffice communication problems. Some other means, such as teleconferencing and videotaping, sound as if they have a lot of potential, but apparently they haven't yet proved to be economically feasible in project work.

Dilution of project control

Problems in controlling money and time plans are multiplied in a design subcontracting situation. It is important that you keep a close watch on costs and schedule early in the work to keep the money and time in line. Again, the owner will hold you responsible for any failure on the part of the subcontractor.

In addition to the subcontract document, you should develop a procedure manual for the control of the subcontracted work. The document should delineate the noncontractual matters that are needed for smooth subcontract operations. For small subcontracts, it can be a few pages, but for larger subcontracts, it should be modeled after the prime project procedure manual. Project procedure manuals will be discussed in detail when we consider project controls.

After discussing all of the potential problems, you may wonder why we go into design subcontracting at all! Believe me, most people do it only after all the other ways to get the job done have been exhausted. On the other hand, I have seen many design subcontracting projects in which the work was properly managed. Sometimes when special expertise is needed, a subcontractor can do the work better, quicker, and cheaper than you can. If the subcontractor's expertise is so broad that you are able to get a lump sum price, your cost control problems can also be much smaller.

I hope this discussion of design subcontracting will make you sufficiently aware of the potential pitfalls to plan them out of existence.

Construction Subcontracting

Most of the problems we discussed under the subject of design subcontracting do not apply in construction work. Such trades as insulation, painting, electrical work, sheet metal, and roofing are normally subcontracted on all construction jobs. Since the subcontractor has to come to the jobsite to perform the work, the problems of communications, cost control, and quality are easier to handle. The prime contractor's field supervisors are normally available on the jobsite full-time to oversee the work.

The prime project control people also supervise the activities of the subcontractors for conformance to schedule, make progress payments, and handle change orders. As with any subcontract, it is critical to have a good set of subcontract documents to control the work. Subcontracting in construction reaches its peak when a construction management format is used. In that case, a construction management team is assigned to the project to totally subcontract all of the work for the owner. That arrangement was discussed earlier as the second case under project field supervisory staffing.

Material Resources Planning

The basis for project material resources planning is the project procurement plan. The document is created by the project purchasing agent (or manager). The report resulting from the evaluation of the procurement situation should include current delivery data on any long-delivery equipment and materials. That is in addition to a survey of current market conditions, price trends, materials availability, and vendor lists. Those data are invaluable in formulating the project material resources plan.

Ensuring that the material resources for the project arrive on time involves these important planning areas:

- Long-delivery equipment
- Special materials and alloys
- Common materials in short supply
- Special construction equipment
- Services and system requirements
- Financial resources

An early review of the project's physical resources must be made to give the critical items special attention. They must be recognized early on to preserve any available float or to keep them from slipping into negative float.

Long-delivery equipment

We generally consider any equipment with a delivery of 10 or more months to be *long-delivery equipment,* and delivery times as long as that will place the equipment on the critical path. Examples of long-delivery equipment include such items as high-horsepower centrifugal compressors, turbogenerator sets, heavy-walled vessels, field-erected boilers, paper-making machines, and rolling mills. You may not find any of those items on your equipment list and therefore assume that you do not have any long-delivery equipment to worry about. That is a false assumption; *every project has long-delivery equipment.* On any project, the items with the longest delivery dates are the long-delivery items. If the schedule is going to be improved, it can be done only by improving the delivery of equipment that falls on the critical path.

Long-delivery items must be given top priority in the schedule, starting with the first operations in the project design. If they are engineered equipment, the data sheets must be generated first and pushed into the procurement phase for bid taking. Each step should be expedited to get the order placed. Continued expediting through the vendor data approval stage also is advisable. The vendor will not start working on the equipment until approved shop drawings are received.

Long-delivery equipment is often preordered by the owner on preliminary sizing criteria that are refined later by change orders as the design progresses. If that method is used, unit prices for changes should be obtained from the vendor with the bid price to have better control over change order costs. A typical example of the prepurchase approach is the preordering of paper machines in the pulp and paper industry.

The importance of following the high-priority system throughout the long-delivery process is most important. I was using that approach on a project which involved several large centrifugal compressors with about 12 months delivery. By working closely with several of the best vendors, we developed our equipment specifications for the units. To save time, we issued them for client approval and bids simultaneously. The client's comments and the bids came in about the same time, and the few minor discrepancies were resolved during the technical bid evaluation.

Unfortunately, the client was not able to follow through with placing the order for about 11 months. As a result, we were without the vendor's data needed to complete several major units on the scale models. The delay caused the construction schedule to slip about 9 months. That meant that our early efforts to mediate the effects of the long-delivery items were wasted. In addition to the scheduling damage, our project morale was badly bruised.

Special materials and alloys

It is a good idea to review all materials specified for the project that might not normally be stocked because they are so special. Lately, many manufacturers have taken special pains to reduce inventories, which makes unusual materials even more scarce.

In the process industries, special alloys fall into that category. Some of the rarer ones include the refractory metals, titanium, Hastelloy G, carbate, and to a lesser degree monel, inconel, and some special stainless steels. Some of these special materials may also require special treatment such as casting, tube bending, welding, stress relieving, and testing, which tends to delay their delivery even further. Since most project managers are not strong in that highly specialized area, it is best to have your technical staff or consultants thoroughly investigate any potential delivery problem areas.

The long-delivery-item reviews should be made early enough that there is still time to deal with the problem. After the bids have come in is too late to learn the bad news!

Problems of that nature can also be found in nonprocess projects. The design might call for a special aggregate in precast wall panels, special window systems, a rare quarried and polished stone, or any one of many others that are likely to fall on the critical path. Early planning for delivery can get such an item off the critical path, and that will pay the dividend of early job completion.

Common materials in short supply

Common materials in short supply often include such mundane items as structural steel, concrete, and reinforcing bars. In a large high-rise building the structural steel is heavy and high in tonnage. Early design and takeoff for placing mill orders for the heavy steel are critical to maintaining the schedule. Each section of the structure must be closely scheduled to have the steel delivered at a time that suits the erection sequence. A large dam project uses huge quantities of reinforcing steel, forms, earth fill, and concrete over long periods of time. These relatively common materials must be planned for and delivered on time to meet the schedule. These commonly used materials sometimes tend to be overlooked on larger projects, so don't pass over them lightly. The advice is especially valid if the project is being built in a remote location.

Special construction equipment

Unlike the other materials, special construction equipment does not enter into the final product, but I like to consider it a physical resource. It is something that must be investigated by the construction people during the design phase. I have seen many cases in which it was cheaper to redesign the plant equipment than to move in costly erection equipment which might otherwise be needed. That is difficult to do on a split design-and-construction project unless the construction people can be brought on board early. In the case of lump sum bidding, that is not possible. The design project manager should suggest bringing in a consultant if a construction problem arises.

All these factors come into play when preconstruction planning sessions are used on the project. It is also a good idea to have the construction people make a "constructibility analysis" near the 50 percent design stage as part of the project review.

Services and systems

Strictly speaking, the project services and systems are not physical resources either, but they must be planned for at about the same time the physical ones are. Planning for them is particularly critical on large projects.

One of the biggest decisions is whether to bring an outside contractor on board to execute the contract. If the answer is yes, you will have to initiate the contractor selection procedure as we discussed in detail in Chapter 2.

The project scheduling, estimating, and cost control systems must be decided on at this time so the necessary manual or computerized project control systems can be selected effectively and put into place. If computers are to be used, the hardware and software resources must be organized.

Even such ordinary resources as the office space for the design team and the site facilities for construction have to be planned early. What form of security system will govern the project operations? How much space do we need and when? How will the drawings be prepared? Is a model of the facility to be built? Is construction to be in a remote location? What special climatic conditions are going to prevail? These are just a few typical questions that must be investigated and answered by the project manager. Some will fall on the critical path and some will not, but the critical items must be discovered early enough to allow for their resolution.

Financial resources

It has always been the owner's responsibility to arrange for the project financing. To a degree, that excludes it from the project manager's responsibilities, but the PM cannot be blind to the fact that the project must be backed by sufficient financial resources. Usually, PMs have the authority to approve all reasonable expenditures needed to execute the project. They also have the responsibility to see that vendors and contractors are being paid for goods and services rendered. Underfunded projects can create many problems for the project manager.

When the project is not adequately financed, I would be very careful about being the final authority in the chain of committing the project funds. I advise you to check your legal position if such a situation should arise.

I was involved with two major international petrochemical projects which were underfinanced, and neither has ever been completed. One was canceled before start of construction, and the other was stopped with construction about 35 percent complete. Fortunately, I was the contractor's project manager in both of these cases, so it was not my responsibility to approve expenditures except for engineering. In the absence of solid financing for the second project, it became increasingly obvious that the work could not go on. I was really just as happy to not be around when the project finally fell apart after such a major commitment of nonexistent funds.

My experiences over the years on a large number of projects have taught me at least two things: (1) Projects can eventually be finished *without* such things as planning, organization, control, computers, leadership, sophisticated project systems, and even talented people. (2) Projects cannot be completed without money!

Summary

This chapter wraps up the planning portion of project activities. Sound planning forms the foundation for everything that comes later. Effective project managers must train themselves to plan all facets of the project as well as their day-to-day work activities.

We must remember, however, that a plan is only a proposed baseline for the execution of a project. It is subject to possible changes along the way. Although it is often necessary to change our plans, I do not recommend making any *radical* changes to your original plan. If your plans were well thought out, you should resist pressures to change them.

References

1. Arthur E. Kerridge and Charles H. Vervalen, *Engineering and Construction Management,* Gulf Publishing Company, Houston, Tex., 1986.

Case Study Instructions

1. Prepare a project personnel resource curve for the design portion of your selected project. Use the breakdown of labor-hours from your home office services budget. Discuss any unusual peaks in the curves and what you would do to eliminate them.
2. Use a ratio of 7 to 1 for field labor-hours to design hours to prepare a construction labor personnel curve for the applicable crafts involved. Be sure to include any major subcontractors. If your curves show any high peaks, investigate them and suggest ways to lower them.
3. On the basis of the contracting plan, describe the size and make-up of the field supervisory staff you would anticipate having on your selected project.
4. Assume that your design staff will expect the following shortages in personnel: Case 1, 5 percent; Case 2, 50 percent; and Case 3, 75 percent. What subcontracting plan would you recommend to your management for making up the shortfall if you were acting as (1) in the role of the owner's central engineering group or (2) as a prime design contractor? Give your reasoning in support of your plans.
5. Review your selected project for expected long-delivery items of any kind. Discuss your recommendations for minimizing their adverse affects on your project's strategic end date and schedule.

Chapter

7

Project Organization

With planning out of the way, we are now ready to get into the organizational part of our project execution philosophy of *planning, organizing, and controlling*. I am sure you understand that we do not perform those three activities in a compartmentalized fashion as I present them in this book. All the activities, in fact, overlap and proceed concurrently. We are continually reviewing all three areas and modifying them as necessary to meet day-to-day operating conditions.

Organizational Overview

Organizing can be defined as "The function of creating in advance of execution the basic conditions that are required for successful achievement of objectives." My first law of organizing is "Design the organization around the work to be done to meet the objectives, not the people available." The law is based on the assumption that you have access to a bank of skilled personnel to perform all the necessary activities. However, we know that rarely happens, and some compromises have to be made in selecting a project team.

The general goal of any organizational structure is to establish the proper relationship among:

- The work to be done
- The people doing the work
- The workplace(s)

Later we will work our way into building some typical project organization charts based on the above principles.

Organizational design

By way of building some background in organizational design, it will be valuable to review some of the basic theory and practice used to design organizations. Most project organizations tend toward the military style of using vertical and horizontal structures. As with the military, we also use line and staff positions.

In recent years, other forms of organization charts have been put forward, but they do not seem to have caught on in the capital projects business. A circular arrangement, such as the one shown in Figure 1.2, is an example of a different type of chart. An advantage of the circular layout is its ability to show the relationship of the project manager to all of the other operating departments of the company. Despite that advantage, it does not seem to be used often.

A vertical structure, as the name implies, places one position (block) over the other as shown in Figure 7.1. The vertical dimension establishes the number of layers in the organization. Keep in mind that every layer introduces a communications filter into your organization. Each filter is a choke point for the necessary flow of vital project information.

In the horizontal dimension shown in Figure 7.2, a number of horizontal positions report to a single block above. The horizontal dimension introduces the term "span of control." In a vertical organization (Figure 7.1), the contacts are one on one; in the horizontal case, they are n on 1. If n becomes too great, supervisor A will not have enough time to devote to each of the supervised blocks. There is no fixed maximum for n, since the ability to supervise depends on the nature and complexity of the contact. The number 6 is often mentioned in management practice as a normal maximum span of control, and it is certainly one I would not exceed without giving the matter a lot of thought. Remember too that the communication filters are added in the horizontal mode as well as the vertical.

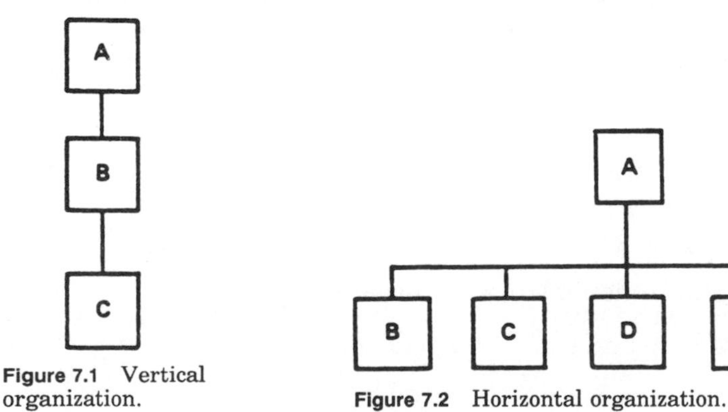

Figure 7.1 Vertical organization.

Figure 7.2 Horizontal organization.

Figure 7.3 compares layers of tall and flat organizational structures. The tall (lower) organization shows the rapid growth in the ratio of intermediate supervisors to workers. The horizontal structure has fewer managers, but each manager has a much broader span of control.

As we can see, the lower format shows a ratio of 15 managers (supervisors) to 48 workers, whereas the upper chart shows only 5 supervisors for 48 workers. Keep in mind that labor-hours are the main part of a project budget and the supervisory hours are relatively non-

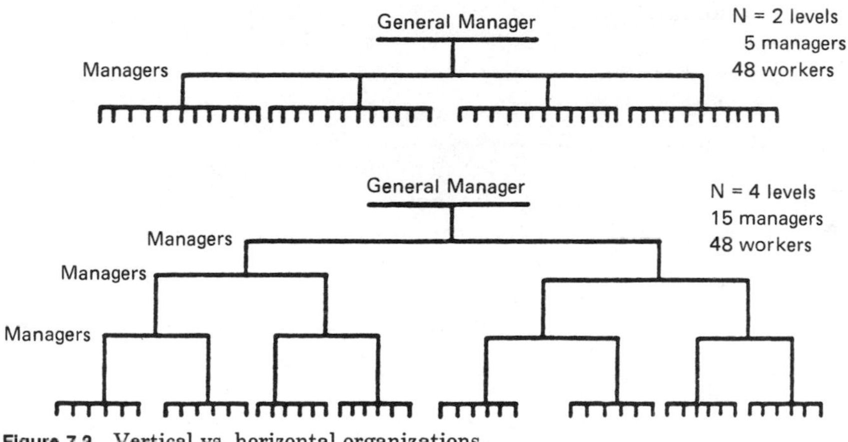

Figure 7.3 Vertical vs. horizontal organizations.

productive. If supervision runs over 10 to 12 percent, your budget is likely to be in trouble.

Obviously, we will want our organization to have an optimum balance of vertical and horizontal dimensions to give us the most effective control of the project work. I prefer to have my project supervisors slightly overloaded to encourage them to extend their capabilities and have an opportunity to grow with the job. Remember that some people exceed their span of control because they are not organizing their time effectively. I do not consider that a sound reason for revising the organizational structure to eliminate the problem. Some on-the-job training or a shift in personnel is definitely a better solution to such an apparent span-of-control problem.

Line and staff functions

The line and staff functions in your organization are best defined by looking at the project goals that we discussed earlier. Our main goal is a quality facility built on time and within budget. To build a quality facility, we must produce a sound design and deliver the project resources in a timely manner. That means that all of the design staff—architects, engineers, designers, checkers, and so on—perform in a line capacity to meet the main project goals.

In a business or manufacturing setting, we normally look on procurement as a staff function. How about our project procurement function: is it line or staff? In the project mode, procurement delivers the project's vital physical resources. That definitely makes it a line function.

How would you classify the functions of scheduling, cost engineering, and project accounting? Although they play a major role in meeting the project's time and money goals, they more closely fit the definition of the staff function. They monitor and report on the performance of but have no direct authority over the various line functions. The authority of the staff groups must be exercised through the project or construction manager.

The analysis of line and staff functions can be extended to analyzing the structures of owner and contracting companies. In owner companies the prime function of the line organization is the production of the product (or service); be it paper, chemicals, electricity, or health services. Therefore, the group responsible for creating the owner's facilities is functioning in a service or staff mode. That is why the owner's operating division (operating as a profit center) has the final say on facility design.

In the case of a contracting organization, the engineering, procurement, and construction groups make up the line organization; they are the profit centers for the contractor's operation. Their prime duty is to provide services to their clients at a profit. That explains why contracting companies have been forced into using a stronger project management approach to improve their project profitability. Project managers in a contractor setting usually have direct access to, and often report directly to, their top management. That is an important aspect to remember when owners and contractors are working together on a project.

Constructing the Organization Chart

Any operating project is worthy of an organization chart showing positions, lines of authority, titles, group relationships, and even the names of the current incumbents. Depending on the size of the project, it may vary from a single letter-size sheet to one or more E-size drawings. When approaching the architecture of your project organization, I hope that you will remember that *simplicity is beautiful* regardless of size!

The main advantage of the organization chart is to show all the project functions and the players in an easily understood format. That makes it easier for new people reporting to the project to get the feel of the team and the work assignments. It is also a good reason to have the chart available early in the project even if all the personnel assignments have not been filled.

My main concern with the efficacy of organization charts is the feeling of compartmentalization they engender. Any chart tends to set up psychological and communication barriers which detract from the team effort needed for successful project operations. A good project manager must work hard at breaking down this barrier-building tendency by playing a strong and charismatic leadership role on the project team. If a cooperative team spirit is not created across all of the blocks on the chart, project performance will suffer.

Since the types of project organizations used in the capital projects business is virtually limitless, it is impossible to show a sample chart for each. The samples discussed here can be modified to suit your particular type of project.

Typical project organization charts

Earlier we said that the purpose of the organization chart is to define the work to be done, the people doing the work, and the location of the work. The work to be done is a function of the project scope. A full-scope project involves design, procurement, and construction activities. We will use that scope to build some typical organization charts.

Organizations involved in a given project include owners (clients), central engineering departments, design consultants, procurement activities, and construction contractors in a wide variety of possible combinations. The chart should clearly define the working relationships and the duties of the personnel within the participating groups.

Remembering that our *overall project goal* is to produce a quality facility that will meet the needs of the owner, it might be a good idea to start at the bottom of the chart with the human resources required to reach that goal effectively. The project execution plan also goes a long way toward shaping the overall organization chart because it defines how the job will be done.[1]

Where the work will be performed also sets some givens in the construction of our chart. The field work is usually remote from the design office, and that calls for a division between design and construction. Procurement can be done in the design office or in the field. Design work can be split between offices and even among specialty design consultants, owners, licensors, subcontractors, and R&D. Each combination affects how we arrange the blocks on the chart.

Keeping the three factors controlling organizational structure in mind should allow you to construct a viable organization chart for any type of project. After all concerned parties have approved the basic structure, you can proceed with reviewing and selecting the key personnel to fill the slots. We will address that important function in more detail later in the chapter.

A typical process project organization chart

The relationships required on a process project organization chart can be fairly complex, especially when an owner, a design contractor, and a third-party constructor are involved. Each of the major players will have a multifaceted organization. Figure 7.4 shows one approach to how the three organizations could interact.

The top group of blocks show the main parts of the owner's team assigned to bring a process facility into being. The central engineering department (staff) has been assigned the job of designing and building a facility for one of its operating divisions (the client). Since the central engineering group does not have sufficient design, procurement, and construction resources, those services are farmed out to contractors.

Central engineering appoints a project manager to be responsible for supervising and coordinating all aspects of creating the new facility. The client's project manager has six major areas of control and/or contact. Fortunately, not all those activities will be peaking at the same time. For example, the work with R&D will peak early in the project, whereas plant start-up will not occur until late. Also, the design and construction activities will peak at different times, so the project manager should be able to effectively keep six balls in the air without undue strain.

The chart clearly shows that the project manager is the focal point for all information and communication between the owner and the various groups responsible for executing the work. That means that the client's

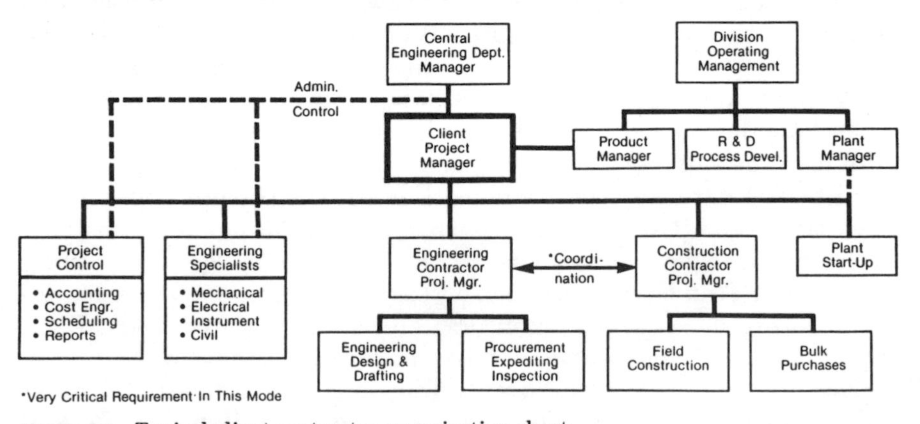

Figure 7.4 Typical client-contractor organization chart.

project manager has to have a well-organized project team to ensure timely flow of all project data, decisions, and approvals through the project team. If that does not happen, a king-size bottleneck that will stifle progress throughout the total organization will occur.

There are two more key information flow filter points on the chart which show up in the engineering contractor and construction contractor project manager slots. Obviously, the three project management slots control the main line functions in the overall project organization. The project control and engineering specialists serving the client's project manager, along with the product manager and R&D people, are supplying key staff support functions to the project. The engineering and construction functions shown in Figure 7.4 will be further detailed in separate charts showing some typical organizations required to support those major activities.

You will note that the only horizontal lines crossing between the vertical sections of the chart are at the upper levels. That brings home my earlier statement about the adverse effects of compartmentalization engendered by most organization charts. The lines of authority show most of the flow up the chart to some point and then back down to the point of execution. We also know that about 80 to 90 percent of the information must flow directly across the chart when a project is moving well. That is what I mean by the project manager's facilitating the flow of project information across the organization, but in a controlled manner.

Sometimes, the cross flow is inhibited by contractual buffers such as the one shown in Figure 7.4 between the design and construction contractors. Neither of the two prime contractors has any contractual obligation to the other except through the owner's representative. That means a specific coordination procedure must be set up to smooth out and control the major flow of project information between them. I have highlighted the critical interface with a double-headed arrow. If that key horizontal communication connection does not exist, or if it breaks down, the project goals will suffer.

Typical design organization chart

The chart shown in Figure 7.5 is a typical engineering-construction contractor chart. It serves as the detailed blowup for the design and procurement portion of Figure 7.4. In that case, the client's organization shown is the same as the one in Figure 7.4. Figure 7.5 could also represent a central engineering department's project organization. In that event, the client's organization would be the owning division's project group responsible for developing the facility.

In any case, the project manager supervises all the functional groups responsible for the line activities such as design, procure-

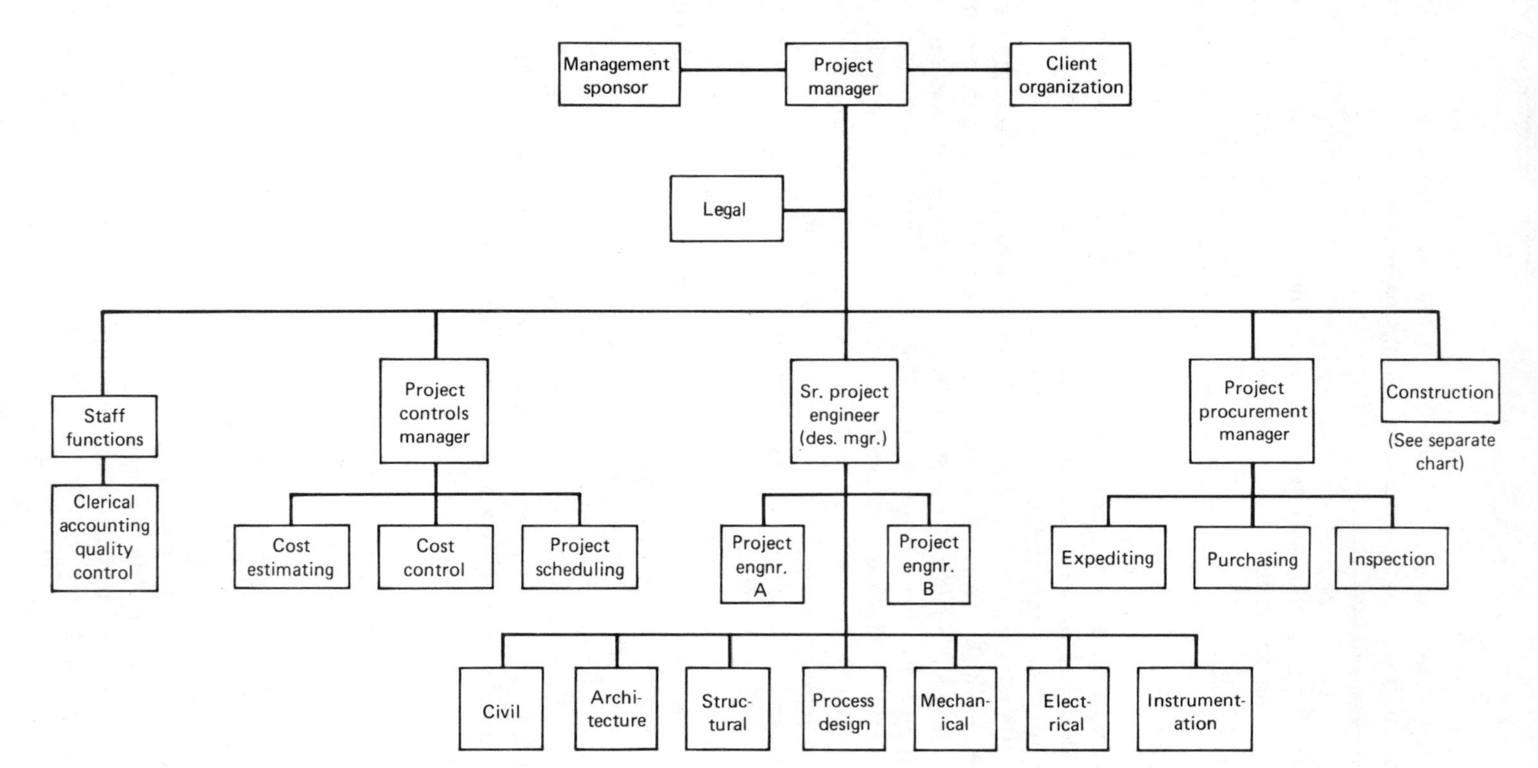

Figure 7.5 Contractor organization chart.

ment, and construction for the client. That is in line with the strong project management environment that exists in the typical contracting organization. One person is totally responsible to the client for project execution and to company management for project profitability.

The block to the left of the project manager shows a "management sponsor." That is a top-management executive who is appointed by the CEO to bring a definitive top-management focus to the project. The sponsor ensures that the project manager is performing up to standard and is getting sufficient top-management review and support in meeting the project goals.

The sponsor also maintains top-level contact with the client's project manager and his or her management to take a proactive stance in client-contractor relations. A good sponsor keeps a finger on the pulse of the project and judiciously steers the project manager away from the rocks of project adversity.

Although it is not usually a problem, the project manager must not let the sponsor take over active control of the project. Smart PMs make good use of their sponsors in handling high-level political matters and getting management support without surrendering control of their projects.

Moving down the chart in Figure 7.5, we find the project control people who are responsible for developing and monitoring the project time and money plans. On large projects, we might find a project control manager heading up the group. On small projects, those people would report directly to the project manager (or engineer) and could be part-time members of the project team.

The group performs only staff functions and has no line authority over the project's design, procurement, or construction operations. It gathers performance and cost data from the line groups and reports the results to the project manager for any necessary corrective action. The project manager has the duty and authority to bring off-target performance back into line.

The line functions are usually grouped together as shown on the bottom line of the chart in Figure 7.5. On larger projects, the design function is directed by a senior project architect, engineer, or design manager. The design leader often has the additional responsibility of acting as deputy project manager during the absence of the project manager. Acting as alternate project manager offers an excellent training ground for future project managers. The arrangement serves the best interests of both the project and the design leader.

The design leader is the first block we come to that requires supervision of a large number of discipline groups. Depending on the type and size of the project, it can run from 5 to 8 groups containing from

40 to 150 design people. Numbers of such magnitude quickly create the possibility of exceeding the design manager's span of control.

The problem is readily solved by adding one or more project engineers to assist with the design supervision. The project illustrated in Figure 7.4 is large enough to afford two project engineers who are assigned to two major geographical areas A and B. Sometimes the division of work between project engineers is made on a discipline or *horizontal* basis. In that case, one project engineer would handle the civil, architectural, and structural disciplines and the other would handle process, mechanical, electrical, and instrumentation for all areas of the project.

I prefer the vertical or geographical approach because it gives the project engineers experience in handling all design disciplines. The arrangement also has the advantage of giving the project engineers broader experience and better on-the-job training. That, in turn, expedites their movement up to design leader and eventually to project manager. Designing the organization chart to promote development of future managers is an important long-term consideration for PMs to keep in mind.

Each design discipline should be headed by a lead person who is responsible for the group's contribution to meeting the project goals. Discipline leaders must be people who have enough management smarts to control the group's productivity and technical output. They are responsible for making the estimates and schedules for producing the finished product. They also monitor the budgets and schedules for the execution of the design or construction work. Comparing them to their military organization counterparts, they are the sergeants who carry out the battle plan in the project trenches.

I have a favorite saying about project managers: "Get out into the trenches to see what is actually happening on your project." It pays to get to know your sergeants and check their performances on a first-hand basis. Discipline lead people are in the natural line of progression for promotion into the ranks of project engineering or department leadership. Improvement in the handling of their project administrative duties will determine just how rapidly promotion will occur.

Construction organization charts

As we have shown in our previous examples, the construction organization is part of the project team reporting to the project manager or director. Therefore, the project manager shows up at the top of the typical construction organization chart shown in Figure 7.6. I want to stress that the project manager has only *functional control* of the field work and *not* the day-to-day construction operations. The project man-

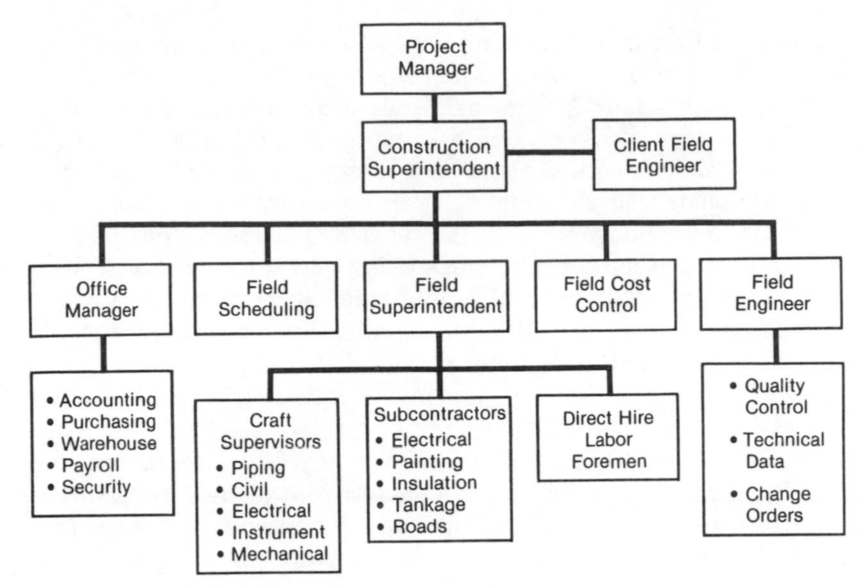

Figure 7.6 Typical construction organization chart.

ager is responsible for meeting the overall project goals of *finishing a quality facility on time and within budget.* The PM's responsibility for construction is to ensure that none of the major project goals are placed in jeopardy during the execution of the field work. Since the field work for an integrated project usually involves the expenditure of 40 to 50 percent of the overall project budget, the construction organization plays a major role in overall project performance.

The leader of the construction team can be known by a variety of titles. Years ago they were called construction or field superintendents. As the management of construction projects grew in scope, technology, and sophistication, the title "construction manager" became more appropriate. Today, the title "field superintendent" is usually used for the person in charge of the actual construction operations under the construction manager. Some heavily construction-oriented companies even use the title "project manager" for the person in charge of the construction operations. In that case, the title "project director" is used for the person in overall charge of design-and-build projects.

The construction manager also has an important contact with the client's organization through the assigned field representative. The client's field engineer is usually assigned to the construction site to protect the client's interests in the field. He or she generally reports back through the office of the owner's project manager. On larger projects (e.g., $50 million and up), the client's field organization can become quite extensive depending on the level of quality, cost, and schedule control that the owner wishes to maintain.

As shown in the chart in Figure 7.6, the construction manager controls the field operations through a group of five or six key line and staff managers. Although the structure has a heavy horizontal component, the construction manager's span of control should not be exceeded if the key managers are well qualified. The chart in Figure 7.6 is predicated on the construction work being performed by the general contractor along with a normal amount of subcontracting. A chart showing a construction management organization for a totally subcontracted job would be entirely different.

The most important line function in the construction organization takes place under the direction of the field superintendent, who manages the largest commitment of human and physical resources on the project. With that sort of major responsibility, the field superintendent must be supported by a large staff of area engineers, craft supervisors, and craft foremen. Those people are responsible for planning the work and organizing the personnel to meet the work schedules. They also coordinate activities among the different crafts and are responsible for the quality of the work. In our military comparison, they are the sergeants directing operations in the trenches where the project battle is being won or lost!

All the other project operations, including the design and procurement efforts, must be directed toward supporting the field operations where the end product is being created. The home office and field operations probably exhibit the worst *compartmentalization syndrome* within the organization. It is a prime concern of the project manager (director) to break down the natural age-old barrier between design and construction so that the overall project goals can be met. Without a doubt, that presents the project manager with the greatest leadership challenge.

The supervision of the construction activities can be subdivided on either a geographical (vertical) or craft (horizontal) basis. The partition might even follow the arrangement used earlier in the design area. On larger projects the division tends to be vertical by areas, whereas on smaller projects the split tends to divide by craft as shown in Figure 7.6.

The comment made earlier about developing future project managers applies to the field organization as well. Future construction managers evolve from the ranks of construction superintendents, area engineers, and craft supervisors. Again a geographical split of duties results in more rapid personnel development because of the broader nature of the responsibilities involved.

The staff functions of project control that start in the home office (on integrated projects) carry over into the field operations. In fact, as the field operations gain momentum and design activity declines, the center for project control functions gravitates to the field. The home office control function then becomes one of overall monitoring and reporting through the project manager's office.

The field scheduling group reports to the construction manager and is responsible for maintaining the construction scheduling activities within the overall project schedule. The key activity of the field scheduler is to lay out the work to be accomplished in the field each week. That is usually done in a weekly scheduling meeting which is attended by all of the key construction supervisors and subcontractors. That is the point at which the CPM schedule is converted to a bar chart format for the detailed control of the execution of each construction activity.

The field cost control group reports directly to the construction manager, and it is responsible for monitoring the construction budget and preparing the field cost report. In conjunction with the field superintendent, it also calculates the percent completion of the field work. By balancing the percent complete against the cost, it calculates the productivity of the field forces. If field productivity lags, it is extremely important to get it back on track immediately. Otherwise, the construction budget will be overrun.

Although it doesn't show on the organization chart, both the field scheduling and cost control groups must have a tieback into their respective home office supervisors. That is necessary to prevent a strong-willed construction manager from dominating the field cost and schedule groups and forcing them to issue overly optimistic cost and schedule reports. Such an idea may seem farfetched, but exactly that situation occurs quite often. Project managers should monitor that area of organizational conflict of interest very closely.

Another important staff function is the field engineering group which directs all of the technical activities in the field. It receives all the technical documentation from the design group and distributes it to the field team. All field revisions and as-built drawing changes are handled by this group, along with any technical interpretation of the drawings by the construction people. The group also handles all of the field survey work needed at the site and sets all of the lines and grades. It serves as the quality control inspector of the on-going construction work and provides liaison with the owner's inspectors.

Field change orders are an important part of the field work that are handled by the field engineering and the cost control groups. Engineering makes the technical input, and cost control furnishes the cost engineering portion for the field change orders. The position of field engineer is another possible springboard into the construction manager's position. With some additional training in the construction area, the field engineer gets a good exposure to construction management.

The office manager is the final staff person to round out the construction manager's staff. He or she reports directly to the construction manager and handles all the administrative duties in the field office. It includes field personnel administration, accounting, field procurement, subcontract administration, site security, first aid, and the field warehouse. That covers seven key areas of administration, which will push the span of control of most office managers to the limit. On a medium-size to large project, the office manager will need a first-class team of section managers to keep the administrative staff on top of its work.

The integrated project organization chart

If we assemble all the examples that we have discussed in this section, we can readily see that even a small project can involve a large number of people. On a relatively small project of $10 million, we could have a design team of 25 people. Combine that with a construction team of 75 and a total of 100 people make contributions to the project. On a project of $50 million, the number becomes 500. Each person's project activities must be channeled into effective work if project expectations are to be met. That points up the importance of designing an effective project organization chart as the first step in building an outstanding project team. Designing the chart is, however, only the first step in building the team.

Selecting and Motivating the Project Team

We said earlier that we first designed the organization to suit the work to be done and then filled the organization from the pool of people available. In the ideal situation, we would like to have two or three candidates presented to us for each key position. Unfortunately, that does not happen very often except in periods of low work load. Even in those periods, it doesn't happen often because the staff has already been reduced to meet the existing low workload condition.

Selecting the project team

If you were an owner who had gone into a contractor's shop to get a job done, you should have picked two or three of the top people during the selection process. You would then work closely with those people to fill the rest of the key slots with capable people. If you are a central engineering or contractor's project manager, you will have to select the people offered by the discipline heads or the manager of the construction department. If you are fortunate enough to have worked with the key people before, you can easily make a decision on the candidates offered. If you do not know the candidates, you should read their résumés, interview them, and check out any available references.

After you have established a candidate's abilities to do the job, you must look at how the person will fit into the team. Does the candidate subscribe to your project management m.o.? Will he or she fit in with the other players and the client? Don't select anyone who you feel has a stronger loyalty to a department head than to you or the project team. Don't, under any circumstances, fill a key slot with anyone who is not a team player! You will only have to change the person later at some cost to the organization.

Having said all the above about selecting only quality people, we also know that pulling it off is very difficult and even a matter of luck. You will have to make some compromises and accept some lesser-quality performers. If possible, it is best to blend them into the lower levels of the organization where they will do less harm.

If you are forced into taking some less qualified people at key levels, back them up with strong people who require less supervision. Also, it may be possible to place them in positions reporting directly to you, which will give you an opportunity to develop them on the job. If certain people show good potential, don't be afraid to take a chance on them.

As the selections are made, add the names to the chart. It may not be possible, or even desirable, to fill all the slots on day 1 because of a slow project buildup. Issue the incomplete chart anyway, because it will help to get the organization working together as well as help to orient new arrivals on the project.

Motivating the project team

The first motivational tool is to establish the project goals and instill them into the minds of the key players. In addition to the general goals of quality, budget, and schedule, some project-specific goals must be formulated and written down. They may involve key milestone dates or budget targets which must be met to earn a bonus. It may be a target to design a higher level of project quality to meet a fixed project financial package. The client may have set some unusually tough aesthetic design standards or some tight environmental goals which have to be met within a tight schedule and budget.

After the specific project goals are selected, they should be discussed by the team members to generate ideas on how they can be met and the role to be played by each member in meeting them. Through those discussions, we can get all of the team members to buy into the goals and become committed to meeting them.

The early project-goal-setting activity is probably the most overlooked and underrated activity in project initiation. Without setting project goals and continually reinforcing them, the likelihood of meeting them is indeed slim. I urge you to remember that this goal-setting philosophy applies equally to the design, procurement, and construction phases of the project!

Preparing job descriptions

The project team job descriptions should not be the standard claptrap found in the departmental files! The format I propose you use is one that is project-specific to each player's accepted goals on the project team. They will not be usable from project to project, so you won't find them on file anywhere.

An interesting approach to preparing the project job descriptions is to have the team members start by writing their own. When they present the drafts to you for review, you are sure to be surprised by just what *they* think their jobs are. The chances of developing good job descriptions are improved when you work out the final drafts together. The final draft must be typed and made a part of both the project organization and the team member files.

A job description must be designed to be a basis for the management by objectives (MBO) program. MBO is a tool that we can borrow from general management to use in managing our projects. If your firm has a companywide MBO program, so much the better. You can slip your project MBO program right into its niche in the overall company system, and your people will already be knowledgeable in its use.

Making the MBO system work

Basically, project managers must initiate MBO programs on their specific projects. The objectives for the program are set by selecting and stating the specific project objectives (project goals) and getting the team members to buy into them. Writing the objectives into the job descriptions gives everyone involved a clear and permanent statement of what is expected. All that remains is setting up the review procedure and schedule of dates to monitor the performance against the objectives. It is up to the project manager, as the project leader, to see that the MBO program is carried out to project completion.

The project manager and the key project leaders must carry the MBO program down through the organization chart to cover all of the project objectives. That means getting down to the cutting edge of the work, be it in design, procurement, project control, administration, construction, or any other key area of the project. Although project managers are ultimately held accountable for project performance, the MBO system forces them to delegate the responsibility for meeting specific project goals and forcing the decision-making process as far down into the organization as possible.

The motivation engendered by delegating responsibility for meeting the goals throughout the organization goes a long way toward building a spirited project team. That sort of project climate makes the practice of leadership much easier while keeping a strong hand on management of the project. Naturally that motivational spirit must be revitalized from time to time during the life of the project. A good time for motivation boosting is during the one-on-one MBO objectives review meetings. Group morale revitalization should take place through well-managed routine project meetings.

Project mobilization

Project mobilization is a critical time in the birth of any project. Everyone is gung-ho to get the project started. The client, the company and department managers, and just about everyone wants to see some dust flying just to feel as though something is going on. Don't let that sort of activity panic you into bringing people onto the project too early. In today's climate, chances are good that your project will be done on a task force basis, which means that all the people who are assigned to your project will be charging full time to it whether they are productive or not.

If the project starts off in a personnel glut situation, people will be forced on you before you are ready to absorb them. The other side of the coin is seen in periods of personnel shortage, then you will be scouring the organization for good people to add to your project. Naturally, each of the conditions requires a different approach.

In a case of oversupply, don't take people on early just to satisfy some department head's desire to cut departmental overhead. Bring your key people on first to assist in the project planning phase as we discussed earlier. It is important to have a good and continuing flow of work available before bringing on the main body of troops. That applies to any phase of the project, be it design, procurement, or construction. If there is a personnel shortage, you may want to consider taking on a few good key people sooner if they are likely to be snapped up by another project. You will have to weigh cost vs. potential benefits. Sometimes the investment will pay off.

The advice given for project mobilization works in reverse when the project is destaffed as it nears completion. People should be returned to their respective departments just as soon as they have completed their assignments. Demobilization is no problem in periods of personnel shortage, but it can be a problem in periods of surplus. In the latter case, the people must go to protect the project budget. Don't park unneeded personnel on your project just to make life easier for other managers.

The most serious project-staffing problems usually arise during periods of personnel shortage. If the work is available, the schedule clock is running, and the project personnel are not forthcoming, you must take effective action to get your necessary staff. Make your staffing needs known to the personnel providers as soon as possible to give the suppliers ample lead time to locate the human resources. Also, make your request in writing and provide copies as high into the organization as necessary to get the desired results.

Once the task force is staffed, it is difficult to make major shifts in loading during the execution phase. People who are released temporarily in the middle of a project often will not be available when they are needed again. That is a nasty problem which can occur during unforeseen project suspensions or slowdowns, and it can have disastrous side effects on project budgets, productivity, and morale. There is no way to manage it out of existence. At best, one can minimize the damage with an intelligent approach to reducing the staff and restaffing when the need arises.

When a major restaffing problem occurs in the middle of a project, all the major players—the owner, contractors, and central engineering groups—must participate in finding the best solution to the problem. The PM must inform interested parties of the short- and long-term effects of early destaffing on the execution of the project master plan. An agreement whereby the project can be reorganized with minimum damage to all of the participating project partners must be reached.

If the need for reorganization was not caused by the project itself, the project manager must take immediate action to protect the original project budget with a claim for relief under the project working agreement. As defender of the project budget, it is no time for the project manager to be fainthearted!

Updating the organization chart

The organization chart should be kept up to date as project personnel changes occur in the course of the job. As we will discuss later, the organization chart and staffing become part of the project procedure manual, which also is subject to continuing revision.

The project organization is a living organism, so it can be expected to change over time. I do not recommend that wholesale reorganizations be attempted in the middle of a project if they can possibly be avoided. Major organizational surgery should be attempted only when it is absolutely necessary to correct catastrophic organizational problems.

Project Modes: Matrix vs. Task Force

It is necessary to discuss the various methods of project execution available to us because it is impossible to tell how the organization is structured just by looking at the organization chart. It is also best to define the terms "task force" and "matrix," which will be used in this discussion.

Matrix organization

A matrix organization is one that functions as a group of specialist departments that handle project work which is brought into the departmental area. Individuals working in the department may be dedicated to work on a specific project for a fixed period of time without doing any other work. A good analogy to this form of operation is a manufacturing operation on a repetitive product.

In effect, the department leader controls the people and the work done in that department. With all of the projects competing for the departmental resources, the department head has the total power over which resources are assigned to which project. It was that feature of the matrix approach that caused the breakdown of effective project performance by the project manager. It was impossible to give the project managers complete responsibility for project performance if they were not also given control of the human resources required to execute the work.

Task force

A task force is a project group that is set up to perform a specific project and that contains all of the home office skills and design disciplines needed to carry out the work. The team is physically located in a self-contained area away from any departmental activities.

Besides the project management group, all of the necessary engineering and design disciplines, the procurement people, project controls group, and the clerical staff necessary to carry out the work will be assigned. All the people are assigned only for the duration of the project; on project completion, they are returned to the discipline department for reassignment. The PM has *functional control* over the assigned personnel only for the duration of the project. Administrative control of the departmental personnel rests with the department manager even during task force assignments.

The main point of difference in the two methods is that the departmental personnel are physically moved to the project task force area. There they are under the functional control of the project manager, and they are not available for assignment to work on other projects without approval of the project manager. The arrangement gives project managers much better control over the human resources assigned to their projects.

Construction groups

Construction organizations are automatically forced into a task force mode by the nature of the work. Since most construction sites are remote from the home office, a project construction group has to be organized to build each specific facility. When that facility has been completed, the team is broken up and sent to other construction sites.

The evolution of the task force mode

Over the years, the capital projects industry has gone almost 100 percent to the task force mode of operation. When I entered the business in 1950, we were still in the early stages of the transition. Most design groups had a strong departmental structure; there was no strong central project management team directing the work except on very large projects. Project engineers or job coordinators guided their project work through the various departments as best they could. Since there was no organized priority system, it was difficult for them to control schedule and costs.

As projects became larger, it was obvious that something had to be done to make project performance—*quality, schedule, and cost*—more predictable. The answer was to find ways to extend to the smaller projects the benefits realized from the task force mode on larger projects. That proved to be easier said than done. During the fifties and sixties, the battle between strong departmental (matrix) format and the project task force format raged back and forth.

In effect, the battle was also drawn along the lines of strong project management vs. strong departmental management. Often the tide of battle swung from one camp to the other depending on whether department or project people made it into the firm's top management! I saw cases in which the method changed back and forth in one year.

Finally, in the seventies, the strong project management (task force) approach won out, and now virtually all medium-size to large firms are task-force-oriented. In today's environment, many firms do not even maintain a designated departmental area for design personnel who are between project assignments. The design department leaders are now staff people reporting directly to the design group head or a vice-president of operations. As we will see later in our discussion, the department heads perform a very important function: maintaining the design discipline staffs at a high level of effectiveness and morale. The discipline leaders are also the technical gurus in their specialized fields of expertise.

How the systems work

Figure 7.7 illustrates how a matrix system functions. The departmental functions are shown across the top of the chart, and various projects requiring human resources are shown down the left side. The number of discipline people needed on each project is shown in the square. If we add the total numbers of people on the project line, we find the number of people on the project. By adding vertically, we get the total department strength.

The lines connecting each block of people to both the department and the project tell us that we are breaking a scientific management commandment by having each project team member report to two "bosses." That dichotomy of leadership is shown graphically in Figure 7.8.

The dichotomy appears to be quite clear as we trace the line of authority down from general management, through the project and department managers and to the people on the project. When we analyze the split of authority between the project and department managers, however, the problem of having two bosses largely disappears.

The supervision of the project people is divided into three major categories of departmental functions, project functions, and joint responsibilities as shown graphically in Figure 7.9. The breakdown of the functions is done in a logical manner to ensure that the project goals of all the key players (as we discussed in Chapter 1) will be attained.

Departmental functions

The department managers continue to take the lead in departmental administrative functions for their work disciplines. The responsibility for adding or deleting qualified departmental staff to match the workload in the project area has a high priority for the department head. In periods of staff fluctuation, it can be a large consumer of the department head's available time.

Maintaining the quality and morale of the discipline staffs at a high

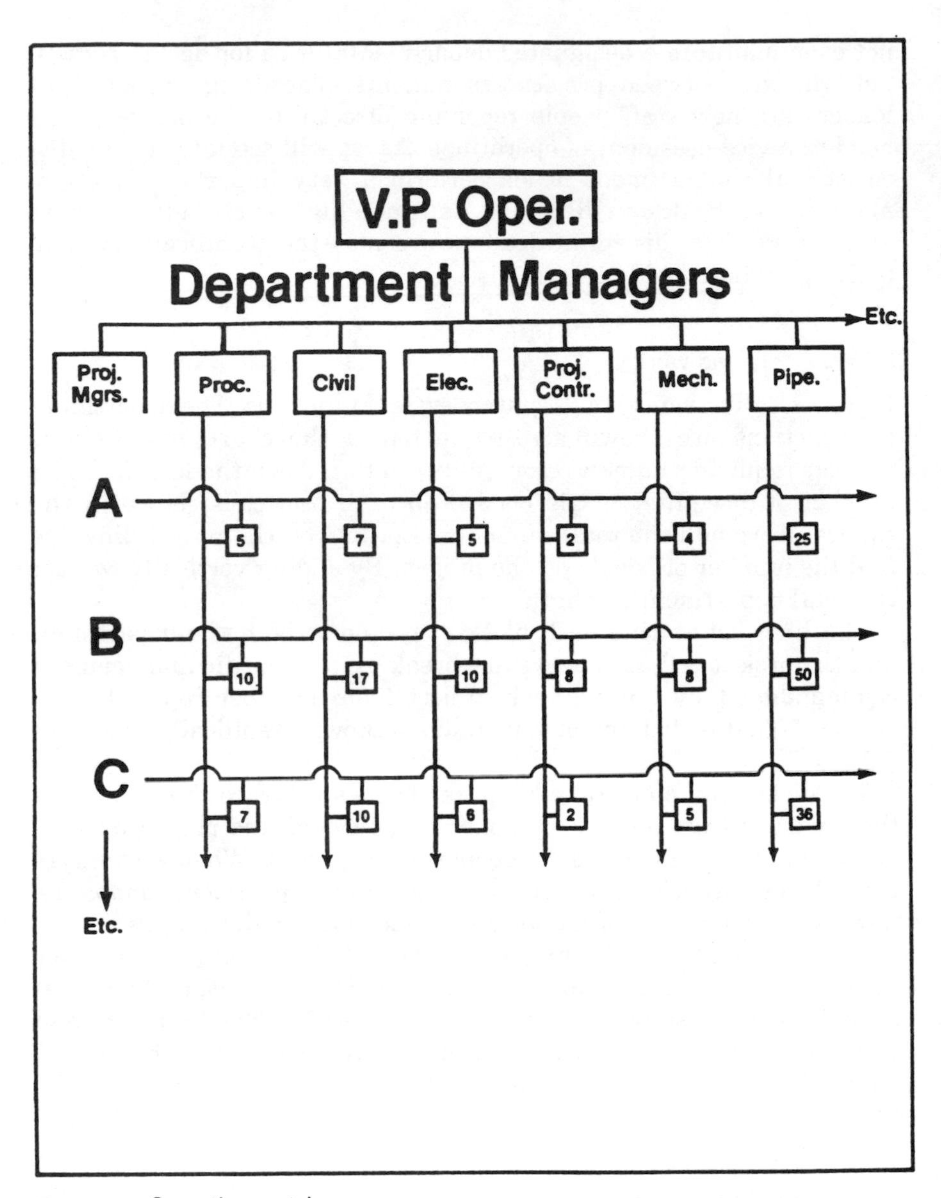

Figure 7.7 Operating matrix.

level makes a major contribution to the overall performance of the design, procurement, and construction operations of the firm. The goal of a top-quality staff is attainable through good hiring and training. The morale is maintained by progressive personnel administration with annual reviews, competitive salary scales, and performance incentives.

Department managers must set and maintain overall quality stan-

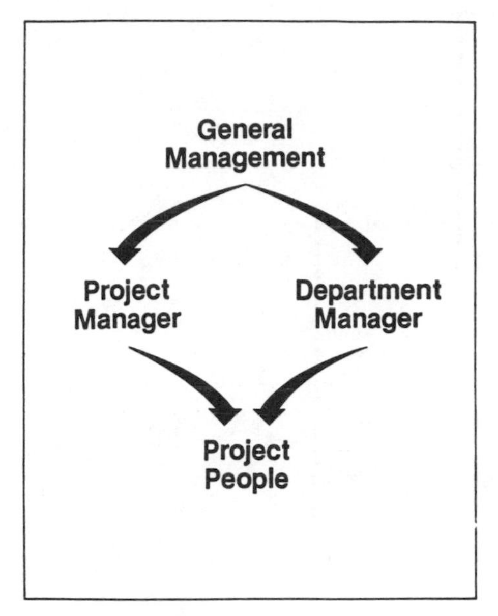

Figure 7.8 Dual supervision.

dards for the department. They must also monitor the standards of quality that are set up and used on each project. The work also involves the development of discipline standards to simplify the work and improve productivity.

Personnel development and training of all department members is the responsibility of the department head. Although the department heads act as principal discipline experts for the firm, they must also develop and refine the skills of all people in the department to the highest levels of performance and promotability. They are responsible for all annual reviews, salary adjustments, promotions, and so on. The only contribution that PMs make to the process is to give fair and thoughtful performance appraisals for the discipline personnel assigned to their projects.

The preparation and administration of departmental budgets for such items as training, vacation, overhead charges, proposals, equipment and services are the responsibility of the discipline manager.

Project functions

All the project functions are controlled by the project staff under the leadership of the project manager. The first project function is the establishment of the project goals, which sets the stage for all the other project functions.

The definition of the project and the scope of work, as defined in the

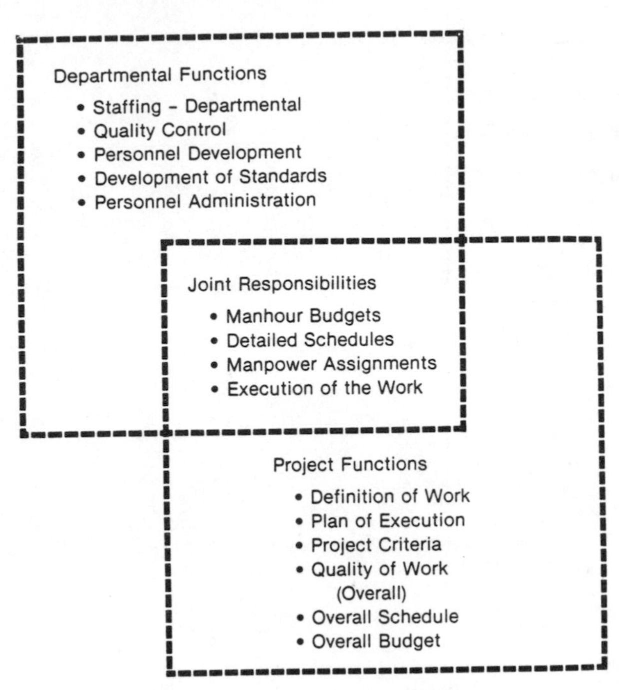

Figure 7.9 Division of project responsibilities.

contract or design book, also is a major project function. It forms the basis for all of the joint responsibilities such as staffing, schedules, and budgets. The master plan for the project execution is a major project function. It lays the groundwork for all the project planning and organizing which we have been discussing up to this point. Establishing the project criteria for capacity, quality, design standards, project execution, and so on, is mainly a project function. Advice from and consultation with discipline experts in this area are certainly warranted.

Setting the quality standards for the overall project work is the responsibility of the project team. There is plenty of opportunity for input and assistance from the discipline leaders in this area. Remember, quality is our number 1 project goal! The other two project goals of meeting schedules and budgets are also primary responsibilities of the project group. Some input from discipline leaders can make a strong contribution to those efforts, but basically the responsibilities are project functions.

Joint responsibilities

Joint responsibilities are the functions listed in the central area of Figure 7.9, where the two responsibility boxes overlap to indicate input from both sides of the departmental-project management equation. Since all these joint responsibilities turn out to be project functions after they are established, the project team can be said to have veto power (through top management) over the final decisions reached.

In fact, top management is prone to override both project and departmental managements in cutting budgets, shortening schedules, and switching key staff people after project level agreement has been reached. Neither the project nor the departmental managers should accept a management veto without going on record with their opinions against the veto. The opinions could come in handy when the condemnation proceedings start at the end of the job!

Personnel budgets and staffing assignments are the two main areas of joint group responsibility. The department and project managers first have to work out the assignments of the discipline group leaders, who, in turn, will prepare the design estimate for the project work. The development of the design estimate was detailed in our discussion of the home office cost estimate preparation in Chapter 5.

The general activities of the design staff assignments and budgets are found in the construction area as well; only the job titles are different. The staffing of the design (or construction) groups is worked out among the design (or construction) manager, the discipline project group leader, and the department head to meet the needs of the project schedule. Generally, only the cases that involve accepting a possible misfit that cannot be resolved at that level will be taken to the project (or construction) manager for resolution.

Department managers should have major inputs to the detailed schedules of the discipline groups. Their many years of experience and their technical expertise are valuable to the discipline scheduling process and also put them in better positions to make their best project-staffing recommendations.

The last joint responsibility—*the execution of the work*—is listed to indicate that everyone directly or indirectly associated with the project has a duty to make his or her best contribution to the project execution. Here we have another classic case of compartmentalization brought on by establishing organization charts. Again, it is up to the project manager to effectively concentrate the input of all department managers into meeting the project goals.

Some Advantages and Disadvantages of the Task Force vs. Matrix Systems

Since the departmental matrix was in existence first, it must have offered some obvious advantages. Recently, the capital projects industry has changed over heavily to a task force mode, which must have offered some important advantages to motivate the industrywide change. As with any list of advantages, there can be a difference of opinion over what is an advantage and what is a disadvantage. That is true in this case also, since some of the arguments appear on both sides of the ledger.

Advantages of the matrix system

Some of the major advantages claimed for the departmental approach are the following:

1. It is more labor-efficient for any size job.
2. It is less costly to operate.
3. It focuses departmental expertise and skills on all projects.
4. It is easier to shift personnel from project to project.
5. The project manager does not waste time on personnel problems.
6. The increased challenge attracts better-qualified people as department leaders.
7. Departmental personnel do not have to physically relocate to and from the task force area.

Although some of those advantages having to do with efficiency and cost do hint at being advantageous to the needs of meeting project goals, most of them lean toward meeting departmental goals. In today's strong project management environment, that is not enough to satisfy top management's desire to meet profitability goals.

Advantages of the task force approach

Most of the following advantages of the task force are much more project goal oriented.

1. It shortens lines of communication.
2. It is more efficient on medium-size and large projects.
3. The project can start faster and finish earlier.
4. It focuses the human resources on the project goals.
5. It makes it easier to generate project team spirit.
6. It is more responsive to meeting project changes and client's needs.

7. Departmental personnel get an opportunity to operate independently and grow with the job.
8. People get more job satisfaction from seeing the results of their work firsthand.
9. It reduces departmental interfaces and natural departmental barriers.

The above list shows that at least half of the advantages are aimed directly at meeting the project goals. The rest are pointed toward such performance-oriented intangibles as improved communications, job satisfaction, team spirit, and morale.

How to resolve the differences

Obviously, it is just as important to transfer the available technical and management expertise from the discipline department to the project task force. Project managers should foster and encourage such participation when it is productively applied. Some hours for the activity should be included in the project budget.

The matter of efficiency in personnel usage works both ways. Task force project managers are generally assigned personnel on a full-time basis, so it behooves them to keep their personnel productively occupied. That means that project staffing must be well planned to ensure a continuing availability of work for the project team members. If it is, the difference in personnel utilization efficiency between the two systems will be almost nil.

The cost and trauma of moving personnel from one area to another are important, but the dividends realized through improved project performance more than offset them. The people in the industry who are unable to adapt to the trauma of change have already found more pleasing work elsewhere.

The trauma of taking department heads out of their line functions and making them strictly staff people has caused some upset in a few organizations. That upset condition has probably caused technical quality to deteriorate somewhat during the transition period. It is easy for department heads to become totally immersed in the administrative side of the business, because they are now handling much larger numbers of people in their departments. A system whereby those valuable people can be rotated into a technical environment from time-to-time would certainly have some advantages.

From the project manager's point of view, the advantages of the task force system are obvious. The project people are gathered all in one place under the watchful eyes of the project staff. The daily nose count requires only a few minutes tour of a compact project area. The pulse of the work is easily gaged from the project manager's office. The project manager is readily accessible to the project team members for almost instant communication. All of those things make project managers' lives easier; how could they fail to be pleased?

Many successful projects have been executed in both environments, so don't fight the system if it is not organized to suit your convenience. Application of the total project management methods set forth in this book can ensure successful project performance in either environment. You need only adapt your methods to suit the system.

Organizational Procedures

Important parts of organizing any project are developing and issuing the project procedures. The documents lay out the ground rules under which the organization will function in executing the work. Each company will have a different standard to which the project procedures are prepared. The procedures should spell out the minimum regulations under which your company management wants its project managers to operate. They can vary from being too simple to offer good project control to being so ponderous that efficient execution is next to impossible. Please remember the KISS principle! As a minimum, the procedures must cover the following major areas which can be involved on your project:

- All contractual obligations
- Basic architectural and engineering design standards
- Communications procedures
- Organization chart
- Change order procedure
- Cost and schedule control systems
- Procurement procedures
- Construction procedures
- Document control
- Project filing system

The Project Procedure Manual

The heart of the project procedures is the project procedure manual (PPM). It is a prime duty of the project manager to see that the document is produced on time and it is working effectively for the life of the project.

Almost every company has a standard table of contents for the PPM which is geared to its type of work. In the event your firm does not have such a standard, I have included a typical one for a design-and-construct project in Figure 7.10. The sample can be expanded or contracted to suit the size and complexity of your particular project. The manual for a small, uncomplicated project can be just a few pages, whereas one for a large project usually runs into one or more volumes.

Review of the Main Sections

The introduction

The introduction should contain a statement of purpose for the project. What is the owner hoping to accomplish with the project? What needs is the project going to fill in the community, industry, or market? The people involved with the work must know what the overall goals are.

Also, it is always well to include a statement that the PPM does not replace the contract and that any conflict between it and the contract will be resolved by the contract.

The project description

A project description gives the location of the project, a site description, an overview of any processes involved, and any other outstanding features of the project. An outline of the scope of work and the services being offered is important to the general knowledge of the team members. It can then be neatly tied into the project objectives, which forms the basis for the project MBO program. All of the goal-oriented groups involved on the project should be covered in this section, including any project team performance incentives. Any work by others involved on the project, including major subcontractors or licensors along with their contributions to the project, should be mentioned here.

Contractual matters

Since the contract is a quasi-confidential document, the key areas affecting project performance should be included in this section. The people who are working on the project but who will not have access to the contract must know how the contract can affect their work. For example, it makes a difference to the project team's performance whether the contract is on a lump sum or a reimbursable basis.

Any requirements for project secrecy or confidentiality must be ad-

1.0 INTRODUCTION
- 1.1 Statement of Purpose
- 1.2 Contract Controlling Statement

2.0 PROJECT DESCRIPTION
- 2.1 Brief Description of Project
- 2.2 Location and Site Description
- 2.3 Scope of Company Services
- 2.4 Work by Others
- 2.5 Project Objectives
- 2.6 Licensing Arrangements

3.0 CONTRACTUAL MATTERS
- 3.1 Type of Contract
- 3.2 Secrecy Requirements
- 3.3 Checklist of Reimbursable and Nonreimbursable Charges
- 3.4 Subcontracts
- 3.5 Special Requirements
- 3.6 Statement of Guarantees and/or Warranties

4.0 PROJECT ORGANIZATION
- 4.1 Project Organization Chart
- 4.2 Brief Job Descriptions
- 4.3 Client Project Organization Chart
- 4.4 Brief Job Descriptions
- 4.5 Organization Chart of Companies or Divisions
- 4.6 Client's Resident Project Team (if applicable)

5.0 PROJECT COORDINATION
- 5.1 Communications Procedures
- 5.2 Communication Systems
- 5.3 Key Personnel Names and Addresses
- 5.4 Correspondence Logging Procedures
- 5.5 Travel and Expense Account Policies
- 5.6 Document Distribution Schedule and Transmittals
- 5.7 Document Approval Procedure
- 5.8 Meetings and Preparation of Meeting Notes
- 5.9 Telephone and Verbal Information Confirmations
- 5.10 Project Language (international projects)
- 5.11 Project Filing System (Appendix D)
- 5.12 Reproduction

Figure 7.10 Table of contents for a project procedure manual.

6.0 DESIGN PROCEDURES
- 6.1 Drawing and Document Formats
- 6.2 Document- and Equipment-Numbering System
- 6.3 Units of Measurement
- 6.4 Applicable Design Standards and Codes
- 6.5 Design Document Schedule and Control
- 6.6 Design Quality Control Procedures
- 6.7 Use of Modeling or Computerized Drafting Technique
- 6.8 Material Requisitions
- 6.9 Technical Bid Evaluation Performance
- 6.10 Vendor Data Procedure
- 6.11 Design Subcontract Procedures
- 6.12 Preparation of Mechanical Catalogs
- 6.13 Others

7.0 PLANNING AND SCHEDULING
- 7.1 Project Master Plan
- 7.2 Project Scheduling Procedures
- 7.3 Preliminary Schedule
- 7.4 Scheduling Control
- 7.5 Earned Value and Progress Status Reports
- 7.6 Special Scheduling Requirements

8.0 PROJECT PROCUREMENT PROCEDURES
- 8.1 Approved Vendors List
- 8.2 Buying Procedure
- 8.3 Expediting and Inspection Services
- 8.4 Shipping Traffic Control Procedures
- 8.5 Vendor Invoice Review, Approval, and Payment

9.0 ESTIMATING
- 9.1 Project Cost Estimating Plan (accuracy and methods)
- 9.2 Appropriations Estimate
- 9.3 Cost Trending and Reporting Procedure
- 9.4 Definitive Estimate
- 9.5 Project Change Order Estimating

10.0 PROJECT CONTROL AND REPORTING
- 10.1 Code of Accounts
- 10.2 Project Budget
- 10.3 Cost Control Procedure (manual or computer)
- 10.4 Project Reporting Procedures
- 10.5 Project Cost Reports: Design, Materials, and Field
- 10.6 Project Progress Reports

Figure 7.10 Table of contents for a project procedure manual. *(Continued)*

10.7 Cash Flow Plan and Report
10.8 Field Progress Reports
10.9 Project Accounting and Auditing Procedures
10.10 Schedule of Reporting Dates
10.11 Invoicing and Payment Procedures
10.12 Computer Charges

11.0 CHANGE ORDER PROCEDURE
11.1 Contractual Requirements
11.2 Change Order Format
11.3 Change Order Log
11.4 Approval Procedure

12.0 COMPUTER SERVICES
12.1 Scope of Hardware Systems
12.2 List of Project Software
12.3 List of Computer Billing Rates

13.0 CONSTRUCTION PROCEDURES
13.1 Construction Contracting Philosophy
13.2 Construction Execution Philosophy
13.3 Field Organization
13.4 Field and Home Office Coordination
13.5 Field Operations Reporting
13.6 Construction Scheduling
13.7 Preparation of Field Operating Procedures
13.8 Construction Progress Photos

Figure 7.10 Table of contents for a project procedure manual. *(Continued)*

dressed in the PPM. All members of the team must conform to the regulations for secrecy agreements and the handling of confidential documents.

Project organization

This is where we keep all of the project organization charts, work descriptions, and any information pertaining to organizations involved with the project. If there are any special organizational interfaces, they should be described in this section.

Project coordination

The main part of the project coordination section covers the communication procedures for the project. The key names and addresses and the correspondence logs are set up to expedite the handling of all project communications. Logging the huge volume of the letters, memos, transmittals, and minutes of meetings generated during the project enables the ready location of vital correspondence later.

Minutes of meetings and confirmation of project information which has been transmitted verbally is critical to maintaining control over the project design work. Often such oral communications result in project scope changes which can have drastic effects on the financial plan.

The document distribution schedule, which sets up who gets copies of correspondence, drawings, specifications, and so on, plays a key role in controlling the project. It establishes the budget for the project reproduction costs, which can be substantial on most jobs. Constant vigil on the part of the project manager is necessary to keep the perennially self-expanding reproduction budget under control.

Document approval procedures are the key to controlling project progress. They should be set up with reasonable but fixed time limits for the approval process. If approval has not been forthcoming when the time period expires, the work should be allowed to proceed without it. Since clients do most of the approving, they are the ones who must agree to such an arrangement.

Having a standard project filing system is a big help in organizing the project. A typical project filing index is shown in Appendix D. As usual, it can be expanded or contracted to suit your individual project's needs. Having a standard project filing system throughout the company makes for better access to project information by all the team members. It is a relatively simple item that pays big dividends in meeting the project goals.

Design procedures

The design procedure plays an important part in any PPM involving design work whether in the home office or in the field. The first part of this section covers such mundane matters as drawing formats and numbering systems. However, when they are not properly thought out to suit the particular type of work being done, the problems will nag you throughout the entire project. That is enough reason for the PM to give them proper attention.

Selecting the applicable project design standards and codes involves legal matters and money, so they must concern the PM. The design firm and the owner are legally bound to meet the minimum code requirements of the area in which the work will be constructed. If codes and standards are improperly selected or applied, expensive rework can result—with disastrous consequences to project performance. As we discussed in the first chapter, the project manager stands at the center of the project team target!

Quality control procedures are established in the PPM for all to read and subscribe to for the duration of the project. The company's reputation is riding on this one, so the procedures must be both results-oriented and cost-effective. This section of the PPM, like most sections, must be constantly monitored for performance.

The ground rules for the critical interface between design and procurement must be covered in this section. Work in this area involves technical bid evaluations and approval of vendor drawings, both of which are critical to project schedule.

Project relations of design and procurement personnel sometimes get edgy because of turf disagreements over vendor contacts. Both parts of the project team must have contact with subcontractors and vendors, so some diplomacy on the part of the project manager is required to keep both groups working effectively toward meeting project goals. Design people should stick to the technical aspects of the buying activities and leave the commercial aspects to procurement.

As was mentioned several times in the section on scheduling, the timely availability of vendor data is the single most critical need for meeting design schedules. Here is where the procedure to make that happen is developed. Make sure the procedure is simple and workable, and follow up on it periodically to make sure that it keeps on working.

If there are going to be any engineering subcontracts, the coordination between the prime and subcontract groups must be worked out here. Don't assume that it will take care of itself, because it won't! Here it is best to assume the worst that can happen and try to develop procedures to minimize any potential hassles before they happen. The method for setting up and monitoring the design document schedule and control system falls into that area. If the method is standard within your organization, just make sure that it will work on your particular project. Sometimes it will require modifications to meet your project's special needs.

Planning and Scheduling

Planning and scheduling comprise a key area which has to be decided upon early in the project. Quite a bit of generalizing has probably gone into it up to this point. Now is the time to crystallize all the prior thinking about scheduling and set down the detailed procedures to be followed for this project. Agreement with the client is critical in this area.

Pay particular attention to the item of establishing an *earned-value system* for reporting project completion in the status reports. A simple cost-effective approach is essential to success in this area.

Procurement procedures

The procurement procedures section lays out the work plan for the procurement and delivery of the physical resources for the project. We are speaking of a procedure to control about 30 to 40 percent of the total project budget, so this area deserves a good deal of project management attention. The starting point is an approved vendors list, an often-overlooked item. If inquiries are sent to ill-chosen vendors, the whole procurement chain will suffer.

As I will state several times in this book, do not for any reason slight the procurement effort on your project, because it plays such an important part in attaining your project goals! I have seen too many project managers mistakenly consider procurement a quasi-clerical function unworthy of their valuable time.

Estimating

Estimating is the foundation of the project financial plan, so it must be well conceived if the money on the project is to be controlled. Many owners do not like spending money on cost estimating because it adds nothing visible to the finished product. That makes the selection of sound estimating procedures to meet a tight budget even more critical. You will have to be creative in developing this section of the procedures to get the best handle on the project cost within the limited estimating funds available.

Project control and reporting

Project control and reporting is generally the largest section in the PPM because there is a lot of ground to cover. It is also a pivotal section because failure here can cause loss of project control, which is sure to result in unmet project expectations. We will be covering most of those subjects in more detail in the next chapter, so I will not dwell on them here. The project manager plays a key role in all of the activities listed, but he or she can also delegate a great deal of the work to project team specialists. In that case, however, the PM becomes the editor of all the material generated by the specialists. It is important to read and check all of the procedures for content, writing style, conflicts, and project goal criteria before releasing them for publication. It will be your first chance to evaluate the ability of your key project staff leaders to communicate!

Particularly important items in this section are the cost control procedures, the project budget, project reporting, and project accounting. As a minimum, they will appear in the PPM for most projects.

Change Order Procedure

The change order procedure could be included in the project controls section, but most people consider it important enough to give it a section by itself. Change orders are the bane of a project's existence. No one connected with the project likes to talk about them, and some even refuse to believe that they exist. Like most other problems, however, they cannot be swept under the rug and they do have to be disposed of before the project closeout.

The legal language for changes in scope used in the contract is usually clear enough and does admit their existence. It is a good starting point for writing a detailed procedure for handling changes. Perhaps the key clause to use here is the one which states: "No work will be started on the change until the parties agree on the scope and cost of the additional work." Since changes happen after the job starts, they usually have a considerable impact on schedules. The above clause gives the project manager some leverage in forcing a decision on acceptance or cancellation of the proposed change because it is holding up the project schedule.

In actual practice, however, the revised work does proceed in order to avoid delaying the schedule. When contractors proceed with unapproved change orders, they are placing themselves at financial risk. I will expand on processing change orders in later chapters.

Computer services

The growth of computers in project work has led us to making computer services a separate section. It could fit in the project controls section for small projects. For larger projects, it rates a section of its own. In any event, money is involved, so give computer services plenty of thought before deciding on the scope that the project will bear.

Construction

On an integrated design-and-build project, the construction section will cover only the generalities of the construction work as they bear on the front-end work of the project. A complete construction procedure manual will be prepared by the construction department (or contractor) when the field work gets organized. Basically, what is required for the PPM now is only the construction information that is needed by the design, procurement, and project control people to prepare estimates, budgets, plans and schedules, receiving procedures, and so on.

Even this brief overview of the PPM table of contents shown in Figure 7.10 should give you some inkling of the broad scope and importance of the document. The development and publication of the PPM is one of the most important responsibilities of the project manager; it is almost as important as organizing an effective project team. Just like the project organization, though, the PPM must be continually monitored, reinforced, and updated as needed to meet the project goals!

Project Procedure Manual Publication

Format

Generally, a loose-leaf format is desirable because the PPM must be constantly updated with changes and varying project conditions. In a contractor-client type of project execution mode, the owner will often use the document that is developed by the contractor in conjunction with the owner. On very large projects, owners may prefer to develop PPMs of their own to cover any special needs they may have.

The best way to expedite development is to make the PPM an agenda item in the client-contractor project kickoff meeting. The interplay of the project procedures between the client and contractor organizations is so complex that more than one face-to-face meeting is likely to be held to get any problem areas resolved.

Another subsidiary tool for use in gathering background data for the PPM is a project design checklist or questionnaire covering the main technical aspects of the project. It should be circulated among all of the key parties to the project for input before the kickoff meeting.

Issuing the PPM

A key factor to remember in issuing your project procedures is to get them published as early in the project as possible. An issue date more than 3 or 4 weeks after project kickoff is too late. Issuing the PPM with "holds" to be cleared up in later issues as the information is finalized is quite normal, so don't try to perfect the first issue.

Early issue of the PPM is an excellent project personnel indoctrinational tool: it gets the new team members up to speed in a hurry. It is essential that they learn the "who, what, when, where, how, and why" of the project without any false starts. That is especially true when a task force approach is being used and this particular group of people may not have worked together as a team before.

Satellite Office Procedures

I have included a reference for satellite office procedures because sometime you may be involved with a project which is being run on a split office basis. We have already noted the possible increase in potential problems when that basis is being used. The only way to minimize those problems is with a good PPM to organize the people and the work.

A good starting point for the satellite office procedure is to use the PPM and tailor it to suit the interoffice operations. The satellite office has to perform the same functions as the prime office, so the systems should be made compatible at the outset. In areas in which the same systems will not fit for some special reason, a workable adaptation must be made. The differences must be minimized to the best degree possible to maximize the opportunities of meeting the project goals.

A special case of the satellite office arrangement occurs when a third-party constructor has a contract directly with the owner. When the design firm has a commitment through the construction stage, a detailed coordination procedure is needed. The division of work for handling material deliveries, design modifications, drawing interpretation, responsibility for start-up, and so on, must be resolved early in the project.

Summary

This chapter gives PMs an insight into organizing the human resources and procedures which are crucial for the successful execution of a capital project. They are the two major factors which must be properly handled to increase the chances of meeting our project goals.

The section dealing with building the project organization involves many human factors that sometimes do not come easily to project managers. Leadership in those areas cannot be delegated to subordinates, so project managers must train themselves to eliminate any weakness in the area.

The area of project procedures is an excellent one for delegation of the many detailed procedures which must be developed. A strong input from the project manager is needed to guarantee a uniform, effective set of working rules designed to make the project run smoothly.

References

1. Arthur E. Kerridge and Charles H. Vervalen, *Engineering & Construction Project Management,* Gulf Publishing Company, Houston, Tex., 1986.

Case Study Instructions

1. Draw up an organization chart for your selected project. Include all key interfaces such as owner, designer, procurement, and construction pertinent to your project.
2. Write job descriptions for the key members of the organization, including all applicable groups.
3. Apply the job descriptions of (2) to a proposed MBO program for your project. Explain how you would set each individual's goals and the schedule for reviewing them.
4. Explain how you would go about filling the key project positions from the human resources available to you from your organization. Explain how you would handle requisitioning the necessary total personnel from your company departments. Do that for a task force and a matrix-type setting.
5. Assume that your management (or client) wanted you to execute your project on a task force instead of a matrix basis. Would you agree or disagree? Give your reasons why. Now assume the opposite case.
6. Prepare the table of contents for the project procedure manual for your selected project. Discuss why you included or deleted specific items from your table of contents.
7. Assume that you were forced to split the design work on your project between two design offices. How would you handle the project procedures necessary for such an operation?

Chapter

8

Project Control

Project control is the pivotal activity that ties all of the previously discussed project management techniques together. Planning and organizing are certainly important in leading us toward meeting our project goals, but effective project control is absolutely essential. We might be a little off target on planning and organizing and get away with it, but we cannot fail even a little bit in control and come out whole.

A definition of control which I think is particularly appropriate for project work is "the work of constraining, coordinating, and regulating action in accordance with plans to meet specific objectives." We have set our objectives of doing a *quality project, on time, and within budget.* We have made our time and financial plans, and we have created an organization to execute them. Now all we have to do is proceed with the constraining, coordinating, and regulating activities to deliver the desired results.

The Control Process

The basic mechanism of the control function is shown in Figure 8.1. The diagram is a good analogy for technical people because we readily recognize it to be like the simple functioning of a wall thermostat. The project control function is the mechanism that keeps the work of the project on target to meet the goals.

Basically, we start the cycle in the upper right corner with actual performance, which is periodically measured. Actual performance is then compared against the planned performance. If there is any deviation (or variance), we analyze the causes. A program of corrective action is formulated and implemented to correct the variance. The cycle is repeated by measuring the revised performance and again com-

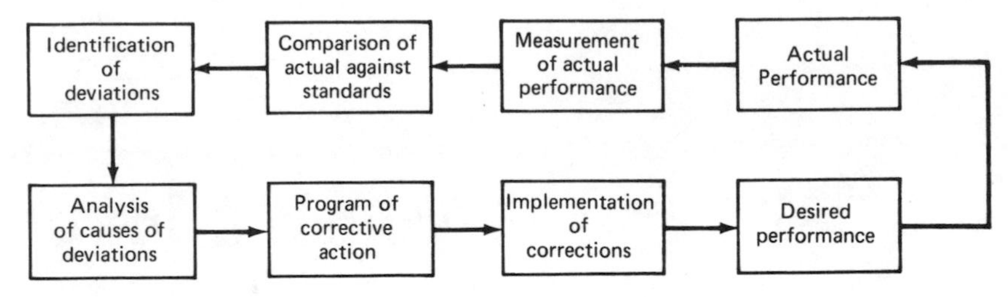

Figure 8.1 The control process.

paring it to standard. The process is repeated until the variance is "tuned out."

Areas of control

Which key project activities must we *constrain, coordinate, and regulate* to reach our project goals? They are the key areas that we developed in the project planning phase, namely:

- The money plan (the project budget)
- The time plan (the project schedule)
- Cash flow projections
- Quality standards
- Material resources
- Productivity

By concentrating on control of these six key areas, we should be able to meet our project goals successfully. Of course, there are a myriad of smaller areas of control, but most of them, except the *human factors,* are related to the above six areas.

Controlling the Money Plan

The cost control system is at the heart of controlling the money plan. Many well-designed cost control systems have been developed over the years to suit the broad spectrum in the capital projects industry. Regardless of the system used, it will take your single-minded devotion throughout the project to make the system work!

Cost control definitions

Cost control, despite its simple name, means a lot of different things to different people. Some often-heard synonyms are cost engineering, cost reporting, value engineering, and cost reduction. None of them alone is equivalent to *cost control.* Let's define a few of the synonyms in order to understand the differences in meanings.

Cost engineering is a generic term which covers the total field of cost estimating, budgeting, and cost control. It is too general a term to use for cost containment.

Cost reporting consists of gathering the cost data and reporting the actual vs. planned results without mentioning the operative word "control."

Value engineering gets closer to cost control because it looks at ways to reduce costs on specific items or activities. It does not look at the total project picture or check the daily performance; it only focuses on specific items in the design area.

Cost reduction also gets closer to cost control and would be fine if it included cost reporting. The result would then be evaluation and containment of costs on a complete project.

As it turns out, true cost control for capital projects involves all of the above activities at various times. To me, cost control means *the purposeful control of all project costs in every way possible.* That means every player on the project team has a part in reducing and controlling costs. The project manager is the leader of the cost containment program and must constantly reinforce that philosophy throughout the life of the project! [1]

Cost control philosophy

A comprehensive philosophy for cost control which I have developed over the years is based on three building blocks:

- Encourage and promote cost-consciousness in the performance of all phases of the work.
- Provide accurate and timely data on cost status and outlook and highlight any unfavorable cost conditions or trends
- Take prompt and effective action to correct problems and provide positive feedback for continuous evaluation of those problem areas.

A major problem area in most cost control systems arises under the second point: "provide accurate and timely data on cost status and outlook." Most good project managers are willing to sacrifice minor differences in the accuracy of the cost data if they can get it in a "timely" manner. Unfortunately, most cost control activity takes on an *accounting* level of accuracy, which often delays publishing the data.

Effective managers are looking for trends in the control of the financial plans, so accuracy within a range of plus or minus 2 to 3 percent is close enough. (Accounting practice could never countenance such inaccuracy in the formal cost report.) Although the age of computers has speeded up the number crunching a lot, the final cost report still takes time to check, print, and deliver. Issue dates of 7 to 10 days after the report cutoff date are not uncommon.

The built-in delay forces the project manager to develop an "unofficial" system to keep a running monthly approximation of the project cost status while awaiting the *official* cost report. That is not an altogether bad situation, since the project manager should learn to develop a "continuing feel" for all facets of job progress even before the formal status reports are issued. If PMs are continually finding *surprises* in their project reports, they are not staying on top of their jobs.

In that regard the project manager comes very close to being an entrepreneur running a small business; successful operators of small businesses are the people who develop a sense of *internalizing* their bookkeeping without waiting for monthly statements from their accountants. It is a matter of learning the ratios and rules of thumb that apply to projects in your particular area of capital projects.

If the project is overstaffed and underutilized, the next cost report will surely show that you are overrunning the personnel budget. That will also apply if the work quality is poor and unbudgeted hours must be spent doing rework. If you wait 4 to 6 weeks for cost information highlighting the problem, you will have lost valuable time for taking corrective action. This *sixth business sense* is one of the chief attributes that separates superior PMs from the average group.

Despite the secondary advantages of untimely cost reports, it is important for the project manager to keep pressure on the project control staff to issue their reports promptly. Examine the report cycle closely to eliminate any delaying factors inherent in the production cycle. The project manager's unofficial gut feeling must be verified regularly to eliminate possible accumulation of any percent error in the internalized system.

Cost control system requirements

To accomplish the cost control philosophy set forth above, the control system must include the basic features listed below to ensure a degree of success:

- A simple but comprehensive code of accounts
- Assignment of specific responsibilities for controlling costs within the project organization
- Use of standard forms and formats based on a standard code of accounts throughout the estimating, procurement, design, construction, and cost control groups

- A sound budget (based on a sound estimate)
- A mechanized system for handling the data on medium-size and large projects

We discussed the need for a standard code of accounts in Chapter 5, which deals with the formulation of the project estimate and financial plan. If you feel rusty on that subject, it would be a good idea to go back and review the advantages of a standard code of accounts.

The assignment of specific responsibilities for cost control should be delineated in the cost control section of the project procedure manual. The description should include the forms to be used, the formats for any reports, assignment of duties, and any other matters needed to create a complete cost control system.

The project manager should also stress the cost control theme when assigning responsibilities in the pertinent job descriptions and the goals in the project MBO program. They are specific assignments made to ensure that the philosophy of *cost consciousness is interjected into all phases of the project work.*

Part II of *Applied Cost Engineering,* by Clark and Lorenzoni,[1] is an excellent in-depth treatise on cost control from the cost engineer's standpoint. It also shows the working relation between the PM and the cost control group on the project. I have included a number of references to this book for more detailed descriptions of various points that I want to stress. Since space limitations preclude reproducing the descriptions here, I recommend that you add this book to your project management library.

The use of the same standard forms and formats throughout all of the project groups eliminates the introduction of transposition errors when dissimilar forms are used. The extra cost incurred by the errors must also be added to the actual transposing cost. The additional work inflates the home office services budget unnecessarily.

If a cost control system is to be effective, it *must* be based on a sound cost estimate and project budget. If the cost figures being controlled are bad to start with, no amount of cost control can make them right. If that situation occurs, the heart will go out of the cost control program and the morale of the cost control team will decline sharply.

If the bad budget is discovered early enough in the job, the project manager should press for a new cost estimate and a revised budget. It will not be gotten without a great deal of turmoil, but it is the only way to properly control the costs for the remainder of the project.

With today's easy access to personal computers and standard software, there is no reason not to have a mechanized system for maintaining the cost data on any small or medium-size project. Using your cost code system along with any standard spreadsheet programs will allow you to set up an inexpensive mechanical system for developing and monitoring the project budget.

If you do not have a mechanized system for a small to medium-size project, do not despair, because a manual system can work just as well. The computer does not add anything to the validity of the numbers; it only massages the numbers faster (we hope!). Somehow, we all feel that numbers generated on a computer are more believable and have an aura of reliability. Nothing could be further from the truth!

A Typical Cost Control System

Figure 8.2 shows how the work flows through the major elements of a typical cost control system. The idea here is to have all of the necessary cost information flowing to the cost control group on a routine basis so it can be assimilated and organized into accurate and timely cost reports.

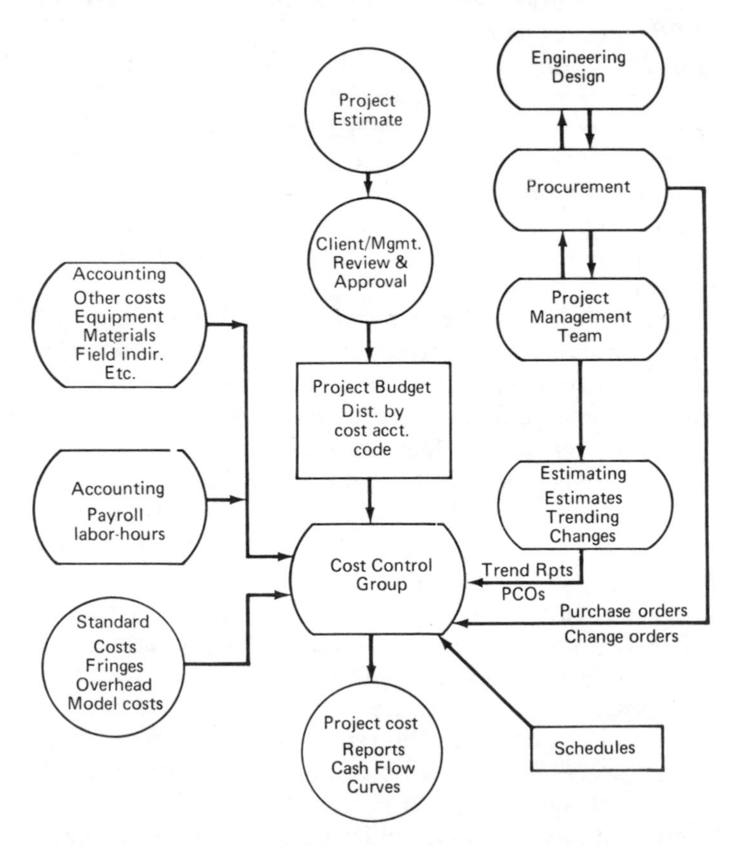

Figure 8.2 Cost control flow diagram.

The process starts with the project estimate passing through client and management reviews before it is converted into the project budget. As we said earlier, the budget is the basic *money plan* for the project which becomes the baseline for the cost control system. The approved budget (the original cost baseline) should never be revised; changes to the *baseline* budget are shown in a modified or *current* budget.

On the left side of the diagram, we show the normal staff departments which feed cost and expenditures into the cost control center on a routine basis. On the right side we show the design, procurement, and estimating functions feeding commitments, project change orders, and estimating data into the cost control center. At the lower right we show that the schedule is regularly issued to the cost control group so it can evaluate the effects of schedule changes on project costs.

At least monthly, the cost control center issues the project cost report and cash flow curves. The idea behind the flow diagram is to have the large volume of routine data flowing in standardized normal channels to ensure that nothing affecting project cost is overlooked. Missing data can result in inaccurate reports, which lead to loss of control of the project money plan. That will result in unmet project expectation for finishing the project under budget.

Staff functions

The staff functions on the left side of Figure 8.2 are recording the monies paid out for the human and nonhuman resources which flow into the project. They are mainly accounting functions which are paying for the commitments made by the operational groups on the right side of the diagram.

In cost control, it is important to differentiate between commitments and expenditures. Commitments are made when we order materials and equipment and have personnel who charge their time to a project. Expenditures occur when the bills for goods and services are paid and the payroll is met. Payrolls convert from commitments to expenditures fairly rapidly, usually in a matter of days or weeks. The time between commitment and expenditure for material and equipment can stretch out for months.

Line functions

The activities of the line functions on the right side of the diagram are more variable and much less routine. The progress of the job can affect schedules, which in turn affect costs. Project change orders, with their adverse effect on budgets and schedules, will always arise. The procurement program is the chief source of the longer-term project commitments which show up in the cost report.

The project team is the center for cost control function on the project. As we said earlier, the project manager, as the project team leader, generates the project cost control philosophy. The project group provides the coordination needed to make the other right side groups perform effectively. All of the corrective cost control action and feedback is processed through the project team.

In addition to coordinating the design and procurement efforts, the project team initiates and processes all the project change orders. It is responsible for getting the change orders estimated, approved, and fed into the cost control center for use in revising the *current* budget.

The project team is also continuously monitoring the cost-trending reports which are maintained between the various project estimates as the design develops and the project progresses. Close coordination between the project team and the project cost engineering group is vital if the project cost control system is to be effective.

The visible products of the system shown in the flow diagram in Figure 8.2 are the monthly cost reports and cash flow curves. They are usually very detailed reports that account for all of the commitments and expenditures on the project. Since the reports are issued to the client and analyzed by the client's cost control people on CPFF projects, they must be accurate. The accuracy requirement is what makes them less timely for the day-to-day needs of the project team's cost control program.

Any off-target trends on cost and cash flow must immediately be made known to the project manager on an informal basis such as through a project memo or a meeting. After the problem has been investigated and some possible solutions have been developed, the client and the company management should also be informed prior to the issuing of the formal cost report. A quick informal notice saves valuable time in solving the problem instead of waiting for the detailed cost report.

It is always better for the PM to get a problem out into the open early, along with some suggested corrective action, rather than wait for the client's cost people to find it in the cost report. The first approach will surely make you look like a more effective project manager than the latter.

How a Cost Control System Really Works

Let us look into some of the details on just how a typical cost control system can work on a capital project. We have already gone through the estimating and budgeting phase, now the project is in the early stages of execution.

The budget is the baseline of the cost control system so we must refresh our memories as to what it contains. In an earlier chapter, we listed the major accounts for a design, procurement, and construction project as follows:

Home office services	10 to 15 percent
Major equipment accounts	25 to 30 percent
Bulk material accounts	15 to 20 percent
Construction and field costs	40 to 45 percent
Contingency and escalation	Variable

I have not included a percentage for the contingency and escalation items, because they vary widely from project to project. They should be determined as part of the budgeting process and added in at the end.

Controlling labor costs

The first and fourth items listed above consist almost 100 percent of human resources costs. Variances in labor budgets can stem from three possible sources as follows:

- Original hourly takeoff error
- Variation in the assumed labor rates
- Variation in actual productivity from standard

Design and construction labor budgets are usually controlled in the same way on a monthly or weekly basis. Any possible error in the original estimate is evaluated by projecting the number of hours needed to complete the work during each reporting period. If overruns start to appear early in the reports, it is a strong indication that the hours in that area may have been underestimated. The PM's early in vestigation into such a symptom is vital. Later in the project, labor-hour estimating errors are harder to detect and prove.

All discipline or group leaders are required to estimate the number of hours to complete the unfinished work in their areas. It is vital that the estimates be based on real facts about how much work has actually been completed and how much is left to do. The estimates to complete must be based on earned value and *physical percent complete and not labor-hours expended.* We will discuss measuring physical percent complete and earned value later in the chapter.

Figures 8.3 and 8.4 are sample computer printouts used for control-

HOME OFFICE COST REPORT

LABOR-HOURS

CONTRACT: S-1234
CLIENT: ABC PRODUCTS
REPORT: NO. 10 PERIOD ENDING: 08/25/87 ISSUE DATE: 09/05/89
TYPE: LUMP SUM PROFESSIONAL SERVICES

LINE	DESCRIPTION	BUDGET		EXPENDED		ESTIMATE		CHANGE THIS PERIOD		CURRENT	PERCENT	
		ORIGINAL	CURRENT	THIS PERIOD	TO DATE	TO COMPLETE	PROJECT FINAL	CURRENT BUDGET	PROJECT FINAL	PROJECT VARIANCE	OF TOTAL	OF EXPEND.
01	PROJ. ADMIN.	13,000	12,800	360	9,190	6,062	15,252			2,272	7.2	60.2
02	BASIC DESIGN											
03	PROCESS ENGRG.	12,300	12,300	186	13,750	2,655	16,405			4,105	7.7	83.8
04	ARCHTECTURAL	1,050	1,050	73	954	148	1,102			52	.5	86.6
05	STRUCTURAL	6,300	6,300	154	4,976	1,641	6,617			317	3.1	75.2
06	CIVIL	3,150	3,150	118	2,085	1,224	3,309			159	1.5	63.0
07	MECH. EQUIPMENT	10,300	10,300	17	3,719	6,922	10,641			611	5.0	34.9
08	MACHINE DESIGN											
09	MATERIAL HANDLING											
10	HVAC/DUST COLL.											
11	PIPING	65,000	64,721	934	33,011	40,582	73,593			8,873	34.7	44.8
12	VESSELS	19,620	19,320	277	13,977	8,275	22,252			2,932	10.5	62.8
13	ELECTRICAL	7,550	7,420	747	7,297	3,301	10,598			3,178	5.0	68.8
14	INSTRUMENTATION	17,850	17,700	1,273	12,231	8,645	20,876			3,176	9.8	58.5
15	COMP.-AIDED DES.	1,340	1,340	88	1,560	309	1,869			529	.8	83.4
16	DESIGN CONTROL	1,960	1,960	68	802	1,174	1,976			16	.9	40.5
17	FIELD RESOLUTION											
18	S-TOTAL DESIGN	159,150	158,270	4,295	103,552	80,938	184,490			26,220	87.1	55.3
19	DESIGN SERVICES	4,000	4,000	180	3,080	2,242	5,500			1,500	2.5	56.0
20	SECRETARIAL	13,150	13,030	368	7,960	5,070	13,030				6.1	61.0
21	COST ENGINEERING	900	900	146	1,381	602	1,983			1,083	.9	69.6
22	SCHEDULING	1,300	1,300	26	1,244	256	1,500			200	.7	82.9
23	S-OTHER ENGRG.	19,350	19,230	720	13,665	8,343	22,013			2,783	10.4	62.0
24	S-ALL ENGRG.	178,500	177,500	4,797	115,791	90,712	206,503			29,003	97.5	56.0
25	PROCUREMENT	400	400	8	284	116	400				.1	71.0
26	HO CONSTR.	2,500	2,500	3	287	2,213	2,500				1.1	11.4
27	ACCOUNTING	600	600	7	492	508	1,000			400	.4	49.2
28	MISCELLANEOUS			20	110	90	200			200		55.0
29	SPECIAL ACCTS.		1,000		987		987			13-	.4	100.0
30	TOTAL ALL HOURS	182,000	182,000	4,835	117,951	93,639	211,590			29,590	100.4	55.7

Figure 8.3 Home office cost report, hours.

HOME OFFICE COST REPORT
SALARIES & OTHER COSTS

CONTRACT: S-1234 --------(ALL AMOUNTS IN THOUSANDS)--------------------------
CLIENT: ABC PRODUCTS
REPORT: NO. 10, PERIOD ENDING: 08/25/87 ISSUE DATE: 09/05/89 PAGE 2
TYPE: LUMP SUM PROFESSIONAL SERVICES

LINE	DESCRIPTION	BUDGET		EXPENDED		ESTIMATE		CHANGE PER		CURR.	PCT.	RATE				
		ORIGINAL	CURRENT	THIS PERIOD	TO DATE	TO COMPLETE	PROJECT FINAL	CURR. BUDG.	PROJECT FINAL	PROJ. VAR.	TO TOTAL	PCT EXP.	BASE EST.	TO DATE	TO COMP.	PROJ. FINAL
31																
32	ENGRG. OTHER COSTS															
33	OVERTIME PREMIUM	19.3	19.3		11.2	8.1	19.3			-	1.3	57.8	0.10	0.09	0.08	0.08
34	TRAVEL & LIVING	121.9	121.9		20.0	101.9	121.9			-	0.7	14.4	0.68	0.17	1.12	0.84
35	SUPPLIES & MISC.	59.5	59.2		13.8	45.4	59.2			-	4.2	23.3	0.33	0.11	0.50	0.28
36	MODEL COSTS	171.9	171.4		24.6	147.3	171.9			-	12.3	14.3	0.98	0.21	1.92	0.43
37	REPRODUCTION	38.1	38.1	0.2	39.4	64.6	100.0			61.9	7.2	35.4	0.21	0.30	0.71	0.48
38	LD. TELE. & TELEX	20.5	20.5		5.8	14.7	20.5			-	1.4	28.3	0.11	0.05	0.16	0.09
39	JOB SHOP FEE COSTS	20.8	20.8	10.0	64.0	20.4	85.0			64.2	6.1	76.0	0.11	0.55	0.22	0.41
30	SUBCONTR. ENGRG.	644.2	644.2		404.7	259.5	664.2			20.0	47.7	60.4	3.60	3.44	2.86	1.21
31	COMPUTER COSTS	294.3	294.3		70.4	70.6	150.0			144.3	10.7	46.9	1.64	0.60	0.87	0.72
32																
33	SUBTOT. ENG. COST	1390.5	1390.2	10.2	650.5	741.5	1392.0			1.8	100.0	46.7	7.79	5.61	8.17	6.74
34	PROCUR.OTHER COSTS															
35	OVERTIME PREMIUM	0.1	0.1			0.1	0.1				5.3		0.20		0.68	0.20
36	TRAVEL & LIVING															
37	SUPPLIES & MISC.	1.4	1.4		0.9	0.5	1.4				94.6	63.8	3.52	3.16	4.39	3.52
38	REPRODUCTIONS															
39	LD TELE. & TELEX															
30																
31	SUBTOT.PROC.COSTS	1.5	1.5		0.9	0.6	1.5				100.0	60.4	3.72	3.16	5.08	3.72
22	HO CONST OTHER COSTS															
23	OVERTIME PREMIUM															
24	TRAVEL & LIVING	52.2	52.2		0.2	52.0	52.2				100.0	0.4	20.88	0.80	23.48	20.88
25	LD TELE. & TELEX															
26	REPRODUCTIONS															
27	MISCELLANEOUS															
28	SUBTOT CON COSTS	52.2	52.2		0.2	52.0	52.2				100.0	0.4	20.88	0.80	23.48	20.88
29	MISCELLANEOUS COSTS	0.7	1.0		0.4	0.6	1.0					40.4				
30	SPECIAL COST ACCT															
31	TOTAL OTHER COSTS	1444.9	1444.9	10.2	652.0	794.6	1446.7			1.8		45.0	7.93	5.52	8.48	6.63
32	TOTAL SALARIES	1893.7	1393.7	54.0	1370.6	1013.6	2384.2			490.5		57.4	10.40	11.62	10.82	11.26
33	TOTAL SAL & OC	3336.6	3338.6	64.2	2022.6	1808.2	3030.9			492.2		57.2	18.34	17.14	19.31	18.10

Figure 8.4 Home office cost report, salaries and other costs.

ling home office costs in the design phase of a typical project. In that case the contract is lump sum for professional services, so cost control is of particular concern to the contractor's project manager. The design estimate was based on the contractor's prior experience on several earlier projects of the same type, and the lump sum price was based on the contractor's estimate.

The original estimate used a home office services estimate form such as the one shown in Figure 5.2. The values were then inserted into the budget as the first column in Figure 8.3, labeled Original Budget. The differences between columns 1 and 2 reflect approved change orders which add or subtract hours from the original budget.

Columns 3 and 4 show the hours expended during the period (1 month) and the total to date, respectively. The hours expended during the period should compare favorably with the project manager's *unofficial* prediction based on the number of discipline people assigned to the project for the reporting period.

The time required to complete the remaining work is evaluated by the discipline leader and entered in column 5 as the estimate to complete (ETC). The computer then adds columns 4 and 5 to get column 6, the Projected Final hours to complete the activity. Any change order hours approved during the reporting period would be shown in column 7 and added to or subtracted from the Current Budget values. Since there were no *approved* changes during that period, the columns are blank.

The computer then compares the Projected Final Hours to complete to the Current Budget to find the variance. In that case, projected overruns show up as positive numbers and underruns as negative numbers. A quick review of the Current Projected Variance column gives the project manager an insight as to where the cost control problems are in the Home Office Services Budget, Figure 8.3. In that case there are overruns in the process, piping, and electrical groups that indicate serious problems. They probably did not show up in one reporting period, so the project manager should already have taken some corrective action.

Although the underruns in civil, mechanical, and instruments help to offset the overrun of the total budget, the hours cannot be switched now; each discipline's budget must stand alone. The overruns may be caused by work done on unapproved change orders, poor productivity, or project delays. The project team has to work with the discipline leader to determine the cause and take corrective action or the design budget will overrun. If so, the result will be a losing project and unmet project expectations.

The last two columns of the report in Figure 8.3 give percents of the discipline hours to the total budgeted hours and the percent of the budget expended. Since the project is about 55 percent expended, there is still time to do something about any overruns. If overruns occur after budgets are 75 percent expended, there is little that can be done to stop the bleeding!

Figure 8.4 is a sample computer printout for the corresponding out-of-pocket expenses for the project. The format is similar to that of the labor-hour budget except that the values are given in dollars instead of hours. The column heads indicating the original and current budgets, expended values, estimate to complete, and projected final cost have the same functions as in the labor-hour budget.

Although expenses are a smaller part of the overall home office budget, they do bear regular checking. Close scrutiny is especially valuable in the larger accounts such as computer utilization, reproduction, travel and living, and overtime premium.

The bottom two lines of Figure 8.4 introduce the salary costs which go with the labor-hours in Figure 8.3. After all, the expenditure of human resources finally is reflected in money. The computer data base contains the individual salary rates for all project people so the total payroll costs can be reflected in the bottom line.

An important rule-of-thumb number, *average project hourly rate,* can be obtained by dividing the total direct payroll cost by the total hours expended. The number is used to check on how your project is loaded from the standpoint of the personnel mix of architects, engineers, designers, drafters, and management people. If your average project hourly rate is not within the normal range for your firm, your project could be suffering from a poor selection of personnel mix required for the work.

Obviously, the average rate is dependent on the type of project being performed. A high-tech study, for example, might require a preponderance of heavyweight (high-priced) specialists to do that type of work. An average design project, however, usually will not stray very far from the norm for the type of project. That sort of ratio must be averaged over a large number of projects and be adjusted, as required, for inflation.

Controlling construction labor budgets

The system for controlling construction labor-hours is very similar to that discussed for the design hours. The only difference is that the work is broken down into more detail by craft for better control. The estimate to complete and the physical progress and earned value systems are used in the same way to calculate the physical percent complete.

Also, the craft labor-hours are budgeted separately from the field supervision or indirect labor-hours. The field supervision and out-of-pocket costs are covered in a field indirect cost budget. The craft labor from the foreman level downward is budgeted to the actual construction operations.

Changes in labor rates

Changes in labor rates are not usually a problem in capital project work, since they are fairly predictable. Home office rates change no more often than twice per year when salary reviews are scheduled. Since cost-of-living raises are subject to the inflation rate, the amount of increase is predictable. The relatively small percent of merit increases usually averages out over the personnel turnover rate within the organization.

The potential increases in cost are factored into the estimate depending on how long the project schedule has to run. Missing an estimate in this area is likely to occur only in periods of unexpected high inflation or job stretch-out.

The field labor rates are even more predictable, since they are often controlled either directly or indirectly by union negotiations with the contractors. Sometimes, the labor agreements are negotiated for several years in advance and provide for fixed increases from year to year. Again, the rates are easily predicted and factored into labor cost estimates and budgets.

Variations in productivity

The area of worker productivity is the most likely one in which labor-hour budgets go astray. Most standard unit labor costs for design or construction do allow for a little variation in the productivity of the work force. The competitive environment in which most capital projects are estimated and performed, however, militates against having a very high safety factor in the area of productivity.

Most managements will not allow padding the standard labor-hours unless an unusually good seller's market exists. Therefore, most *standard hours* are assumed to be at 100 percent efficiency. If the project team's productivity (efficiency) drops much below 100 percent, the labor budget is going to be overrun.

When we think about loss of productivity, our first thought is that the problem lies in poor quality and performance of our project personnel. That is logical enough, since even project managers are human. However, it is also unlikely that *all* (or even a high percent of the people) are of poor quality and not performing up to standard.

We like to overlook the possibility that *management factors* can contribute significantly to poor labor productivity. It is only human nature to find the fault in someone else. In the design area we must look at improper supervision, personnel shortages, poor operating systems, incomplete design data, project delays, poor communications, and so on as management's possible contribution to low productivity.

The above list includes only a few of the possible hundreds of management failures that can adversely affect productivity. The most obvious one which has been brought into sharper focus by the advent of the task force mode of operation is the continuous flow of design data and approvals once the task force has been assembled. The flow of information *must be continuous* to maintain good productivity. Personnel cannot be moved on and off the project at will to match an uneven flow of project information without killing task force productivity.

Many of the same management failures, such as poor supervision and work planning, missing or late design information, inaccurate design documents, and missing or late materials, also affect field productivity. To the list we can add some items peculiar to the construction environment: unusually bad weather, strikes, unexpected obstructions, labor shortages, and remote site access.

On international projects we can list even more typical productivity killers such as cultural differences, political unrest, onerous government regulations, and lack of tools and equipment—to name a few. A case in point in that area was the fall of the Shah of Iran, which caught by surprise many foreign companies operating there. Later, I read some articles that blamed project losses on the failure of the PM to foretell the collapse of the regime! Given that sort of thinking, project managers should not be surprised at being blamed for anything at all that goes wrong on or off their projects.

Underrunning the labor-hours

Some people do not consider it a problem when the fortuitous circumstance of underrunning the labor hours arises, but it does deserve some comment on how to handle it. Remember, one of Murphy's laws is that "the amount of money expended will always rise to the amount allotted."

Some group leaders (like many politicians) think that the hours budgeted to them at the start of the project are theirs to spend no matter what. They can waste the hours in a variety of ways such as getting plain sloppy with the work, keeping unneeded people on the project, and doing work outside the scope. Project managers must be ever alert to those situations and preserve any underruns to improve overall project cost performance or to offset overruns elsewhere in the project. Since project managers are judged only on overall project performance and not on individual discipline performance, they have to impress on all project leaders the vital motto: Preserve underruns wherever they occur!

Summary of labor cost control

The total cost of human resources for an integrated project may run as high as 35 percent, including field labor. On a design-only project, the design labor cost will be at least 60 percent of the design fee; in either case, control of the labor costs is an important part of project cost control.

Periodic reviews, estimates to complete, and timely cost reports are critical to good results from any cost control system. The most likely factor that causes labor budgets to overrun is poor productivity. Poor productivity more often results from ineffective project management than from unqualified workers. The project manager must find and eradicate the causes of poor management practices that contribute to low productivity. Labor underruns must be guarded and nurtured through the closeout of the project.

Controlling Material Resource Costs

The physical resources on a given project can run from 40 to 60 percent of total installed project cost, which makes it a budgetary force to be reckoned with. The human element is present to a much smaller degree in that part of the budget. We do not have to deal with personnel productivity factors when we are dealing with equipment and materials. Prices are controlled by market forces and people participating in the timely delivery of the physical resources, so controlling material cost is not entirely devoid of human influences.

The overall philosophy for controlling the physical resources budget is much the same as for controlling the human resources. We start with an estimate of the physical resources cost and convert it into a budget which becomes our baseline for buying them. As the project progresses, we check the actual delivered cost against the estimated cost of each item. The estimate of the cost to complete the delivery of each physical resource occurs in the same way as the control of human resources. Any differences between predicted and planned are reflected in a variance column for the project team's use to control equipment and material costs.

Reviewing a cost report

The best way to get an overview of cost control of the materials budget is to review a typical capital project budget as shown in Figures 8.5

and 8.6. The sample budget that I have selected is from an actual project which was handled in a computerized format because of its size. Manual budgets work the same way except that they are posted and calculated by people, which makes the manual work a bit onerous on larger projects.

Actually, the total budget for the project consisted of about 100 pages of printouts, but I have selected the two most critical ones which show the total costs. Figure 8.5 lists a summary of the major equipment and materials accounts. Each of the listed accounts is a roll-up of the detailed listing for all equipment items and further breakdowns of the bulk materials. If the summary numbers for any account show a cost control problem, the detailed listing should be checked to find the source of the trouble.

The header section of the report is repeated on every page; it shows the title, project, location, and reporting period. The title of the report indicates that this is a summary sheet of all areas, which means that the body of the report lists all materials and equipment for each area of the project. Sectionalizing the report is handy when a project is subdivided into areas for individual supervisors, who can readily use their specific sections of the report to control costs in their assigned areas.

In the column on the far left side are the account code numbers, which are from the standard cost codes used on the project. They are the stem accounts used in the detailed breakdown for equipment. The first of the column heads to the right of the description column is Original Budget. As was said earlier, the column should never be revised unless it has become necessary to reestimate the project. Any movement of equipment and materials among the accounts is handled with the column headed Trans(fers).

The next column reflects revision to the project in the form of approved change order, which show up in the cost report only after they have been officially approved. The computer combines the values in the change columns with the Original Budget numbers which then results in the Current Budget column. All the subsequent calculations are based on the current budget number as the latest baseline.

The next series of columns contain the "action items" which report the financial activities as they occur on the project. The first two columns cover the *commitment activities* for the report period and the project to date. Commitments are the values of purchase orders and subcontracts that will spend project funds at future billing dates. No cash changes hands in either of the two columns.

The Cost to Date column shows the actual funds spent against each account. As invoices are received and checks are drawn against the

CONTRACT:
CLIENT :
PROJECT :
LOCATION:

*** COST PROGRESS REPORT ***
MATERIAL STATUS REPORT
SUMMARY ALL AREAS

REPORT DATE: JULY 27,1989
PERIOD ENDING: JULY 24, 1989
REPORT NO.: 005
ALL FIGURES IN THOUSANDS
PAGE OF

DESCRIPTION	ORIGINAL BUDGET	TRANS	APPR EXTRA	CURRENT BUDGET	COMMIT PERIOD	COMMIT TO DATE	COST TO DATE	EST TO COMPLETE	PROJ COST	PROJ VAR	X COMMIT
MAJOR EQUIPMENT	2582	0	0	2582	0	0	0	2998	2996	414	0.
0111 TANKS	138	0	0	138	101	172	0	198	370	232	46.5
0112 VESSELS AND DRUMS	1363	0	0	1363	857	1447	0	773	2220	857	65.2
0113 TOWERS AND REACTORS	6646	0	0	6646	0	4030	0	3641	7671	1025	52.5
0114 INTERNALS	1435	0	0	1435	0	0	0	1519	1519	84	0.
0115 HEAT EXCHANGES	4460	0	0	4460	-25	1072	0	4018	5090	630	21.1
0117 COMPRESSORS/DRIVERS	3490	0	0	3490	0	2235	0	356	2591	-899	86.3
0118 PUMPS AND DRIVERS	1674	0	0	1674	4	4	0	1745	1749	75	.2
0119 SPECIAL EQUIPMENT	4958	0	0	4958	634	636	0	3871	4507	-451	14.1
MAJOR EQUIP ESCAL	1246	0	0	1246	0	0	0	998	998	-248	0.
*TOTAL EQUIPMENT	27992	0	0	27992	1571	9596	0	20115	29711	1719	32.3
0121 A/G PIPE	2285	0	0	2285	0	0	0	4685	4685	2400	0.
0122 A/G FLANGES	1281	0	0	1281	0	0	0	1226	1226	-55	0.
0123 A/G FITTINGS	1715	0	0	1715	0	0	0	1695	1695	-20	0.
0124 U/G PIPING MATERIAL	957	0	0	.957	0	0	0	957	957	0	0.
0125 SHOP FABRICATION	2379	0	0	2379	0	0	0	2379	2379	0	0.
0126 A/G VALVES	2661	0	0	2661	0	0	0	2137	2137	-524	0.
0127 PIPING SPECIALTY	1326	0	0	1326	0	0	0	1340	1340	14	0.
0131 FDNS/STRUCT/PVG MAT	1016	0	0	1016	0	0	0	1095	1095	79	0.
0138 STRUCTURAL STEEL	2095	0	0	2095	0	0	0	2131	2131	36	0.
0140 MAJ ELECT EQUIP	3286	0	0	3286	0	314	0	3033	3347	61	9.4
0141 ELECT MATL/DEVICES	1583	0	0	1583	0	0	0	1822	1822	239	0.
0150 INSTR/CONTROL DEV	4020	0	0	4020	0	0	0	4249	4249	229	0.
0151 INSTR VALVES	1242	0	0	1242	0	0	0	1242	1242	0	0.
0152 INSTR/BULK MATL	160	0	0	160	0	0	0	160	160	0	0.
0165 BULK FRIEGHT	390	0	0	390	0	0	0	390	390	0	0.
0166 VENDOR REP	192	0	0	192	0	0	0	192	192	0	0.
0199 STEAM PIPELINE	5040	0	0	5040	0	0	0	0	0	-5040	0.
BULK MATERIAL	6957	0	0	6957	0	0	0	7360	7360	403	0.
BULK MATL ESCAL	3659	0	0	3659	0	0	0	3765	3765	106	0.
*TOTAL BULKS	42244	0	0	42244	0	314	0	39858	40172	-2072	.8
**TOTAL DIRECT MATERIAL	70236	0	0	70236	1571	9910	0	59973	69883	-353	14.2

Figure 8.5 Cost progress report, equipment and materials.

CONTRACT
CLIENT

PROJECT
LOCATION

** COST PROGRESS REPORT **
MONTHLY JOB PROGRESS REPORT
SUMMARY ALL AREAS

REPORT DATE: JULY 27, 1989
PERIOD ENDING: JULY 24, 1989
REPORT NO.: 005
ALL FIGURES IN THOUSANDS
PAGE OF

DESCRIPTION	ORIGINAL BUDGET	TRANS	APPR EXTRA	CURRENT BUDGET	COMMIT PERIOD	COMMIT TO DATE	COST TO DATE	EST TO COMPLETE	PROJ COST	PROJ VAR	% COMMIT
**TOTAL DIRECT MATERIAL	70236	0	0	70236	1571	9910	0	59973	69883	-353	14.2
**TOTAL SUBCONTRACTS	37633	0	0	37633	0	0	0	26448	26448	-11185	0.
**TOTAL DIRECT LABOR	22203	0	0	22203	0	0	0	17035	17035	-5168	0.
***TOTAL DIRECT COST	130072	0	0	130072	1571	9910	0	103456	113366	-16706	8.7
**TOTAL FIELD INDIRECTS	18797	0	0	18797	0	0	0	16146	16146	-2651	0.
**TOTAL PRO-SERVICES	12933	0	461	13394	551	2733	2733	12833	15566	2172	17.6
**TOTAL OTHER COSTS	6099	0	0	6099	0	0	0	5618	5618	-481	0.
***TOTAL INDIRECT COST	37829	0	461	38290	551	2733	2733	34597	37330	-960	7.3
**TOTAL OVERHEAD	5417	0	173	5590	271	1059	1059	5136	6195	650	17.1
***TOTAL	173835	0	643	174469	2339	13702	3792	143831	157533	-16936	8.7
**TOTAL ESCALATION	0	0	0	0	0	0	0	0	0	0	0.
**TOTAL CONTINGENCY	14001	0	0	14001	0	0	0	13194	13194	-807	0.
****GRAND TOTAL	187836	0	634	188470	2339	13702	3792	157025	170727	-17743	8.0

Figure 8.6 Cost progress report, summary.

project account, the amounts show up in the column representing the actual cash flow for the month and the project to date.

The Est(imate) to Complete column represents the cost of bringing that account to completion in accordance with the project scope. In equipment and material accounts it represents the money which has yet to be committed for unordered physical resources or subcontracts. After all project materials and equipment are committed under fixed price orders, a small sum should be held in the Est(imate) to Complete column to cover any unexpected costs that might arise.

Each estimated cost-to-complete value is added to corresponding value in the Commit(ted) to Date column to get the revised expected total cost for the account. The resulting number is the most current Pro(jected final) Cost for that account and, when all the accounts are totaled, for the total project.

The new projected final cost is then compared by computer to the current budget figure, and the difference is shown in the Proj(ected) Var(iance) column as a positive or negative number. A positive number is an overrun, and a negative one is an underrun. This is one time when being negative is beautiful! That rarely happens, but a minus sign on the variance column total does reflect a favorable cost trend for the project.

Project management would be beautifully simple if all we had to do was look at one number, but it doesn't work that way. Even if the total variance number is negative, we must analyze all of the accounts to see how it got that way. A negative value is definitely a good sign, but here it is no cause for complacency. Perhaps the number is not really factual; perhaps it should not be so low or high; or perhaps we could even improve on it. Proving to yourself that the number is really a *correct statement of the cost picture* is the only way to avoid nasty surprises at the end of the project.

Is there a cost control problem in Figure 8.5?

Figure 8.5 does, in fact, show an example of how reading only the bottom line of a cost report can get one into trouble. The bottom line of the variance column shows an underrun of $353,000, which indicates that the total material and equipment account is under budget and therefore in good shape. In looking at the numbers in the Project Var(iance) column, however, it is obvious that something is wrong. The equipment account shows an overrun of $1,719,000 alone, certainly not a good showing. The bulk material account shows an underrun of $2,072,000, which is fine. But, how can that be when we are overrunning aboveground piping by $2,400,000?

Account number 0199 for a Steam Pipeline shows an underrun of $5,040,000, which turns out to be the culprit. The item was recently deleted from the project as being unnecessary, but it is still being carried as an underrun instead of being shown as a negative change order in column 3. Carrying the account in the wrong column has created a false sense of well-being by turning a $4.687 million overrun into a $353,000 underrun. Although $4,680,000 represents only a 7 percent overrun of a $70,236,000 Total Direct Material account, the overrun does represent 33 percent of the Total Contingency account of $14,000,000 shown in Figure 8.6.

It is still early in the project, with only 14.2 percent of materials committed, so there is cause for concern about cost control in the physical resources portion of the project. A detailed analysis of the materials account should be made to better evaluate the final projected outcome of the material accounts.

Cost control summary, all areas

Figure 8.6 shows a typical budget summary sheet for the same project. Let us examine the major accounts to see if we can spot any problem areas from the numbers shown. From the Total Pro-Services (home office costs) numbers, we can see that the design is about 17.6 percent completed, so the design is just starting to peak. The total home office services is the sum of Pro-Services, overhead, and fee totaling $21,826,000. That number divided by the total installed cost (TIC) of $187,836,000 gives a design cost–to–TIC ratio of 11.6 percent, which is in line for a petrochemical project.

On the other hand, the Pro-Services account is already showing a projected overrun of $2,172,000, or a 17 percent increase. The higher value is still within the 11.6 percent rule-of-thumb number relating to the TIC, which is favorable. However, predicting a 17 percent increase in design cost that early in the project, without any change orders listed, does look suspicious. It could indicate that the design cost was underestimated or that change orders are not being processed properly because the increase has come so early in the project. In any event, a thorough review of the home office services account seems to be in order.

We have already discussed the false indication of an underrun in direct materials caused by the deleted steam line. The next account in Figure 8.6 is construction subcontracts, which looks weird at this stage of the project. The account shows a projected underrun of $11,185,000 on a base budget of $37,633,000 or 30 percent. Since no subcontracts have been committed as yet, it is difficult to imagine that such a prediction could be made without having a reduction in the scope of work. The same comment applies to the Total Direct Labor account, which shows an underrun of $5,168,000, or 23 percent, before one labor-hour has been expended. All of them add up to a nifty underrun of $16,706,000 when only one account has been 14.2 percent committed. If we drop down to the Total line, only $13,702,000 has been committed, which is really only 0.07 percent of the funds budgeted.

Those are only a few points about the cost report that we developed from analyzing just two summary pages. I am sure that there are plenty more suspicious areas hidden away in the backup to the summary pages. Having wide swings of that sort in some of the accounts so early in the project indicates a possibility of faulty estimating or someone painting an overly rosy picture in the cost control report. Analyzing a project cost report is not unlike reading a company's financial statements: Is it making money? Is it solvent? What are its cash reserves?

Most project managers have technical backgrounds and very little accounting and financial training. Therefore, it is vital for project managers to get the knack of using rule-of-thumb ratios to analyze and interpret a cost report. If you don't come by that knack naturally, you should take a few financial and accounting courses to develop your skills in that area.

Escalation and contingency

The project team elected to handle the escalation account differently than I recommended. The escalation was distributed in lump sums to the major accounts as shown in the last lines of the Major Equipment and Bulk Material accounts in Figure 8.5. The escalation on equipment is only 5 percent, so it appears that we are in an era of low inflation. The Major Equipment Escalation account shows an underrun of $246,000 in the face of a projected overrun in that account of $1,719,000, which seems a bit optimistic at such an early stage of the buying program. Apparently the overrun is not due to price increases.

The Bulk Material Escalation allotment is slightly higher, at about 9 percent, which is not unusual considering that the project took place in the late 1970s. The odd thing about the account is that it projects a $106,000 overrun in the face of a total account underrun of $2,072,000. That is a further indication that the underrun on bulk material is bogus!

The Total Contingency account is shown as a lump sum of $14,001,000 on the next to last line in Figure 8.6. That is about 8 percent of the TIC, a fairly conservative figure even for a project of such large size. It indicates a degree of confidence in the estimate at this early phase of the project. In addition, the cost report is forecasting a reduction of $807,000 at this point—another sign of optimism.

In earlier chapters we discussed holding escalation and contingency in separate reserve accounts until needed. That is being done in the budget even though the escalation has been parceled out to the major material accounts. Assigning fixed amounts to the major accounts is probably a good practice, since it allows more flexibility in putting the money where it is most needed. Money should never be assigned to individual account items, as we mentioned earlier, until a real need arises.

The contingency account could have been handled in the same way as the escalation account with some slight advantage. There must be some parts of the estimate that are felt to be more accurate or less accurate than others. By breaking down the contingency fund, more money could be assigned to the weaker areas of the budget. In either case, the funds can be drawn from the main contingency account as required.

It is normal to reduce the contingency and escalation funds as the commitment percentage increases and the exposed portion of the unspent budget becomes smaller. Any changes in those accounts requires the approval of the owner and the company management. As project manager, it is best to manage those funds with a lot of thought before recommending any revisions. The complexion of a budget can change overnight, so maintaining a conservative balance for as long as you can makes good sense. You will be even a bigger hero if there are healthy balances at the end of the project!

Manual vs. computerized budgets

The personal computer revolution has pretty well taken the argument out of the discussion about manual vs. computerized budgets. I used to recommend using manual methods on smaller projects because the cost of putting the figures in the mainframe computer was too great. That heavy expense was due to both software development costs and hardware charges.

Now, however, we can put a PC with commercial software on a small project at moderate cost. In fact, the PC and software are often already there. All that is needed is to assign a clerical person the job of inputting the data. The only caveat is to make sure that the person is trained and does the budget posting regularly enough to maintain current figures.

As we said earlier, if a PC is not available on the project and there is no chance of getting one, don't develop an inferiority complex. You can do the job manually with equally good results if the budget is good and you manage the funds properly. The format for the manual budget is the same as for the computerized sample in Figures 8.5 and 8.6.

Specific Areas for Cost Control

We have been addressing the subject of cost control on capital projects in a general way; now it is time to get specific. The number of specific areas is considerable, and we will touch on only some of the more fertile areas for cost reduction. It is a general rule of cost control that the front end of a project is the area most likely to offer cost-reduction opportunities. After a project passes the 60 percent design milestone, the likelihood of attaining cost reductions becomes much smaller. The same thing is true of the field work; it is the preconstruction planning and the setting up of the field operations that offer the best areas for cost reduction. We then go into a stronger *cost control mode* later in the project to ensure that the budget does not get out of control.

The design area

The overall cost parameters of the project are set during the basic design area of any project. The initial project scope establishes the size and function of the facility. Many specific questions involving major impacts on project cost must be resolved. The feasibility of various approaches to the project must be tested—usually along economic lines.

In architectural projects, the basic uses of space and traffic flow are established. The function of the facility must be dissected and each part must be questioned and studied to assure an optimized design. Architects call this formative stage of a project *programming*. William Pena developed a book, *Problem Seeking,*[2] which defines the underlying principles of programming in these five steps:

1. Establish *goals.*
2. Collect and analyze *facts.*
3. Uncover and test *concepts.*
4. Determine *needs.*
5. State the *problem.*

The book goes on to delineate a complete programming system for establishing the basic design for any type of project. An organized approach to establishing the basic design parameters for the project can get the cost control program off on the right footing.

In a process plant, the basic process design package must be developed in much the same way. Heat and material balances are made. Product purity specifications are set. Basic equipment is sized. All those functions can have a dramatic effect on final facility cost.

In subsequent design steps also there are many decisions affecting cost. Can expensive or redundant equipment be eliminated? Are we using the optimal construction materials? Can energy be conserved with better heat exchange? Any of those design factors can yield cost savings in capital and operating costs if properly handled.

After the basic and schematic design phases are completed and we enter the construction document phase, design factors affecting cost continue to play major roles. Here we must keep a sharp eye out for a natural tendency to gold-plate the project. That can be in the form of redundant systems, unnecessary controls, overdesign, and other aspects which do not improve the overall mission of the facility. Beware of the belt-and-suspenders syndrome!

Detailed design can yield more substantial cost reductions in such areas as more efficient plot planning, better building and equipment layout, and better systems integration. The civil, electrical, mechanical, and instrumentation areas also offer fertile fields for cost reductions. Specific areas to look at are unnecessary earth moving, paving, drainage, foundation design, main power supply, area classification, and excessive code requirements in general.

It is difficult for project managers to ferret out all the design problem areas on their own because they may not have the time or expertise to do so. What they can do, however, is alert their project engineers and design team leaders to search for cost-cutting ideas in those vital areas. The cost-cutting ideas are then referred to the PM for screening and deciding which ones should be considered for implementation. Since many of those ideas will represent major changes, the potential savings must also be weighed against their effect on project schedule. The time value of money must always be considered in implementing a major change that affects project schedule.

Home office costs

The home office budget usually represents only 6 to 12 percent of the total installed project cost. Although it is a relatively small part of the project cost, it does get expended under the close scrutiny of the home office management. In design-only projects, the home office budget represents the total budget on which the firm's profit is based. Those factors make the home office budget worthy of cost control attention from the project manager.

In some larger engineering-construction company operations, the design phase is sometimes offered as a "loss leader item" to get a highly profitable construction project. In that case, the design estimate will be on the low side and provide little chance to yield a profit. The design group's performance will still be judged as a separate profit center, so the design project manager has an added challenge to eke out a positive result. As we discussed earlier in this chapter, the main items of cost in the home office budget are:

- Design personnel hours

- Average labor-hourly rate
- Computer costs (including CADD)
- Other out-of-pocket costs

The design hours must be monitored at least monthly on large projects and more often (i.e., biweekly) on smaller projects. Using some sort of home office cost report, as shown in Figures 8.3 and 8.4, is vital to successful cost control. This is where a running estimate of the labor-hours is essential to keep tabs on how the time is being spent during the report period. Keep in touch with the discipline leaders to get a feel for how they really think their discipline budgets are faring. Often they will give you an indication of a problem area even before it turns up in the home office cost report.

Getting out into the trenches is so important in the discipline area of cost control. By reviewing the actual work to complete with the discipline leader, you will be better able to spot overly optimistic or pessimistic estimates to complete in the cost report.

Some digging into the out-of-pocket expenses also is very productive. Since everything is charged to your project by way of a contract number, it is possible to get another project's expenses charged to your project when a number gets transposed. It is not necessary to make an accounting audit of the out-of-pocket costs; just spot-check for anything out of the ordinary.

An early response to off-target results is vital to successful cost control. If you spot an adverse trend developing, get into the matter immediately so you can respond with an effective plan for corrective action. As I will often say in this book, as leader of the project it is up to you to implement the project control systems and direct any corrective action. Naturally, it is also wise to use the advice and counsel of your project staff.

Procurement cost control

Buying the large amount of physical resources and subcontracts required for capital projects is an important area of cost control which is often overlooked. Some project managers tend to look on purchasing as a clerical or staff function without much bearing on the outcome of the project. Nothing could be further from the truth! A savings of 5 to 10 percent of the equipment, materials, and subcontract budget can represent a sizable savings to the project. Pick your procurement people with the same care as you do your technical staff.

Spend some time to develop a sound procurement plan to ensure timely and efficient delivery of vendor data to the design group and the physical resources to the field. As we saw earlier, good productivity by human resources in design and construction depends heavily on having the project resources available when they are needed. An indirect cost reduction results from an effective purchasing program.

Effective cost control in procurement is founded on good procurement procedures, which should embrace the following major factors:

- A good approved-vendor list
- Ethical bidding practices
- Sound negotiating techniques
- Change order controls
- Control of open-ended orders and subcontracts
- Control of procurement, expediting, and inspection costs
- Maintenance of purchasing status reports

Procurement's goal is to buy the specified quality of goods and services for the project at the best possible prices. The first four items on the list contribute the most toward reaching that goal.

The suppliers on the vendor list must be capable of supplying the right kind of goods at competitive prices to meet the schedule. Therefore, they must be carefully screened to ensure that they make what you want to buy and have the capacity to deliver it when you need it. A fully loaded shop will not be able to deliver and will not be as competitive as a hungry one.

Don't load up the approved-vendor list with more bidders than are needed to ensure competitive prices. Any marginal suppliers should be dropped, along with any having full order books. Having more bids than necessary adds to the time needed for technical and commercial bid tabulating without improving the desired result. Good practice calls for having a minimum of three bidders and a maximum of six. It is well to go with the higher number during periods when "no bids" are prevalent to ensure having at least three qualified bidders to choose from.

Ethical bidding practices have been established over the years in most western countries to get the best offer the first time from reliable bidders. If bidders know their prices are going to be peddled to the competition, they will not give their best price the first time around. The minor cost reductions that result from using so-called auctioneering tactics are more than likely to be offset by time lost in the schedule.

That theory on ethical bidding does not preclude the need for negotiation in the procurement process. Sound negotiating techniques can reduce costs on large orders and subcontracts. The negotiating process often brings out some of the more obscure aspects of the procurement. Is the supplier really meeting scope and specifications? Are all of the requirements written into the specs really needed? Is it possible to get a better warranty or delivery at no additional cost? Those are only a few hints of the possible fruits of the negotiating process.

In most cases, the negotiating team is made up of procurement and design personnel to ensure that the technical and commercial points are covered as they arise. Team negotiation is always better because the ball can be passed among the members to suit the occasion.

Minimizing change order costs in the first place is the best line of defense to use, which means that the quality and definition of design must be stressed before the procurement process begins. Since change orders are inevitable, their cost is best controlled by obtaining unit prices for potential changes during the bidding process while the sense of competition is still present. When you are already locked into a vendor, your position for negotiating the cost of extras is weakened.

A good example of protecting oneself on change orders is the practice of getting unit nozzle prices when bidding vessels, or volume price breaks when buying structural steel, rebar, and concrete. Having the prenegotiated prices for changes eliminates a lot of haggling over change order costs which can consume a sizable piece of your procurement budget.

Open-ended purchase orders or subcontracts invite cost control problems so they should be used only when they are absolutely necessary. If you are forced into the position of using open-ended agreements, be sure to put check-and-balance systems in place for early warning of problems. The checks can be in the form of funding limitations so flags will go up to indicate that more commitment is needed. At that time, progress and productivity can be reviewed to predict what the final number is likely to be. If something like that is not done, you will no doubt get a nasty surprise at the end of the job just when you thought that the financial goals had been attained. Make a point of having a checklist of all the open-ended commitments you have in order to ensure effective control.

A rule of thumb for estimating the overall procurement cost is about 5 percent of the total material, equipment, and subcontract cost. That breaks down to 2.5 to 3 percent for buying, 1.5 percent for expediting, and 1 percent for inspection.

Purchasing hours and costs are reported as part of the home office cost report, which was discussed earlier. The hours spent should be compared to the performance of the procurement commitment curve. The commitment curve is the S that results when we plot the projected value of orders placed against elapsed time. The procurement curve lags the design progress curve and leads the cash flow curve.

I have included controlling and reporting purchase order status under cost control because lack of that information can adversely affect the project. Keeping abreast of the procurement status of tagged equipment is relatively simple, but controlling bulk material orders in detail is something else. Computerized systems for tracking a project's physical resources have been developed by many firms. The systems are really spreadsheets which schedule each procurement activity for all of the equipment and materials. As procurement progresses, the report is updated to keep the information current. The project manager should monitor the material status report, in conjunction with the procurement commitment curve and the home office cost report, to measure the cost-effectiveness of the procurement program.

Many of the early computerized procurement-tracking programs were developed to handle a great deal of detail which required the use of a mainframe computer. Needless to say, the programs proved to be too costly for the benefits derived. With the advent of the PC and standard spreadsheet software, a cost-effective procurement-tracking program can be developed for even a relatively small project. In either case, though, I advise against going into too much detail in the bulk material area.

From this discussion of procurement activities, it again becomes obvious that the most money can be saved with good planning in the early stages of the project.

Construction costs

Since construction spends the lion's share of the project budget, we would expect it to be a fertile area for cost reduction and control. As we look at the possible areas listed below, we can see that the best time to save is early in the project.

- Constructibility analysis
- Planning temporary facilities and utilities
- Planning heavy lifts and associated equipment
- Assigning laydown and working areas on site
- Organizing the field supervisory staff
- Establishing the construction philosophy
- Field mobilization and demobilization
- Field personnel leveling and field productivity

In many cases it is difficult to get the construction team into the project early enough to realize some of the potential cost savings. In split projects for which the design is completed before the construction is bid it is virtually impossible. This feature of split contracting is one of the major drawbacks in that type of project execution. However, it is possible to get the constructor's involvement early when a negotiated third-party constructor arrangement is used.

The constructibility analysis is relatively new; it has been developed and promoted in recent years by some of the large contracting firms. The advantage to the contractor is readily apparent, but the potential cost reduction to the owner is also quite real.

The idea is to bring a few highly qualified construction specialists into the project during the design phase. The specialists review the design as it develops and offer suggestions which will make the facility easier to construct and thereby reduce construction costs. Although the people are fairly expensive, they do not have to be used full-time to reap the advantages. The best situation is have them carry right through to become part of the construction team. The knowledge gained during the design phase makes them especially valuable in planning the early field operations.

Temporary construction facilities are another area that must be analyzed for cost savings. The facilities should be adequate to do the job and suitable for the local climate and labor conditions. Temporary utilities should have adequate capacity to safely carry the expected construction loads until the regular facility utilities come on line. Frequent breakdowns of overloaded utility systems during construction can be very costly and are intolerable on a cost-conscious project. Pennies saved up front can cost big dollars at the peak of construction!

Gold plating the field facilities should also be avoided for obvious reasons. Make sure that the field facilities budget is spent in the right places. It may, however, be necessary to add some special touches if local conditions are poor and some morale building is needed.

Early planning of the heavy lifts for the project is normally an offshoot of the constructibility analysis. The equipment for the heavy lifts is bound to be expensive and hard to find, so it must be scheduled early and monitored for efficient use. On international projects, it may even have to be imported. Ensuring that the design group includes the proper lifting lugs into the plant equipment is equally important. Well-planned lifts are the safest lifts!

Custom-designed construction equipment such as special form systems, lift-slab equipment, and concrete batch plants must be planned early. They are key "management" contributors to improved field productivity, which is the number 1 cost reduction area in field work.

Laying out the construction site and assigning the laydown and work areas for the subcontractors is an important component in controlling field costs. The modern field site must be planned like an industrial plant to ensure smooth work flow and high productivity. An overly remote construction parking lot, for example, can cost thousands of dollars in direct and indirect costs!

Organizing the field supervisory staff is the largest single cost-sensitive item in the field indirect cost budget. Since many of those people travel from site to site on large projects, the salary, living, and relocation allowances tend to make the hourly rate higher than normal. My philosophy for any organization is to be slightly understaffed for best efficiency. Idle hands in the field organization can cause even more problems than they do in the home office.

Selecting the most effective contracting plan can affect project costs dramatically. Certain situations favor the use of an integrated contract approach over a negotiated third-party constructor arrangement, and vice versa. Other avenues to explore are going union or open shop, direct hire or construction management with subcontractors, and acting as your own general contractor. All those options should be investigated to determine the most feasible contracting approach. The decisions cannot be made solely on the basis of cost.

Monitoring and improving field labor productivity is the single most important source of cost improvement during the construction phase. If the contract is cost-plus, the owner has the greatest stake in maintaining above average productivity. If it is lump sum, the contractor has the greatest stake in field productivity. In either case, the PM should have the greatest interest of all!

The field cost engineers and the construction supervisors must evaluate the field productivity for all crafts at least weekly. Any unfavorable trends must be dealt with immediately to prevent erosion of the field labor budget. Field labor leveling can make a significant contribution to the overall field labor productivity by not overloading the individual work areas. The leveling process must be carefully handled so the end date will not be adversely affected by making use of the float in the schedule.

All the cost reduction areas in construction are the direct concern of the construction project manager. If the construction manager is reporting to an overall project manager (or project director), that person is also responsible to see that the field is aware of the cost reduction goals. Frequent visits to the construction site are the only way to ensure that the goals are being met satisfactorily. Reading the field reports alone will not give the project director the necessary gut feeling of what is really going on in the field.

If you are not comfortable visiting the field to check on such matters, don't feel badly. Most construction managers do not feel obligated to make home office people feel at home. The only solution is to learn as much as possible about construction so you can read the situation in the field yourself. Accepting some field assignments during your formative years is the best way to get a proper construction indoctrination. Lacking that, the only way to get educated is by making frequent field visits and asking dumb questions until you get it right in your own mind.

The project management team

The project team is the watch dog over the whole cost control and reduction operation. It has an impressive list of cost areas to cover. A few of the major ones are:

- Adherence to specific project scope
- Strict application of project procedures, control systems, and budgets
- Approval of purchases and bid tabulations
- Monitoring of project change orders, internal and external
- Monitoring of blanket orders and subcontracts
- Overall design improvement
- Monitoring of and controlling home office costs
- Monitoring of field operations relating to quality, cost, and schedule

The PMs have the greatest responsibility to enforce staying within the project scope. No one should be better qualified, since they were instrumental in writing the original project scope. The philosophy of staying within the project scope must be transmitted to the rest of the project team by the project manager. I recommend using a missionary approach to impart this theme on a regular basis.

Strict adherence to the project procedures and systems can also improve cost performance. That is based on the logical assumption that the procedures and systems were well thought out before they were put into use. If any procedures do turn out to be faulty and are adversely affecting project performance and cost, by all means change them for the better.

There are many free spirits in all areas of the capital projects business who find standard procedures and systems unsuited to their lifestyle. Project disorganization is costly and a destroyer of budgets, schedules, and cost control. As project manager you cannot afford to let the project run amok!

In addition to reviewing the budget and cost reports, the PM should sign off on all bid tabulations for equipment, materials, and subcontracts. That puts an immediate screen in effect to catch any out-of-budget purchases before they are made. It is not advisable to delegate this important function to others who might not point out potential problem areas to you. Checking the values of the out-of-budget orders will give you a running mental tab on the winners and losers in your physical resources budget. You will need only the budget report to confirm what you already know.

Contract change orders are another province of cost control that should be handled by the project manager. Others on your staff must be taught to watch for change orders and to prepare the detailed forms for approval, but it is the project manager who is responsible for getting the orders approved. A budget that is not adjusted for scope changes is a budget that will soon be out of control.

The other items on the list have already been discussed in detail. They are listed in the project team's list of responsibilities because the *buck stops with the project manager*. It is impossible, however, for the PM to cover all of the cost control alone. The field is just too broad and too specialized for one person to handle on a good-size project. Use your staff as your eyes and ears and for advice, but the final decision on the corrective action has to be yours.

Controlling Cash Flow

Forecasting and controlling the project cash flow is important to several project groups. They are mainly the financial people, but we are all interested in knowing that enough cash will be available when the bills come due.

The owner's needs

Information on the cash flow for a large project is vital to the owner's financial people. Capital projects often involve budgets in the millions and hundreds of millions of dollars. In either case, we are speaking of large sums of money in interest costs. Large construction loans are structured in a myriad of ways, but they generally follow a pattern of commitment, drawdown, and repayment phases.

During the commitment phase, the interest rate will be in the range of 1 to 1.5 percent. In that case the bankers have only agreed to have the money ready when needed by the owner. The money is actually loaned short-term somewhere else until the funds are drawn on.

As the project progresses and cash is needed to make progress payments, the owner draws against the loan commitment funds. At that point the loan rate increases to the negotiated rate because the borrower now has full use of the money. If the owner draws down amounts greater than are needed for actual expenses, the surplus funds are banked at an interest rate as nearly equal to the loan rate as the owner can get.

The repayment phase starts when the project is completed and running and is generating cash flow to repay the loan. The interest charges on the reserve and drawdown funds have been accumulating during the project execution period. The interest costs are a significant part of the owner's project financial plan, which is badly upset should the project finish late.

An accurate and current cash flow projection is extremely important to the project financial planners. Any wide swings from the plan should be brought to the owner's attention as soon as a problem is recognized. The project manager should evaluate the cash flow, along with the other key project financial controls, every month.

The contractor's needs

Other important project participants with an interest in the cash flow are any contractors who may be involved. Like any good business person, the contractor wants to run the project with the owner's money. At today's high interest rates, the contracting company cannot afford to have its cash tied up in the project. That is especially true if the contract is set up to run all of the projects materials and equipment procurement through the contractor's books, which happens when contractors write the orders on their purchase order forms and pay the vendors' invoices.

For example, let us assume that the contractor is carrying an average invoice balance of $3 million against the owner over a period of one year. Borrowing that money at 10 percent interest would cost $300,000 interest per year, or $25,000 per month. If contractors used their cash reserves to carry the invoice account, those interest expenses would represent lost income from their bank accounts.

The situation is easily resolved by means of a zero balance bank account set up just for the project. Each month the contractor tells the owner how much cash outflow is expected that month. The owner then deposits that amount in the zero balance account. The contractor writes the checks for the approved vendors' invoices out of that account, essentially drawing the account down to zero.

Zero bank accounts do not pay the depositor any interest, so the bank makes its profit from the short-term interest it earns on the average balance in the account. If the account is not drawn down to zero in any month, the amount left over is deducted from the cash projection for the next month. The project cash flow curve is used as the basis for making the requests for funds each month. The actual amount transferred and spent is plotted on the cash flow curve each month to become the *actual* cash flow curve.

Cash flow curves

A typical cash flow curve follows the traditional S shape with a slow start and finish and a straight center section. Cash flow is sometimes shown in the form of a table or as a vertical bar chart. A typical cash flow chart is shown in Figure 8.7.

The cash flow projection is made up of three major cost components as follows:

- Home office services progress payments
- Payments for goods and services made through procurement
- Direct and indirect construction costs such as labor, materials, subcontracts, and field indirect costs

The cash flow curve lags the design progress curve because the contractor normally finances the home office costs for up to 30 days. That may not be the case if the project is being done in-house, when design costs show up on the curve immediately. The cash flow also lags the procurement commitment curve because most payments are made on delivery of the goods. On some international projects the delay may be shortened because of the custom of making vendor-supplier progress payments starting with a down payment on order placement.

The cost engineers use the projected design, procurement, and construction S curves to draw the projected cash flow curve. Like most S curves, it is subject to some slippage to the right caused by schedule delays, late deliveries, and so on. Therefore, the cash flow projection must be recycled quarterly and the forecast updated as required.

Cash flow on lump sum projects

The cash flow on lump sum projects requires the project manager's special attention. It is vital to ensure that the progress payments are keeping up with the cash outflow. On most lump sum projects, progress payments are set up on a schedule geared to certain project milestones. In that case the contractor will try to load more cash flow into the front end of the project to create a forward-loaded payment schedule.

One reason to do so is to offset the payment losses due to the normal 10 percent retention on the contractor's billings. The owner's purpose for the retention is to keep the contractor interested in completing the

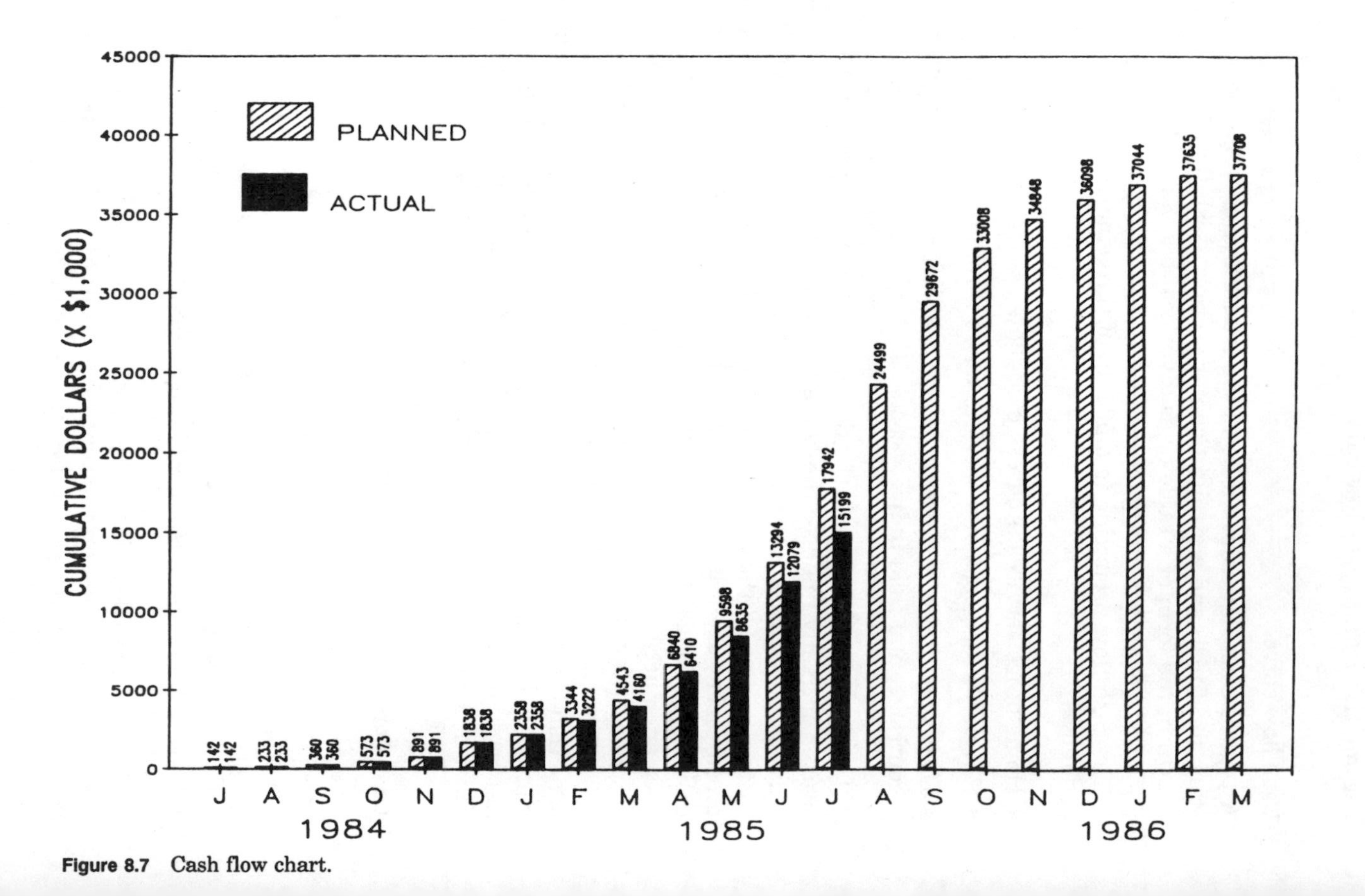

Figure 8.7 Cash flow chart.

project on schedule—especially the final cleanup items. Retention is another financial burden that contractors wish to avoid. It results in another case of lost interest on the funds retained by the owner.

One way to avoid the loss, and still satisfy the owner's needs, is to convert the retention funds into a performance bond until the project is completed. A performance bond costs the contractor only about 1.5 percent as opposed to 10 percent in lost interest. Also, clauses can be introduced into the contract to reduce the amount retained as the project moves into its later stages and the owner's exposure lessens.

Project invoicing

It is normal practice for project managers to review all invoices before they are mailed to the client. That is not so much to check the arithmetic as it is to check for possible errors in the formatting that is required by contract and project procedures. A minor error can be a cause for not processing the invoice, which means another delay in the cash flow.

It is the job of the client's project manager to review all the incoming invoices before passing them on to accounting for payment. The review also ensures conformance to project procedures rather than arithmetic. It is important to keep the invoice procedure moving on schedule to have the best possible contractor-client relationship. If minor errors are found, only the part in error should be held up for resolution and payment in the next invoice.

On most domestic projects the cash usually flows smoothly, so collecting the money is not a major problem, but on international projects it is often another matter! Most foreign work operates under an irrevocable letter of credit to the contractor's bank for the amount of the contract. That ensures that the funds to pay for the project are available in the contractor's country, so it becomes much more difficult for the owner to renege on legitimate progress payments. If the contractor can prove the work was done, the bank will release the payment.

Cash flow summary

Although cash flow is often overlooked in the plethora of other demands on the project manager's time, it is too important to be passed over lightly. The top managements of all of the companies involved in the project are going to be watching the cash flow. The needs of each group must be satisfied if you are to be rated as a top-performing project manager.

A realistic cash flow plan, followed by effective procedures for controlling and reporting it, will go a long way toward improving the project's cash management. The main thing to remember is not to surprise any of the cash flow participants with major changes at the last minute. If cash flow problems start to surface, get them out into the open quickly so they can be resolved.

Schedule Control

We sometimes hear schedule control referred to as *time control* in relation to making the *time plan* for the project work. Actually, time cannot be controlled; it marches on relentlessly in fixed amounts without any yet known means of controlling it. Time is, in fact, quite inelastic.

The project schedule is a plan of work per unit time, so it is somewhat flexible. By speeding up the task (doing it in less time), we can make up for past or future lost time and still make the predetermined scheduled date. Therefore, the work is *elastic* and must be controlled to the *inelastic* time scale. I realize that this discussion sounds very basic, but you will be surprised at the number of project managers who think that they can create more time to make the schedule.

Monitoring the schedule

If we are to control the schedule, we have to monitor progress (or lack of it) against the time scale. We broke the work down into specific tasks (work activities) as a convenient way to check elements of our work against time. The average status of all those work activities is our measure of overall physical progress on the total project.

We must also consider that each work activity has to be *weighted* for its percent of the total project. The weight is determined by the value of the human and/or physical resources used to accomplish the task. The weighted value of the activity times the physical percent completion tells us how much *earned value* that activity is contributing to our percent complete. Conversely, the earned value divided by the total budget for that activity will give us the physical percent complete for that activity.

Measuring physical progress

Again it is necessary to point out the fallacy of using labor-hours expended to calculate the physical percent of completion. If you are not operating at exactly 100 percent efficiency, the percent of completion will be in error. That is why we stress the necessity of estimating physical progress and not expended progress.

The physical percent complete must be tied to the *earned value* of the work accomplished to date. The earned value of the work activity is measured by breaking the task down into a logical system of checkpoints and assigning a percent of completion up to that point.

A good example to use is the preparation of a typical design drawing which can be broken down as follows:

Operation	Progress, %
General conceptual layout	35
Drafting completed	85
Checking	90
Correct checking marks	95
Issue for construction	100

The system is applied to each drawing on the project along with the labor-hours assigned. As each operation is completed, the earned amount is transferred into the physical percent complete column and shows up as earned value against the schedule. Some systems allow an estimate made of the status of completion between the milestones, and some allow only progress to be reported as the milestone is reached. The former method probably gives a more accurate picture of the total earned value, but it also opens the door to section leader's fudging the intermediate percent complete.

The same approach can be used for all of the other work activities in the design area such as engineering, spec writing, bid evaluation, and shop drawing review. The indirect costs, such as project management, project controls, purchasing, and secretarial and clerical, are prorated as a percent of completion off the time line of the schedule.

The cost-effectiveness of any earned value system must be evaluated in relation to the benefits provided. On a large project the amount of clerical work can become horrendous, even when computerized methods are used. On smaller projects, the problem becomes worse because limited funds are allotted for such controls.

Much of the data gathering takes place during the monthly status evaluation as the group leader evaluates the percent complete of each document and the group's physical percent complete. The cost people can then take those figures and review them for accuracy before converting them to the overall project physical percent complete.

The same technique can be used in the construction area by breaking down each trade into manageable work activities. A ready example is the installation of foundations which is something usually required on all capital projects. A typical breakdown can be made as follows:

Operation	Progress, %
Layout	5
Excavation	20
Forming	50
Set rebar and anchor bolts	75
Pour concrete	90
Strip forms and backfill	100

The same comments about the design phase apply to the field work. The field engineer or quantity surveyor evaluates the status of each operation being worked on during the reporting period and arrives at the physical percent complete. The field cost engineer then collects all of the input and calculates the physical percent complete for the whole field operation.

That system of cross checking physical progress between the line and staff groups makes it more difficult to hide overly optimistic forecasts or to make errors in the estimate to complete. That is all done before the schedulers get into the act and plot the actual progress against that which was planned.

Planned vs. actual physical progress

Once the physical progress for the period is calculated, it becomes a simple matter to plot the new value on the planned progress S curve as shown in Figure 8.8. The curve is derived from the original project schedule by plotting planned percent complete and personnel hours over the elapsed time for the project. The resulting S curves are shown as solid lines in Figure 8.8. The actual physical progress and hours are then plotted monthly as dashed lines on the same graph (Figure 8.8).

If the plot for the actual completion falls on the solid line, the project is on schedule. If the actual progress curve falls below the solid line (as in Figure 8.8), the project is running behind schedule. That could be due to an understaffed project because the actual hours expended curve is also below planned.

Should we be fortunate enough to have the actual progress line fall above the solid line, we are ahead of schedule. That is still not a time for complacency, however, because progress can easily fall below the line in the following period!

If there are no extenuating circumstances such as unrecorded change orders, lack of owner's performance, and force majeure forcing the actual progress curve below the planned progress curve, there must be another serious problem that is causing the adverse trend. The failure to maintain schedule then lies at the door of the project team, and the PM must take urgent steps to rectify the problem.

First, determine the cause of the problem. Is it due to such general

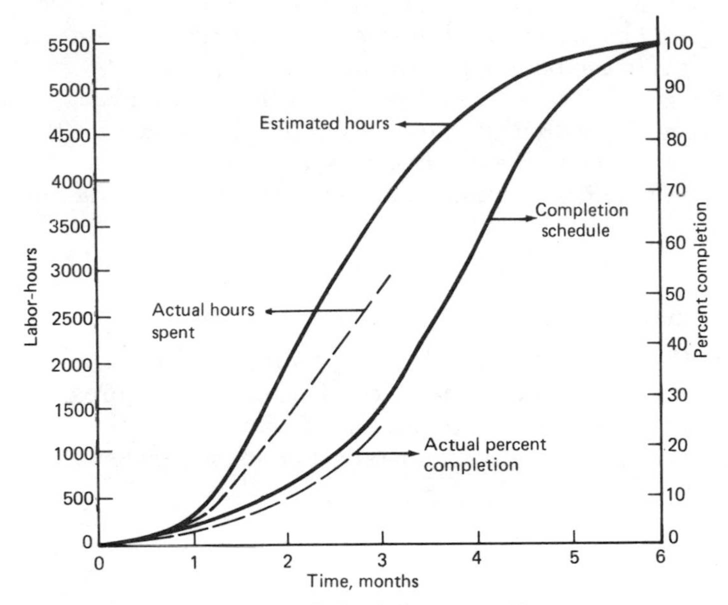

Figure 8.8 Planned vs. actual physical progress S curve.

causes as staffing problems, poor productivity, low morale, and lack of leadership? Or is it due to failure of a specific group to perform up to standard? In most cases, the problem will involve more than one cause, so the suspected areas must be prioritized to attack the worst ones first.

Although I stress immediate response to substandard performance, be careful not to overreact. That can sometimes exacerbate the problem instead of solving it. Discussing the problem with key staff members is a good approach for getting all the pertinent input on the subject. After weighing the input from the staff, you, as project manager, should then make the decision on the course of action to be taken. Hopefully, you will be able to get the key staff people to buy into your decision. The latter point is important to ensure that the selected solution has a chance to succeed.

If we study the situation presented in Figure 8.8, we can learn why it is so difficult to get a job back on schedule once it slips below the planned curve. After the project has passed the buildup part of the curve (the lower arc of the S), it enters the straight line portion when the peak staff is on board.

On the straight part of the curve, most projects make about 7 to 10 percent progress per month. The period includes the time when the productivity is at the highest level. It is at its lowest during the curved portions of the S, both at the beginning and at the end of the project. It is inherently difficult to get monthly progress above 10 percent over the several months needed to regain the lost time and return the actual progress to the planned curve.

I take that negative view just to let you know that getting back on schedule isn't easy! Nevertheless, we do have to try to make up the lost time. We have to look to the unfinished work for the solution. The time lost in the past is history, and no amount of wishful thinking can bring it back!

Ways to improve the schedule

We can control the future, as we said, by speeding up the activities that are predicted to finish late. The discussion refers to all areas of the project such as design, procurement, and construction.

There are a few ways to improve your rate of progress when you are behind schedule.

- Improve productivity
- Increase staff
- Work overtime
- Reduce the work
- Subcontract part of the work

Improving productivity is the best and cheapest way to increase the speed of doing the work, either by using better-qualified workers or by improving work methods. Because it takes time it generally will not work when a quick fix is needed, but it is a good cure for the longer term.

Provided there is ample room to work and qualified people are available, adding more people is probably the next best way to make up time. If neither room nor work for the new people is available, productivity is likely to fall off to give a net gain of zero. Rarely do we find qualified people sitting around awaiting assignment to our project, so this solution is not suited to a quick fix either.

Overtime work is the next option, and it is the one most used. The system can increase the available staff hours on a project by 15 to 30 percent immediately. The added hours come from people already trained in the work, so learning curve losses are nil. The disadvantage of overtime arises after its extended use. After prolonged overtime work, productivity has been proved to fall off. The people put in the time but do the same work as in an 8-hour day.

Construction people usually work more overtime than do design people. In fact, overtime is often used as a ploy to attract workers during periods of skilled labor shortage. I recommend that practice be avoided because the high premium for construction pay and the resulting loss of efficiency can combine to ruin the field labor budget.[4]

The cost of premium time payments must also be weighed against the value of the anticipated time-saving benefits. Many home office people fall into an overtime-exempt category, so their overtime does not carry a premium rate. Drafters and CAD operators usually fall into a premium time category, as do all contract design people and construction craft labor.

All things considered, using scheduled overtime offers the most advantages for making up lost time for getting back on schedule. The increased cost due to premium time and the potential efficiency loss over long periods of overtime have to be taken into account. If the productivity is already poor, however, using overtime at premium rates is only throwing good money after bad!

Reducing the work to regain schedule is another route to consider. It is applicable only in rare cases when some work operation can be eliminated to save time. If those items were easy to find, they shouldn't have been in the work in the first place, so beware when using this approach to the problem.

Sometimes we try to shift work from the design area to the field. For example, we could do the piping isometrics and material takeoff in the field to reduce the home office work. That is merely shifting the work to another location to make the home office budget look good. The arrangement can work well on remodeling jobs, where doing the work in the field is more efficient when field measurements are required. The shift can also be made to improve schedule rather than for budgetary reasons.

If the schedule is slipping because of an inability to staff the project from your own sources, you may want to consider subcontracting a portion of the work. Earlier we discussed some of the trials and tribulations of subcontracting design work. It is important to research the method properly before being committed to it. Since it is a contractual matter, the client and company management have to be brought into it to approve any subcontracting plan.

Any of the latter work reduction ideas must be well thought out before trying them. They are usually recommended only as a last resort, and they are often a short step from the frying pan to the fire.

Tips on schedule control

When monitoring the schedule, it is best to keep close tabs on the 20 percent or so activities that are on the critical path. If the critical items slip, they can make some of the near-critical activities late. If the critical path items rise much above 20 percent, the project will have to be rescheduled.

Be especially careful at the beginning of the project because getting behind schedule then will probably dog your project performance the rest of the way. Quick detection and reaction are important. Schedule problems hardly ever get better by themselves. However, be careful not to overreact and do something that is counterproductive.

Monitoring procurement commitments

On integrated projects, especially in the early stages, procurement activity has a profound impact on the project schedule. The design and construction downstream activities are closely tied to the availability of vendor data and delivery of the physical resources.

That statement does not apply to straight architectural projects when the construction is bid lump sum after completion of the design. In that case, the design is done in a broad fashion to suit a variety of equipment types which might satisfy the specifications. Also, the contractors are usually responsible for procuring all of their materials and equipment, so they are in control of their own destiny on schedule.

The strictly lump sum split contract arrangement does lose the advantage of early ordering of long-delivery equipment to fast-track the schedule. In turn, that places greater strain on the contractor's procurement schedule.

The tool used to gauge the performance of your procurement team is the purchasing commitment curve. As with any activity with buildup, peak, and build-down phases, the result is the familiar S curve as shown in Figure 8.9. The value of the materials and equipment to be purchased is plotted against the time allowed for the procurement. The curve should show an early upturn because of the early placement of the long-delivery equipment orders, which are usually big-ticket items. Most bulk materials are purchased later in the project after design is about 30 percent complete.

The commitment dollar value is posted to the curve only when the written purchase order is actually issued. It is bad practice to count verbal orders as commitments. Pressure must be maintained on procurement to get the written purchase orders out, because vendors do not perform nearly as well on verbal orders.

If the commitments fall below the line, the project manager must investigate the cause as soon as the reason is detected. Procrastination will eventually affect the schedule across the whole project. The problem could be due to delays in the technical side in preparation of specifications and requisitions, making bid tabulations, and so on. Delays are also possible on the purchasing side because of late issue of

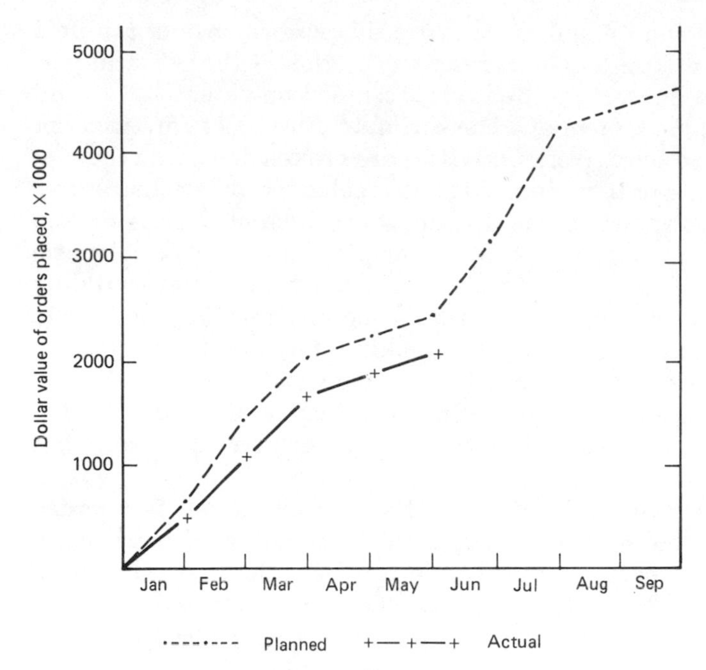

Figure 8.9 Procurement commitment curve.

bids, poor vendor response, adverse market conditions, or a variety of other reasons.

After the buying activity has been completed, procurement activity will continue in the expediting and inspection mode until all of the equipment and materials are delivered to the field. Progress in that area is hard to chart and must be gauged through the expediting reports and reported late deliveries from the field. The PM must regularly monitor the expediting reports to keep his or her finger on the pulse of deliveries starting with the delivery of the vendor data.

When equipment and material deliveries are really critical, I have often held the expediting review meetings in the field. That shortens the communications lines between the field forces and the expediting people to zero. I usually scheduled the meeting to correspond with my field trips so I could participate in the meetings. That close proximity to the field work seems to give the expediters a real feel for the urgency of making the deliveries on time.

Monitoring construction progress

Keeping track of the field work on larger projects is done in very much the same way as for the design work. The field cost engineers are responsible for setting up and monitoring the performance of the field forces. The scheduling people use the calculation of the physical percent complete in forecasting the project completion date.

The earned value system and the estimate to complete are used for each field work activity. Work activities are broken down into the specific crafts such as civil, piping, rigging, steel erection, electrical, and painting by geographical areas. In turn, each item of work is broken down into its major components. For example, foundations can be broken down into values for each operation such as excavation, building forms, setting rebar, pouring concrete, concrete finishing, and stripping forms. In some cases, however, breaking the work down into such detail is not considered economically feasible within the field indirect cost budget. In that case, an all-inclusive value for each cubic yard of foundations placed can be used with slightly less accurate results but lower cost.

The actual work completed on all of the field operations is recorded weekly and reported monthly along with the actual labor-hours used. The latter are compared to the standard hours allotted to the work, and the resulting productivity is plotted on a curve for each craft.

The productivity rate is calculated by the following formula:[1]

$$\text{Productivity, \%} = \frac{\text{budgeted hours}}{\text{actual hours}}$$

The calculation is made for each craft in each area weekly to monitor the productivity trends. Any unexplained adverse trends must be evaluated and acted on quickly to stave off time and money overruns. The productivity rate does vary over the life of the project as shown in Figure 8.10[1] so don't be too upset when the productivity starts below 1.0.

The productivity starts low because of start-up problems and the time needed to learn the local conditions. It peaks just before the peak in the labor curve and then starts to fall off. Finishing below the average in the final stages of the project is virtually guaranteed because of the nature of the cleanup operations at the end of the job.

The objective of a good field performance is to finish the field work at a cumulative efficiency of 1.0 or better. That means that we have to have a period of 1.0 + productivity during the peak of the work to offset the losses at the start and finish. If the productivity never gets above 1.0, the budget and schedule are bound to be exceeded and the project expectations will not be met.

The field scheduling people work closely with the cost control people in calculating the physical percent complete for the field work. The schedulers plot the monthly (or weekly) percent complete on the projected S curve [3] to compare scheduled completion with actual comple-

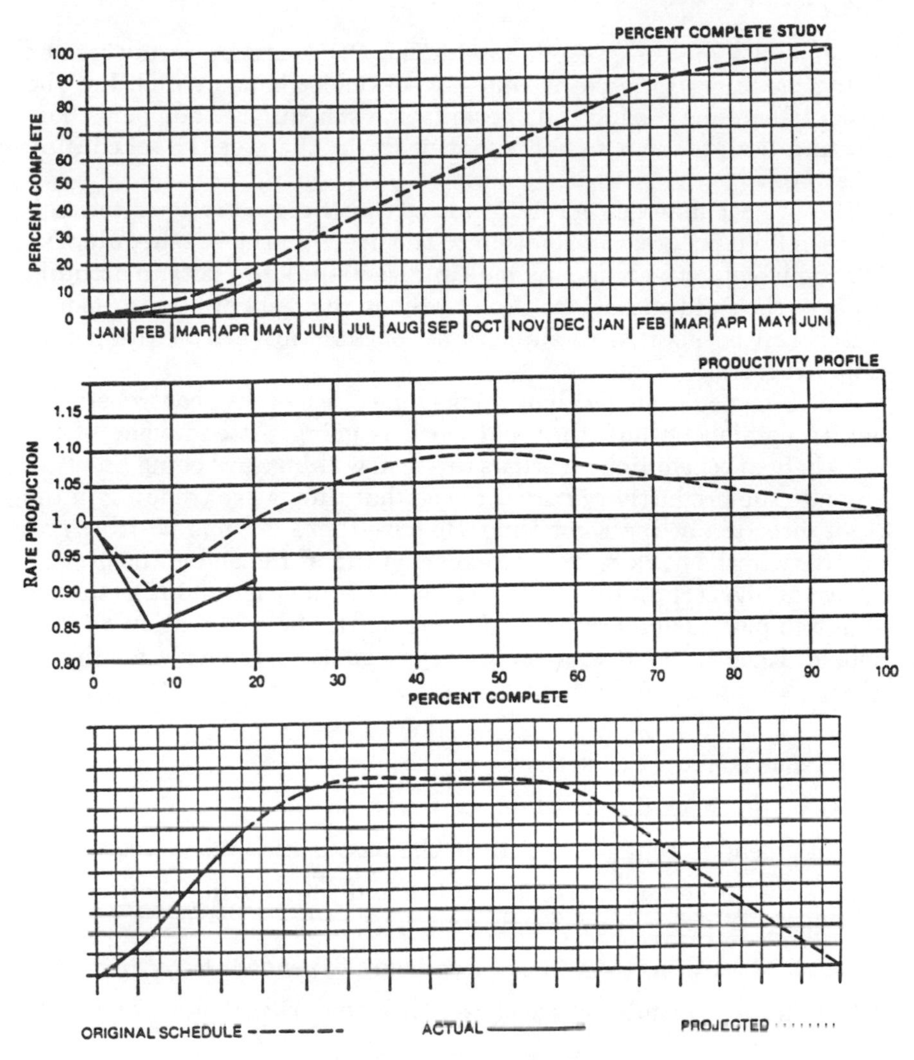

Progress/Productivity Report

Figure 8.10 Productivity curve.

tion. Any failure to meet the projected schedule must be explained and recommendations for improving performance must be made.

The actual scheduling of the field work is generally done manually on a weekly basis by using the overall CPM schedule as a control document. The tasks for all of the crafts and subcontractors for that week are laid out in a field supervisor's planning meeting. The meeting is chaired by the construction manager and the field superintendent. Any of the prior week's missed targets are reviewed together with the assignments for the present week. The CPM can be sorted by the field schedulers to make 2- and 4-week look-aheads which list the tasks to be done during that period. The schedulers then sort them out and assign them to the supervisors in the weekly scheduling meeting.

The project manager's role in all of that field activity is to monitor the field progress in relation to the overall project schedule. Doing the job effectively means making frequent site visits to actually observe the field operations. Don't rely on the field progress reports, which can be slanted to support an optimistic view of the actual field progress.

This is a good time to call on your home office project control staff to get an opinion on how the field work is going. Attend some of the weekly field scheduling meetings to see how things are being handled. Dig into the productivity curves to see that things are trending in the right direction at the right time. Be sure there is some positive productivity in the bank before the expected run at the end of the project. Above all, be certain that your own house is in order and that the design and purchasing arms are supporting the field operation as they should. Keep in mind your basic project goals to deliver a facility *as specified, on time, and within budget!*

A schedule control checklist

- Cycle and check the schedule regularly.
- Look ahead to control the future; the past is history.
- Give the three major areas of EPC equal time during their critical periods.
- Check *physical progress,* not labor-hours expended.
- Check the schedule by exception; use the critical item and look-ahead sorts.
- Take early action to correct slippage.
- Make corrective decisions logically; don't overreact.
- Use overtime wisely.

Equipment and Material Control

On a major EPC project, we may be handling from 500 to 1000 tagged equipment items, along with large quantities of bulk materials in the piping, electrical, civil, and other trades. The logistics of getting all that material to the construction site on time represents a major administrative undertaking.

Controlling major equipment

Keeping tabs on the tagged equipment is relatively straightforward, beginning with the equipment list which is developed as part of the basic design data package. Specifications for the procurement of the items are written and sent to procurement for purchase. Bids are taken and analyzed, and a vendor is selected for each item. That initiates the vendor data cycle and the manufacture of the goods. The equipment is manufactured, inspected, and shipped to the field, where it is received and stored until it is installed. A field receiving report is prepared and sent to the home office so the vendor's invoice can be paid.

To simplify that process, like items are grouped to make combined purchase orders. For example, all the centrifugal pumps can be gathered into one order, all pressure vessels in another, and so on. Each item can be listed on the status report, but grouping them greatly accelerates the procurement process.

The start and finish dates for each of the equipment control activities have to be scheduled to ensure that the equipment will arrive on schedule for installation. In earlier chapters, we discussed ways of monitoring the delivery of the equipment by use of spreadsheet programs to record the progress of each activity.

As each step is accomplished, it is logged into the system, which is used to generate a weekly material status report. The report is used by the project manager, the schedulers, and the expediters to monitor the progress of the procurement program. In the early stages, the accent is on delivery of vendor data; later, it is on the equipment itself.

Bulk material control

The bulk materials are handled similarly except that a material takeoff (MTO) stage is required to know how much material to order. On an integrated EPC project, a preliminary order of bulk material is made for up to 50 percent of the material required. It is usually based on the favorable prices for the total quantities projected to be required. Here we combine smart scheduling with thoughtful cost control! Another example of early bulk ordering is the early takeoff of the structural steel for placing a mill order before the detailed design is completed. It's another of the many time- and money-saving ideas to keep in mind when planning your project materials management program.

As the design progresses and detailed MTOs are done, the final order quantities are set and additional materials are ordered against the previously agreed-upon favorable bulk material prices. Generally, an excess of about 5 to 10 percent over the takeoff amount for bulk materials such as piping and electrical fittings is ordered to cover losses and rework in the field. Minor problems disposing of surplus materials at the end of the project are better than having material shortages during peak field productivity.

Naturally, some common sense is needed in ordering the surplus materials; much more care must be taken in ordering surplus exotic materials which are prohibitively expensive and are made only to special order. The situation can also set up a time trap because of the long delivery time for most exotic materials. If you do come up short, waiting for delivery of critical items needed to complete the project can ruin your schedule.

All those points support my thesis that project managers must realize the importance of planning, organizing, and controlling their project material resources. A well-managed materials program can save time and money at every turn and make a valuable contribution to meeting the project goals.

Controlling Quality

International competition has made quality control a primary concern in industry worldwide, and capital projects are not exceptions. In addition to maintaining a company's reputation, the incidence of design quality lawsuits has also been on the increase. There is certainly no cause for complacency about quality control in our business!

The major areas of quality control in the capital projects arena are:

- The project design basis
- Design and engineering
- Equipment and materials
- Field construction
- Final inspection and acceptance

We will examine each of the areas in detail, along with the project manager's role in quality control.

The owner's design basis

Quality control cuts across many organizational lines of the project, but it begins with the owner's design basis for the facility. That is where the baseline for quality of the facility is established. Some of the many external and internal factors affecting the design basis are as follows:

- Market forces
- Expected market life of the product or service
- Product, service, and facility economics
- Financing capability
- Operational safety
- Environmental and zoning factors
- Government regulations
- Company or agency policies
- Maintenance, services, and operating costs

Each of those factors can have a profound effect on the cost and economics of the facility, which will in turn affect the quality that the owning entity establishes in its design basis. The owner must specify the economic life, which, along with the financing limitations, establishes the desired quality level for the facility. The design team then uses the applicable code requirements and designs to a quality standard within the limits of the project financing.

Design quality control

The design group, whether captive or contractor, will develop the owner's design basis into the conceptual design within the limits set by the owner. In that phase the preliminary plans and specifications which finalize the scope and quality of the facility are developed. Design and feasibility studies, along with cost estimates, are made to arrive at the optimum design within the design basis.

During this stage, an important job for the owner's project manager is to satisfy his or her management that the quality established in the design basis is being adhered to. Open communication between the owner and design project manager during the developmental stage is vital to successful implementation of the design basis.

When the conceptual design review meeting with the owner results in approval of the design basis, the detail design phase begins. From here on control of quality passes to the design team with only limited review and approval of key documents by the owner to assure conformance to the design basis and the contract.

The design team is responsible for its own design quality, which means that accuracy and consistency among the calculations, drawings, specifications, and all project documents is essential. It is the design leader's obligation to assure that the design quality control systems are in place and are functioning. The best reason for sound quality control during design is to minimize costly changes and corrections in the field. Correcting field work is expensive by itself, and it has a traumatic effect on the schedule.

If the design team has inherently weak quality control procedures, it behooves the project manager to bring pressure to bear on the discipline leadership in the design departments to strengthen the procedures for those areas. Poor design quality can lead to failure to meet any of the major project goals. The project team should also review and sign off on all design documents for conformance to the design basis and the contract. That is especially important on the documents that the owner's project team also is approving.

For some years now, leading contract firms have been using a system of internal project audits. They are random spot checks of certain documents selected by the auditing group. That is fine as a general overview of the firm's quality, but is not a replacement for thorough design-checking procedures.

Equipment and materials quality control

Quality assurance for the project's physical resources is generally left to the owner's or contractor's inspection department. That level of quality control prior to delivering the goods to the field is normally found only on process and high-technology types of projects which have a large amount of custom-designed equipment. To avoid costly rework during construction, it is vital to confirm the quality before specialized equipment is shipped.

That part of quality control often functions as part of the procurement or engineering department. Inspectors make shop visits at critical points during the fabrication cycle to monitor the quality of the fabricator's work. Final testing and inspection are made prior to acceptance for shipment. Conformance to the applicable drawings, codes, and material requirements is certified and made part of the equipment file.

The project manager's role in the quality control activity is to monitor the inspection reports and act on the areas in which quality control comes into conflict with meeting schedules. Conflicts require a delicate balance in the compromises which have to be made to protect the best interests of the owner and the project schedule.

It is not considered cost-effective to inspect such off-the-shelf items as pipe, valves, fittings, and conduit and wire in developed countries. If those items are being produced by third-world suppliers, a quality inspection may be necessary to assure that standards of quality are being met. Material certificates certifying the chemical analysis and metallurgy of those types of goods are often required to satisfy code requirements. All that documentation is filed along with the final test reports, which are turned over to the owners for their plant records.

The project managers for the owner and the design-procurement team should work out the scope of the materials testing program in order to assess the program's cost. That is usually established as part of the procurement services scope. The degree of material quality control is a function of the owner's policies, code regulations, and supplier reliability. Owners usually reserve the right to attend any inspections that have been delegated to others, but they rarely exercise the option.

Construction quality control

Construction quality control is perhaps the most involved and complex process on capital projects. That is to be expected, because the major portion of the budget is spent in that area. Also, working in the open with a diverse number of crafts involved adds to the field quality control work.

The construction team establishes its quality control program to the degree that the finished work will finally be accepted by the owner without any costly rework. In A&E types of projects the owner often hires the design firm as the final authority on acceptable quality. Most quality standards are established in specifications such as those for concrete testing, steel erection tolerances, building code inspections, and equipment standards.

On larger A&E projects construction quality inspectors will be on the site full-time to observe daily operations. On smaller projects the inspections may be intermittent, usually at critical stages of the work. Field inspection reports should be filed on a regular basis as a record of the quality assurance program for the project.

On integrated EPC projects involving major engineering works and process projects, the quality of the field work is monitored by the contractor under the watchful eye of the owner's resident construction team. You may remember that that group showed up on our field organization chart in Chapter 7.

The field engineer is responsible for field quality control operations reporting directly to the construction manager. The arrangement raises a question: Is it sound quality control practice to have the field organization inspect its own work? To avoid that situation, an owner will often insist on setting up its own quality control team to monitor the quality of the construction work. The only real disadvantage of that approach is the additional cost.

In any case, the field engineer arranges for the services of local field testing laboratories to perform the soils, concrete, radiography, and other testing required by the specifications. The field engineer is also responsible for inspecting concrete pouring, steel erection, electric grounding, welder qualification, and setting the lines and grades for the project. All test results and reports are maintained in a quality control file and turned over to the owner at the end of the project.

The project manager's role in the field quality control is to monitor the overall performance to ensure that the program meets the project goals. The owner and contractor project managers coordinate their efforts in that regard through their respective field teams and mutually resolve any problems which rise to their level for action.

The ultimate goal of any construction quality control program for all the parties involved in the project is to have a smooth final inspection and acceptance of the facility as described in the next section. Outstanding examples of how *not* to control field quality are some of the nuclear projects which had to be virtually torn down and rebuilt to get them accepted.

Final facility checkout and acceptance

The final checkout of the completed facility actually starts somewhat before mechanical completion. As construction reaches the 90 percent complete stage, it enters the "punchlist phase." The construction team makes its own inspections to determine what is left for it to finish before it can say the project is complete.

By that time the owner is really anxious to take over the facility and start its operations. To that end, it will often accept the facility on the conditional basis that all of the work will be finished before the contractor leaves the site. Depending on the type of project, that is known by a variety of names. In A&E work it may be known as "substantial completion"; in other areas it is called "mechanical completion." In any case, it is the point at which the owner takes over the *care, custody, and control* of the facility for its "beneficial occupancy." In other words, it is the point at which the owner moves in and starts operating the facility. That is often reached on a unitized basis in a process plant when the owner accepts the utility systems first—followed by successive process units later.

The contractor may still have people in the area finishing up such details as painting, insulation, caulking, and other minor finishing work. At that time the construction and design team, in conjunction with the owner's representatives, starts a series of final inspections which delineate what has to be done to bring the project to completion. The result is a detailed check to see that facility conforms completely to the plans and specifications as well as the contract.

The final check also applies to the design portion of the project to ensure that all the contractual requirements for as-built drawings, mechanical catalogs, code requirements, and so on have been turned over to the owner. It is the project manager's responsibility to ensure that all contractual requirements have been complied with. All loose ends will have to be cleared up before the project manager can get a final letter of acceptance for the facility. At that time, all retained funds should be released and all bills paid to ensure a lien-free environment in which the owner can operate the facility.

On process plants the final checkout can be much more involved because the quality check has to go all the way back to the engineering flow diagrams to ensure that the plant will operate safely and properly. Testing and adjusting the piping and instrumentation systems require detailed inspection and calibration before the systems can be accepted as operational. At the same time, the owner brings in its operating personnel who use the turnover period for on-the-job training.

If the process project involves a contractor's process design, final performance tests will also have to be run. They should be attempted only after the plant is completely operational and has reached a steady-state condition. When the tests are properly completed and documented, the plant acceptance can be signed off by the owner.

Project managers are often required to take up residence in the field for several months to handle the details of the complex acceptance process. Very often, everyone's nerves become frayed during this stressful time. The project manager must exert strong leadership to bring the final checkout and acceptance process to a successful conclusion.

That is the time for the final accounting to determine whether all of the project's goals and expectations have been met. Have we really delivered the facility that was contracted for—was it constructed as specified, on time, and under budget? If the answer is yes (or even close to yes), it is a happy time for all concerned. If the answer is no, it is up to the project manager to put the best spin possible on a substandard outcome. Special attention should be given to the project history to try to avoid making the same mistakes next time.

Summary of quality control

Quality control involves all project activities from the conceptual stage to project turnover. Substandard quality is not acceptable at any price. Everyone working on the project—owner, project team, design organization, procurement group, and construction forces—plays a major role in quality control. It is the owner's role to establish the quality desired and ensure it is delivered. The project managers are responsible for implementing the quality control processes for the project and for leading and monitoring the quality control effort.

Operating the Controls

As a means of summarizing the far-flung requirements of project control, I have prepared a list of recommended guidelines for handling them. The best designed control system in the world is not going to lead to a successful project unless we operate the controls correctly.

Establish priorities

The number 1 requirement for effective project management is establishing priorities for your work activities. The most urgent activities should go to the head of your things-to-do-list! Don't handle the activities in the order in which they arrive in your in box. Remember, the project organization is functioning on the basis of what comes from your out box.

Sort the whole contents of your in box at least daily in addition to the day-to-day operations of your project. Make sure the items related to project quality, cost, and schedule receive top priority. Also, items related to project and contractor-client relations should receive high priority. Be responsive to project people problems. To a slightly lesser degree, the items arising from your company management should also be treated as priority items.

All the priority items have to be worked in with the routinely scheduled project activities such as project reviews, monthly progress reports, and field trips, which also are making heavy demands on your valuable time. I have found that keeping a written priority or daily things-to-do list is valuable in managing your time. It is a simple way to prevent overlooking or forgetting a high-priority item now and then.

Control by exception

Using control-by-exception techniques is the only way to stay on top of the plethora of daily project activities. That is true even if you have properly delegated as much as possible to other project team members.

Many control systems, such as CPM schedules, control budgets, and quality and material control reports, are designed to highlight the exceptional conditions. Make sure to check the exceptional items first, and then go into the normal items as time permits.

Keep looking ahead

Running a project is like piloting a ship. It is nice to know the depth right under your keel, but it is more important to know what dangers lie immediately ahead. Keep your sonar turned on!

Take advantage of the control systems which have look-ahead capability, such as periodic CPM look-ahead sorts and estimated-to-complete projections. Keep an eye on the project performance curves for design, procurement, and cash flow to note any trends away from the projected norms.

Check the actual work of the groups performing the work, whether it is design, procurement, or construction that is charted on the curves. Is there a sudden drop in personnel when the group should be at maximum production? How is productivity holding up? Checks should be made by visiting the groups performing the work and physically observing the work in progress. Also, talk to the group leaders, you will be amazed at what they will tell you verbally that they would not dream of putting into their written reports! Take a supervisor to lunch; visit the discipline managers; keep tuned to the project grapevine to get a real feel for how things are going. Check all of the intelligence you have gathered to ensure it is correct, and then act on it accordingly!

Reaction time

The function of any control system is to have a rapid reaction time to off-target items. Corrective action should be started immediately. The longer you delay, the worse the condition is likely to become and the longer the time to get back on target!

Once time is lost from a schedule or money from a budget, it is very difficult to get even again. Also, you will have lost one of your opportunities to beat the schedule or budget. Do not postpone taking action to correct an activity because it may result in unpleasantness to someone else or yourself. When you feel that the need for action is real, get started on it!

Single-mindedness

"Single-mindedness" is a long word, but it is in the dictionary. It is a noun and means "the ability of a person to have or show a single aim or purpose." In the case of project managers, their common aim or purpose is to have successful projects and to attain their projects' goals.

A lot of people along the way will try to blunt your single-mindedness and divert you from meeting your goals. One reason is that some people may not share your goals; they may even have a personal dislike for you. Others may be incompetent or just plain lazy. All those reasons must be brushed aside and not allowed to deter you from reaching your intended goals. There will be times when a suitable compromise may be the only solution, but never sell out completely!

If you have reached an impasse through your normal channels, don't be afraid to go over the opposition's head to meet your project goals. Remember, top management gave you a mission to accomplish the project and you accepted it. I have found most top managements to be very supportive of their project managers when an impasse arises on the project. The first place to start in a case like that is the management sponsor for your project.

A final word of caution: Single-mindedness is good, but the way you present it to get your project needs satisfied is equally important. We will talk about the fine points of avoiding the bull-in-the-china-shop syndrome later in the book.

Some common mistakes in operating the controls

In addition to common mistakes in operating the controls are failing to plan, organize, and control all of your project activities. A few more that I have found to occur fairly often are these:

- Failure to maintain good project and client-contractor relations. We will discuss that high-priority item later under human factors.
- Overestimating the cost of the design or the construction of the project, thus causing project cancellation. The same kind of thing can be said about underestimating the project and thus causing it to be uneconomical.
- Failure to get a good procurement program organized.
- Failure to properly document meetings, project changes, telephone calls, and other important project happenings. If the job runs into trouble later on, you will need the documentation to defend your position against a variety of possible criticisms.
- Failure to hold project review meetings with client and management to properly check progress and coordinate the work. A corollary to that is to hold too many project meetings, which waste valuable time.
- Failure to control meetings, which makes them nonproductive. I

will detail a procedure to avoid that in the chapter on communications.

- Failure to exercise project management control over the field activities. More money and time are spent here than in any other part of the project. Try to develop some field experience somewhere along your career path.

Most of those tips start with the word "failure," and that is what you are going to have if you do not avoid the pitfalls listed in this section on project controls. Keep in mind that it takes about three or four good projects to wipe out the reputation gained on one bad one!

References

1. Forrest D. Clark and A. B. Lorenzoni, *Applied Cost Engineering,* 2d ed., Marcel Decker, New York, 1985.
2. William Pena, *Problem Seeking: An Architectural Programming Primer,* 3d ed., AIA Press, Washington, D.C., 1987.
3. Arthur E. Kerridge and Charles H. Vervalen, *Engineering and Construction Project Management,* Gulf Publishing Company, Houston, Tex., 1986.
4. *Scheduled Overtime Effect on Construction Projects,* The Business Roundtable, 200 Park Avenue, New York, 1982.

Case Study Instructions

1. Prepare an outline of the project wide cost control program that you feel would best suit all phases of your selected project. Also, list the goals of the program and how you would present your cost control philosophy to the key members of your project team.
2. What are the most attractive areas of cost reduction on your project? How would you go about investigating and screening them for presentation to your client?
3. Assume that you have full procurement responsibility on your project if the scope does not already include it. What key features would you look for in your project procurement agent's proposed procurement program?
4. Outline your proposed schedule control program in conformance with the type of schedule selected earlier for your project. Assume the schedule S curve shows you are 5 percent behind design schedule at the 50 percent complete point. What is your program for getting the design work back on schedule? Would your solution be any different if the procurement commitment curve were also 5 percent behind?

5. Your construction manager has a labor survey report indicating a potential labor shortage during your peak construction effort. He wants to start out working five 10-hour days to attract better-qualified craft labor. Would you support or try to defeat the proposal? State your reasons either way.
6. Your project design productivity is below par at the 50 percent complete point. What percent productivity should you be showing at this point? What do you propose to do about it? How are you measuring earned value and/or physical progress?

   Assume the same scenario for the construction effort and answer the same questions.
7. How do you plan to control and report the procurement and delivery status of equipment and bulk materials? Sketch an outline format for the type of report you would use.
8. Outline a comprehensive quality control program for all phases of your selected project. Make a detailed outline for the facility turnover sequence and procedure you recommend for your project.
9. Make a list of areas in which you would be especially careful in operating the controls on your project. Expand on any areas of your project which you think are particularly critical. Are there any areas in operating the controls where you have had problems in the past? How will you try to improve in those areas?

Chapter

9

Project Engineering

No discussion of project management would be complete without including the role of the project engineer or the project architect. Frequently the route to the project manager title is through the position of senior project engineer. Hence, it is important to understand how the project engineering side of the business relates to project management. This chapter is intended to address that linkage, and not to give a complete guide to the practice of project engineering. A whole book would be required to cover the subject in that depth. Several such books are listed among the references at the end of this chapter.

Many project engineers and architects are performing project management duties without benefit of the project manager title. Since use of the title "project manager" has come more into vogue recently, the difference between the two titles is not quite as blurred as it was years ago.

The basic duties of project engineers and project architects involved in capital projects are generally very similar, except for some minor nuances that are introduced because of differences in project environments. To avoid the repetition of the ponderous term "project engineer-architect," I propose to use the term "project engineering" in its generic sense. The material covered in this chapter is readily applicable to the *project design* function for all types of engineering, architecture, proposals, studies, or any other technical endeavor.

Project engineers usually function in two basic modes in the capital projects area. One mode is as project engineer in complete charge of the project when he or she is also functioning as project manager. When operating in that mode, the project engineer assumes all of the project management duties and responsibilities covered in this book. And that is in addition to the detailed discussion of duties in this chapter only! Since the average project engineer is *not* Superman, that type of assignment can work only when handling one or more small projects.

The second mode is serving as a senior project engineer responsible for the design activities on a medium-size to large project. The discussion in this chapter will be based on that mode of operation, and it also applies to mode 1 on a smaller scale.

The Project Engineering Environment

The basic difference between project management and project engineering is that the project engineer still has a major technical input to the project. On average, the split of a project engineer's duties are about 50 percent administrative and 50 percent technical even when acting as project manager on smaller projects.

When a project engineer is operating independently on one small project or a series of small projects, he or she is usually handling a project or projects in the range of $50,000 to $500,000 total budget. Often, the project is to remodel or revamp an industrial, commercial, or institutional facility in a plant engineering setting. If the project is executed by using a captive design group, it is usually done on a matrix basis in competition for resources with other ongoing work. The construction is often done by the facility's maintenance department or is farmed out to local contractors.

Without a dedicated project team, the lead project engineer has to function like a one-man band in performing all of the project management functions such as scheduling, cost control, and purchasing. In my opinion, that sort of project environment is much more demanding than that of a project manager or engineer dedicated to a single major project. Generally, project engineers in multiproject environments are so busy putting out fires that they do not have enough time to plan, organize, and control their work properly. I sometimes think that they don't even have enough time to figure out ways to improve their operating procedures.

Those who prove that they can survive in that sort of environment certainly should be good candidates for becoming project managers. Their main problem in handling the transition to project manager is giving up the technical part of their old duties. I have often advised young project engineers moving up to project management to get rid of all the technical files that they have accumulated over the years. That way, they will not be tempted to look into their old files for information to solve technical problems which should be delegated to the project engineers. If they cannot bring themselves to shed the old technical duties, they are bound to fail at their new project management environment which now consists of virtually 100 percent *management* work.

The other environment for the project engineer, of course, is serving as the technical leader and coordinator on the project team performing a large project. That could be in either a central engineering department or a design contractor environment. Such an assignment would be filling the lead project engineer slot as shown in the organization chart in Figure 7.5.

On large projects, assistant project engineers reporting to the lead project engineer are usually assigned to coordinate specific units or portions of the project. In that mode they perform mostly technical work with a minimum of administrative duties.

Sometimes, they actually perform design work such as requisitioning equipment, evaluating bids, supervising layout and plot plan studies, and writing specifications. In that position, junior project engineers should understudy the lead project engineer to learn about coordinating a whole project. Junior PEs should also learn all the management techniques they can to go along with the technical experience being absorbed in their roles as assistants.

In a construction environment, the project engineer could be serving as field engineer or field project engineer responsible for the technical content of a construction project. The career growth path from there would be more likely to go through construction project management to project director in the home office. Construction firms also have project engineers in the home office coordinating the estimating, bidding, engineering, and procurement activities to support the field operations.

The Administrative Side of Project Engineering

The senior project engineer or project architect is the person to whom the project manager should delegate the execution of the technical portion of the project work. That means that the PM wants the lead project engineer assigned to the project early so the LPE can participate in master planning and staffing of the project.

Project initiation phase

Among the project engineer's early duties are establishing the scope of the design work, making the design estimate, developing the design portion of the project procedure manual, and developing the design team's organization chart and personnel loading curves. All of that makes for an extremely busy 4 to 6 weeks as soon as the LPE comes on board!

If the early duties are not properly planned, organized, and controlled, neither the project nor the project engineer will ever get off the ground toward a successful project. Having the project engineer involved during the project's formative stages is absolutely essential to getting the critical front-end phase off to a smooth start.

Concurrently with the above administrative functions, the project manager must initiate the flow of the project design data which is a primary responsibility of the LPE position. That means playing the lead role in preparing the design data questionnaires or participating in the project programming sessions. Establishing the design basis is second in importance only to defining the scope of work.

On a large project, where the early duties can easily exceed the span of control for the lead project engineer, there should be one or more project engineers to share the load. In that case, the lead project engineer must delegate some of his or her duties to the assistant project engineers.

On a small project, the LPE coordinates and directs the activities of a 10- to 25-person design team which is assigned to the project. On larger projects, the personnel load can easily climb to 100 to 150 people including all of the design disciplines involved. In either case, the PE works through a discipline leader in charge of each design group. Coordinating the goal setting and scheduling the *deliverables* for such a large group of design professionals gives the project engineer plenty of opportunities to refine leadership skills in a hurry.

In addition to all those project relations, the lead project engineer is the prime contact with the client on design matters. That means getting the project design basis established, as well as ensuring a continuing flow of design data feeding the design team. The activity is done through a continual series of meetings, telephone calls, correspondence, and so on which take large bites out of the PE's available time.

Preparing the design estimate and budget

As leader of the project design team, the LPE is responsible for developing the design cost estimate. When the estimate is approved, it will become the design control budget. The working document for preparing the estimate is the design portion of a home office services estimate form similar to those shown in Figure 5.3.

To get all of the disciplines on a sound footing for the estimate, the project engineer will fill out a project services checklist (PSC) similar to the one in Appendix B. The completed PSC should be reviewed with the project manager and other key project staff personnel before issuing it to the discipline leaders for preparing the personnel estimates.

Along with the PSC, the project engineer collects any existing design data from any reliable source. Anything that will assist those who prepare and refine the design estimate should be included in the package. A word of caution though: Do not include any material that is likely to create a "fright factor." That will only lead to ultraconservative estimates by the discipline leaders.

I have found that a discipline leader who has made an estimate will defend it to the death. It is much more effective to keep the original estimate under control than to fight for having it lowered to more reasonable levels. Very often, the defense is so strong that the estimate winds up being chopped arbitrarily by management. Also, the chop is often intentionally deep just to drive home the point that the estimate was too high. As usual, two wrongs *definitely* do not make a right and both sides lose.

That example is a good reason for the project design leaders to take a strong stand when reviewing the proposed design estimate before it gets to management. One thing to look for is the duplication or overloading of contingency factors beyond those warranted by the state of the design data. Project engineers must satisfy themselves that the estimates are sound before passing them up the line.

Design schedule input

The project engineers have the prime responsibility to supervise the development of the design portion of the project schedule. Just as with the budgetary activities discussed above, they are responsible for keeping the design team on the design schedule which they helped to develop.

The project engineer has the prime responsibility for inputting the design data delivery dates which lie at the heart of the design schedule. The discipline leaders cannot schedule their deliverables until the key dates are established. The project engineer then integrates the discipline schedules with the master plan for producing the design deliverables in accordance with the milestone dates in the overall project schedule.

The project scheduler produces the detailed schedule in close cooperation with the project engineer and the design discipline group leaders. Any interdisciplinary scheduling squabbles should be resolved by the project engineer in conjunction with the project manager if necessary. The coordinating activity includes resolving any interdepartmental problems involving procurement and construction, when applicable.

The project engineers usually receive copies of the 2-week look-ahead schedule sort which lists the activities due to be completed that week. On some projects, the responsible project engineer's name is listed on the 2-week look-ahead bulletin board listing critical milestone dates. Many scheduling departments are now using the project due date bulletin board to call attention to important milestone dates which must be met to keep the project on schedule. The technique has proved to be effective when the project manager reviews the 2-week look-ahead list with the project engineers on a regular basis. If the dates are starting to slip, the project engineer and the project manager must find out why and then do something about it.

Naturally, no one can keep work on schedule by looking at only a 2-week look-ahead schedule. The project engineer has to monitor the monthly physical progress report to ensure that all the disciplines are staying on the planned schedule. Good "entrepreneurial" project engineers should learn how to develop a gut feeling about the physical progress before the monthly schedule report is issued. They should be making plans on how to get the late disciplines back on schedule before the progress report blows the whistle on them.

Planning and mobilizing the design team

Now that the scope of work, the schedule, and the labor-hours have been set, the personnel required to execute the design have to be brought on board. The project engineer works in conjunction with the discipline leaders to arrange for the proper number of design personnel to meet the project's design goals.

It is important that the project engineer (and the project manager) support the discipline leaders in obtaining the necessary design personnel from the department heads. In times of people shortages, all pressure needed to staff the project must be brought to bear to ensure there are sufficient personnel to maintain schedule and meet quality standards.

For efficient personnel usage, it is equally important that design personnel are not brought on board too early. As coordinator of design information flow, the project engineer is in the best position to judge the proper timing for building up the design team. The people are actually brought into the project by the discipline group leader. The project engineer monitors personnel buildup and assists the discipline leaders in obtaining the right quantity and type of people for the project. The project engineer also brings the project manager to bear on the problem when that is required.

The project engineer must spend some time monitoring the flow of vendor data through the project design groups. The vendor data clerk should issue the late vendor data list to the project engineer daily. If late data reports become too prevalent, it is time to get into the vendor data approval system and shake it up.

The project engineer's intimate knowledge of the work status in the discipline design groups places him or her in an excellent position to advise the project manager about demobilizing design groups as their parts of the project wind down. Don't be softhearted in the demobilization process at the end of the project. A person whose contribution to the project is finished must go back to the discipline department immediately. Finding a new assignment for that person is the department head's job, not the project engineer's. As the design work deputy of the project manager, it is the project engineer's duty to finish the design work within the budget.

Serving as deputy project manager

The duties of the project engineer mirror those of the project manager to such a degree that using him or her as the deputy project manager is logical. That may or may not show up on the organization chart, but making the temporary assignment takes only a project memo.

Usually, serving as deputy PM can work for only short stints because of the pressures of the project engineer's normal duties. If project management work is especially heavy at that time, the project engineering work is bound to suffer. If the project engineer is supported by good assistant PEs, some of his or her normal duties could be delegated down the line temporarily.

Ordinarily, while acting in the deputy PM capacity, the project engineer will handle only the necessary routine items and put off any major decisions until the project manager's return. If any major emergencies arise, the project engineer can always discuss any proposed actions with the project's management sponsor.

An aspiring project engineer should welcome the temporary duty as an opportunity to get the feel of running the whole show and expanding his or her personal horizon. The only risk lies in going too far out on a limb when handling a major unplanned emergency without consulting management. Otherwise, exercising good judgment and common sense should see you through safely. But having both duties for a short time is very likely to exceed your span of control. Use that as an opportunity to test your ability to delegate some work to subordinates. Just remember to keep cool and act like a professional!

The Technical Side of Project Engineering

The project engineers play a key role in the technical development of the design right from day 1. They must be thoroughly knowledgeable in the technical scope of work. They must know the basis of design and participate in the early problem solving sessions in which the basis of design is established. Each should play an important part in writing the final scope of work on the technical side.

For that reason, project engineers must have in-depth training, knowledge, and experience in the type of work involved in the facility. They generally gain it by having worked up through the ranks of one or more of the design discipline groups. More than likely, they have spent some time as discipline group leaders themselves, so they are in a good position to coordinate and direct design teams. To handle the multifaceted nature of their new positions, newly appointed PEs must broaden themselves quickly into all technical areas which are strange to them.

The PE as design data coordinator

The project engineers are the clearing house for the transfer of all design data from the client to the design team. That applies to either an in-house project or a contracting environment. In the development of the design documents, design discipline leaders are in daily contact with their counterparts on the client's design team. However, if official design data are modified or additional information is transmitted, the work must be done officially through the project engineer or the project manager. The project engineer then issues the new information to *all* parties having a need to know. If the change affects project scope, the project engineer should initiate a change order in accordance with project procedures.

Having such a centralized clearing house for the design data is the only way a project's design basis can be kept on track and its scope can be kept under control. On any project daily questions will arise between client and contractor and among the disciplines over design data interpretation. Those that cannot be readily solved at that level are brought to the project engineer for resolution. Any design problems that cannot be solved at the project engineer level should be referred to the project manager for final resolution.

Monitoring the technical scope of work

The project manager depends on the project engineer to supervise the scope of the technical work and that is virtually the total scope of PE work on design-only projects. On integrated design, procure, and build projects, however, the design portion is considerably less percentage-wise.

It is up to the project engineer, as supervisor of the design disciplines, to see that there is no significant work done outside the project scope. Having prime responsibility for maintaining both budget and schedule is enough reason for the PE to keep nonscope activities out of the design agenda. Any such activity should become the immediate subject of a change order. That is the fastest way to determine if the person requesting the change is serious.

The project engineer and quality control

The project engineer usually signs off on all working documents produced by the project's design team before passing the documents up line for review by the project manager. It behooves the project engineer to have the documents in the best possible condition before releasing them through the PM for client approval.

The design quality control program must be in place early in the project to avoid producing substandard documents. As the project's design leader, the project engineer should be the main driving force behind getting an effective quality control program installed and working. The extent of the program has a direct effect on the design labor budget because all checking hours must be accounted for.

The proposed design quality control program must be approved by the project manager and the discipline department managers involved. If possible, it is best to stay with the quality programs that are already used within the company and adapt them to any special situations on the project.

If a good quality control system is in effect and is enforced during document preparation, the project engineer's final approval becomes much easier. A quick review for conformance to the project's engineering standards and the design data is all that's needed. If a lot of errors are found, the documents have to go back for extensive rework, with disastrous consequences to productivity and schedule. Important secondary effects of poor quality are poor client relations and low project morale.

Passing all the design documents through one control point allows the project engineer to ensure that the design team's product has a high degree of uniformity and professionalism. Specifications, requisitions, process descriptions, operating instructions, and so on must be clearly and concisely written to get the best results from the documents. Drawings must be clear, neat, and well presented to maximize efficiency in the field. A poor design product can wreak havoc on the budget and the project schedule.

There is no way a good project engineer can treat the final approval of design documents as a perfunctory duty. If shoddy products are passing over the PE's desk, the quality control program is in dire need of an immediate overhaul! Unfortunately, the approval stage occurs at the end of the design cycle when all of the schedule float has been used up in other operations. If you see this scenario looming in your future, it is in your best interest to at least spot check the product quality before it arrives on your desk.

Any firm (or project) that knowingly ignores quality control is heading for trouble and eventual extinction. That can place a heavy burden on the project engineer if good quality control is not practiced and QC costs are not budgeted in the beginning. If you cannot change that sort of QC environment in your firm, you may want to consider changing your environment.

Many design firms today are taking a leaf from the U.S. automaker's book and establishing quality control departments. Those groups have specific charters to establish sound quality control policies and procedures and promote quality work at the point of document creation. A project engineer's life in such a quality-conscious environment can be a lot easier!

The procurement interface

The technical interface between design and procurement, when those services are included in the scope, falls within the duties of the project engineer. As I have mentioned several times in this book, the relations between procurement and design sometimes get a little testy because of territorial disputes. As the leader of the design group, it is up to the project engineer to see to it that the interdepartmental relations do not get out of control to the point of poor performance. A smoothly functioning interface can make a strong contribution to improving schedules and budgets.

A good working relationship with procurement also is essential for the processing of that ever-valuable design commodity—vendor data! The faster equipment orders are placed, the faster the critical vendor data are available to the designers. The processing of vendor data is so simple that it should be routine, but it's truly amazing how many times the system becomes unglued and needs some prompt reaction to fix it.

Participation in the procurement process also offers project engineers an excellent opportunity to exercise some judgment in cost control and cost reduction. The technical bid evaluation stage is a good place to find items which might have been specified or designed too conservatively and are therefore too costly. That is the last chance for the design people to reduce costs on materials and equipment.

Good project engineers must engender a cost-conscious approach in the people performing the detailed technical bid evaluations. The bid evaluators should be trained to bring any potential cost reductions to the project engineer's attention for possible further investigation and action.

The relationship between design and procurement is normally under the watchful eye of a good project manager, so effective project engineers owe it to themselves to stay out in front in that area. It would be difficult for a PM to give a high performance rating to a project engineer who didn't exercise full control of that critical area.

The construction interface

Although the project manager has overall responsibility for the performance of the construction work on integrated projects, the project engineer is the focal point in the vital design-field interface. The end product of most capital projects is the finished facility, which we have said is to be completed as specified.

To reach that major project goal, there must be complete cooperation between design and construction. The most logical channel for the necessary design document interpretation, correction of design errors, and solution to field design problems is between the field engineer and the project engineer. Project engineers often take up residence in the field after the design work is completed. In those cases, they often play major roles in the final facility checkout and start-up.

During the design phase and after field opening, the field engineer uses the project engineer as the main contact for resolution of design-construction problems. The project engineer is in the best position to sort out the problems and bring the best technical brains to bear on a solution. Field problems that can have a significant effect on cost and schedule should immediately be brought to the attention of the project manager. If there is a continuum of small problems in certain areas, the project manager should also be alerted to assist in correcting the situation.

Frequent trips to the field by the design project engineer are invaluable for spotting potential design problems, potential interferences, and the like even before they occur. As the one person on the project with the most comprehensive and intimate view of the overall design, the project engineer is the only one who brings that unique input to the field work. Good project managers will also develop that sort of design overview capability, but not to the depth of the project engineer who has lived with the design from day 1.

Personnel development for growing project engineers is a particularly valuable aspect in handling field interface problems. That is the best time to get the necessary field experience to broaden the PEs' backgrounds for moving into project management. Project engineers who aspire to be project managers will be wise to keep their eyes open to other facets of the construction environment while carrying out their design duties in the field. That should be from the standpoint of observation only, and not from engaging firsthand in the political arena at the construction site.

Communications and the project engineer

From the foregoing discussion, you can readily see that communications is a fundamental need in the project engineering field. Chapter 10, which is devoted to project communications, is fully applicable to the practice of project engineering. The project engineer must be as adept at communicating as the project manager is.

The project engineer's communications repertoire runs the whole gamut from project correspondence, reports, minutes of meetings, chairing meetings, and verbal presentations to performance reviews. A project engineer cannot survive without effective, across-the-board communications skills. Regardless of the strength of your technical skills, you will never progress in the field of project engineering without top communicating ability. If you feel that you are weak in any areas, you must gain or polish your communications skills without delay.

The best way to do that is through professional writing and public speaking courses followed by an ongoing program of honing the skills and critiquing them to a fine edge through practice. You must become a purveyor of your ideas to your peers and upper management to have your valuable technical contributions as a project engineer accepted. Chapter 10 will give you more details on conducting a program to expand your communicating abilities.

Human Factors in Project Engineering

Although the project engineer's milieu does not cut across quite as many lines of communication in the company as the project manager's, there are enough of them to bring some human factors into play. Chapter 11 discusses the human factors encountered in project management, which are very similar to those for project engineering. All of the points covered in that chapter are equally applicable to project engineers, so there is no need to repeat them here. There are, however, a few points worth mentioning.

The project engineer–project manager relationship

It does not take long to realize that the selection of the lead project engineer is about as critical as the selection of a project manager. If the pair does not work well together, it will be next to impossible to meet the project goals. In critical times (like project initiation), it's vital that the two people develop a real feel for each other's modus operandus in attacking the project. That sort of rapport, along with strong feelings of mutual respect, can go a long way toward producing a successful project.

It is almost impossible to overstress the need for having a strong bond between project manager and project engineer. Project managers must also serve in the role of mentor to aspiring project engineers to groom them for future project management roles. Delegating more than normal significant management duties to project engineers helps the PEs grow with the job. One way to do that is to have the PE perform as your deputy. Having someone who has both your confidence and the necessary skills available can be a distinct advantage when vacation time rolls around!

Client relations

Remember, every project has a client whether or not the project people and the client are working for the same company. The project engineer has a great deal of client contact in the areas of transferring the design basis and issuing the design documents. Most client comments on the design documents are handled by the project engineer and the discipline leader.

Good client relations are essential on both sides of the project engineering equation if the project is to be successful. The rules of play in that area are identical to those for project managers discussed in Chapter 11. If those rules are followed by the project engineers, successful client relations should result, and they will make a large contribution to the project's success.

Project relations

We have already touched on the procurement and construction areas, in which the project engineer plays a key role in meeting the project goals. A mature and professional approach to project relations in the above areas, along with those in the design team, is absolutely vital.

Project and client relations can sometimes come into conflict with one another at the project engineering level where some hard decisions have to be made. Generally, the client relations will have preference but not at the expense of project relations. A balance has to be struck so neither side is declared the loser.

Leadership

Strong leadership is absolutely essential to becoming an effective project engineer. Most project managers delegate a good deal of authority to the lead project engineer. It is up to the project engineer to exercise that authority through sound leadership.

Leadership is just as important on small projects as on larger ones. On small projects there are fewer personnel, but the leader is much closer to the people being led. When everyone is in the trenches fighting the battle side by side, it is sometimes difficult to portray a strong leadership role. However, the need to instill the project mission into the smaller group is often more critical than into a larger one. On a small project there is less margin for error!

Review the list of qualities for leadership in Chapter 11 occasionally to see how your leadership style stacks up to what is required in your particular project engineering environment. Try to strengthen the areas in which you feel that improvement would help you perform your job more effectively. The best leadership training is practicing leadership on the job, so don't be afraid to try different techniques to sharpen your skills.

References

1. Ernest E. Ludwig, *Applied Project Management for the Process Industries,* 2d ed., Gulf Publishing, Houston, Tx., 1987.
2. Arthur E. Kerridge and Charles H. Vervalen, *Engineering and Construction Management,* Gulf Publishing Company, Houston, Tex., 1986.
3. Richard E. Westney, *Managing the Engineering and Construction of Small Projects,* Marcel Dekker, New York, 1985.
4. Dan Mackie, *Engineering Management of Capital Projects,* McGraw-Hill Ryerson, Toronto, Ont., and New York, 1984.
5. Howard F. Rase and M. H. Barrow, *Project Engineering of Process Plants,* Wiley, New York, 1957.

Case Study Instructions

This case study will require a role change from that of project manager to senior (lead) project engineer for your selected project.

1. Prepare the design cost estimate for all the disciplines you expect to use for your selected project. Include your best estimate for out-of-pocket costs for the design group. Suggestion: Use the home office estimate format shown in Figure 5.3.
2. The project manager has asked you to prepare the design cost control program input for the Project Procedure Manual. Outline the key features of your plan and the accounting data you will need to make it work.
3. Prepare a plan for transferring the project design data from the client to the design team for your selected project.
   a. How will the basic package be handled?
   b. What will you do to prepare for the project kickoff meetings?
   c. How will revisions to the design basis be handled?
4. Refer to Chapter 4, Project Scheduling, and prepare a logic diagram for the design portion of your selected project. Assume some logical delivery dates for the design data transfer discussed in (2) and some elapsed times for delivery of the design work. Don't forget to work within the planned completion dates for the overall schedule.
5. Prepare an outline of the quality control procedures you propose to use on the design portion of the work. Cover all the areas from document inception to issuing for construction.
6. The project buyer has complained that the mechanical equipment group under your control has been contacting several vendors with design changes which have affected price. The mechanical people complain that procurement is taking too long to process the revised information. How do you propose to resolve this problem?

Chapter

10

Project Communications

Successful project execution is virtually impossible unless you have an effective communication system. Good communication skills are basic to becoming a successful project manager. Effective communication is the lubricant that keeps project machinery running smoothly.

Figure 1.2 highlights the project manager's key role in project communications. It shows the many communication channels encountered in a complex project organization. If communications break down, each of those channels is a potential barrier to information flow. The only way to control potential information barriers is through effective communication.

Most project management communication deals with conflict resolution. That means project managers must maximize their communicating abilities. If you are lacking in that critical area, you should enhance your writing, public speaking, visual presentations, and listening skills through training and practice. I will offer specific suggestions in that area as we discuss the various types of project communications.

Why do communications fail?

There are myriad reasons why communications fail in all parts of our lives. A few of the more common reasons why business communications fail are the following:

- Not having a clear goal in mind
- Staying in a negative mode
- Concentrating on your own thoughts to the exclusion of the other person's ideas
- Not establishing rapport

- Assuming that others have the same information on the subject that you have
- Being impatient—not hearing the other party out
- Mistaking interpretations for facts
- Failure to analyze and handle resistance
- Being ashamed to admit you don't understand
- Overabundance of ego

Communication is defined as "1. a transmitting, 2. a giving or exchanging of information, and so on as by talk or writing." By comparing the common reasons for failure with the definition, it is easy to analyze the failures. I am sure that each of us can go down the list and pick out three or four failures that apply to us. They are the ones we must improve on if we are to become successful communicators.

One of my techniques is to mentally transpose myself into the intended receivers of my communications. Then I read the message while using their knowledge of the situation. If I can still understand what I said, the communication just might work. If I cannot understand it, I rework the message. If you practice that technique, you can even use it in oral communications as well as written ones.

If we don't establish rapport in communications, anything we say is likely to fall on fallow ground. The easiest way to effect rapport in your project communications is to build your credibility as a professional manager. People will receive your messages more readily if they respect your opinion. All you have to do is keep an open mind to build *full rapport.*

My personal number 1 communications problem is being patient while receiving communications. After many years of trying to improve, I still have not perfected the art. If you have the same problem, don't give up too easily. It can be done!

Handling resistance in the mind of the message receiver also requires patience. If putting your idea across is important to you, analyze the resistance and cancel it. You must remember that factor when *selling* your idea *upward* or *downward* in the project organization.

The last two items on the problem list often go together. Never let your ego get in the way of admitting that you don't understand something. None of us will ever learn all there is to know about managing a project. Showing that you are big enough to admit ignorance now and then will increase your stature, not decrease it. Another way to handle that problem is to play the message back to the sender the way you understood it. That avoids the open admission of not understanding the message.

Those are a few basic points for improving your communications. Do take advantage of any communications training courses that will help overcome any weak areas. Above all, remember to practice, practice, practice!

Communication systems

In addition to improving their communicating abilities, PMs must keep abreast of the latest communications hardware. In the past few years a spate of new communications tools has hit the market. These new tools multiply your communications skills through hardware alone. Fax now allows us to send documents anywhere in the world with the speed of placing a telephone call.

Multicity telephone conferences are now commonplace. Videoconferencing is also becoming cost-effective as its costs come down and travel costs rise. Project managers should check out and evaluate these new techniques and take full advantage of those suited to the project. Calculate the cost/benefit ratio for these new tools and propose them in the project cost estimate.

Project Communications

Project managers are the prime spokespersons for their respective organizations. That duty places them in the position of either creating or monitoring most project communications. To set a good example, project managers must make sure that all project communications are handled professionally. They should review all correspondence and at least edit it to a consistent professional standard.

Typically, project communications fall into the following key categories:

1. Project correspondence
2. Audio-visual presentations
3. Project reporting
4. Meetings
5. Listening

As we discuss each of these areas, we will find that list covers almost every form of communications there is.

Project correspondence

The primary project communication link is between the client and the contractor project managers. All major project correspondence should pass through that channel. That is the only effective way to control the project design, scope, and other contractual obligations. The contract usually sets up that communication channel by calling for official notices to pass between the project managers. Remember, that applies whether you are contracting within the company or externally.

Project managers should delegate some specialized outside correspondence to other key staff members to speed up communications. However, any matters concerning scope, cost, schedule, and so on must pass through the project manager channel to be official. That is the only way to avoid conflicting instructions and project chaos.

Project correspondence serves another important function in addition to sending messages. All the correspondence should find its way into the project files to document the project. When problems arise later in the job, it often becomes necessary to build a paper trail to find out what happened.

The project secretary should assign a number to and record each document in a correspondence log. That makes it is easier to locate correspondence without thumbing through the bulky files directly. That applies to incoming and outgoing letters, memos, document transmittals, and minutes of meetings. Maintaining those logs on a spreadsheet is a good PC computer application.

Audio-visual presentations

Project managers often make audio-visual presentations during the project. As we said in Chapter 2, the first one could be the dog-and-pony show used to sell the project. Later presentations occur during project and management review meetings, feasibility reports, and PR talks to citizens groups, among many others.

In preparing for a presentation it is vital to select the single most important idea or thought you want to convey to your audience. Marshall all of your arguments, visual aids, and statements in support of your main theme.

Start your preparation with an outline of your theme. Base it on the most positive way to present your arguments from the introduction to the closing punch line. Work the outline over to refine it into a clear, concise, and effective presentation of your thoughts supporting your ideas. If you are working with a group, bounce your ideas off the other members as the outline develops. Write the copy, and prepare the supporting visual aids supporting the idea. Remember, the clearly spoken word, reinforced with a strong visual image, increases the understanding and retention of the idea.

Rehearse the presentation to get your timing worked out and polish off any rough edges. Make the dry run in front of a nonsympathetic group acting in a devil's advocate role to critique your effort. Have the members ask all of the difficult questions they can think of so you will be ready to handle them in the actual presentation.

Be sure that your visual aids are of good quality and are easily legible from any point in the presentation room. Check out the presentation facilities to be sure that the layout and equipment will work well for your materials. *How to Run Better Business Meetings* published by the 3M Company is an excellent guide for arranging the physical facilities for effective presentations.[1]

Management consultants in presentation training have developed the training into a fine art over the years. Usually, these courses will take a management group into seclusion for several days to indoctrinate its members in the art. They teach the attendees the basic techniques of creating such visual media as slides, flip charts, overhead viewgraphs, and videos. They develop story boards incorporating the spoken and visual parts of the presentation. Through the benefits of videotape and a critique by the rest of the attendees, the presenter receives pointers on how to improve. That is an excellent way to gain basic skills in the presentation arts. Later, you can fine-tune them in actual practice. You should take advantage of presentation training at the first opportunity.

Project reporting

Another powerful project communications tool is project reporting. Creating project reports forces the project staff to review all aspects of project activity at least once a month. Setting the standard for high-quality project reports is an important role for project managers.

Producing a concise and well-presented monthly progress report is an excellent way for the project manager to set the standard. It is also a good time to review the other project control reports and check how the project is going. The monthly progress report is a summary of how well the project team is moving toward meeting its goals.

The main purpose of the progress report is to inform the key interest groups on how the work is progressing. A secondary purpose is to keep a running history of important project activities. These key groups are interested in the progress report:

- The client's organization, local and main offices
- The contractor's top management
- Key people on the project

The operative words in that discussion is *to inform* key people of important project activities. A rambling poorly written progress report will not inform the reader. The progress report must also be factual, and results-oriented and report any problem areas to present a true picture of project status. It should be written in a positive, direct style to instill confidence in the reader that the project team is in control of the project.

The writing style must be forceful and direct. Stay away from the passive voice and weak verbs. Keep your sentences short and use a new paragraph when the subject changes. There are several new style manuals and computer programs on the market today aimed at improving our writing. One that I used in writing this book is *RightWriter®*, which works with most word processors. It won't write the report for you, but it can train you to writing in a more forceful style.

The monthly report format should be consistent from month to month so readers don't waste time hunting for the key indicators each month. Most well-managed companies today have developed standard report formats to suit their types of projects. Using a standard format saves project managers from having to develop a new format for each project. Even with standard formats, some latitude is possible to meet particular client requirements in the report.

Figure 10.1 shows a typical progress report format that I have used over the years. It happens to be for a process plant, but you can readily adapt it to any type of facility. Compare it with your company's present format to see if it can contribute to doing a better job of reporting project performance.

A major report format decision is whether or not to include complete copies of the project control reports in the monthly progress report. That may be the way to go on small projects, but it makes for a bulky and unwieldy report on large projects. In the latter case, a summary section for the major detailed reports highlighting any accomplishments and problem areas produces a more readable report. The client's control people are the only ones interested in perusing the detailed project control reports. It is their duty to pass any unusual conditions on to their project manager for action.

The client and contractor managements receive and read the monthly progress report, so it should not contain any confidential contractor information. For example, a client should not expect to see the contractor's detailed cost information on a lump sum project. Also, don't discuss any confidential personnel matters in the report. Prepare an addendum covering any of these confidential company matters as an attachment to your company management's copy.

Project reports should be factual. Painting a rosy picture by ignoring problems only leads to more serious problems later. Bringing prob-

PROJECT PROGRESS REPORT
TABLE OF CONTENTS

0.0 TITLE PAGE OR COVER SHEET

Project name and number
Client and contractor's names and logos
Report number and period covered

1.0 TABLE OF CONTENTS

2.0 MANAGEMENT SUMMARY

A brief precis or abstract of project activities

3.0 OVERVIEW OF PROJECT OPERATIONS

Narrative report by disciplines

3.1 Project management
3.2 Design disciplines
3.2.1 Process engineering
3.2.2 Detail design
3.2.3 Subcontract engineering
3.3 Procurement
3.4 Project controls
3.4.1 Scheduling
3.4.2 Cost control
3.4.3 Cost estimating
3.4.4 Material control
3.5 Construction

Summary of construction activities

4.0 PROGRESS CURVES OR CHARTS

4.1 Design activities
4.2 Procurement commitments
4.3 Construction progress
4.4 Personnel loading by discipline or craft

5.0 SUMMARY OF PROJECT CONTROL REPORTS

5.1 Cost control
5.2 Schedule
5.3 Cash flow
5.4 Change order register

6.0 WORK TO BE DONE NEXT MONTH

6.1 Design activities
6.2 Procurement
6.3 Construction

7.0 PROBLEM AREAS

7.1 Information needs
7.2 Decisions required
7.3 Other problems
7.4 Suggested solutions
7.5 Indicate unresolved items from last month

8.0 CONSTRUCTION PHOTOS

9.0 DRAWING AND SPECIFICATION LISTS (optional)

Figure 10.1 Table of contents for monthly progress report.

lems out into the open as they arise, along with possible solutions, is the best way to handle the situation. That approach shows a positive attitude in the face of an adverse situation. Problem areas require immediate attention from both project managers to mount a quick and concerted attack to develop the best possible solution.

Don't include a lot of dull and tedious statistics in the report. Graphs and charts present data in a more interesting and readable format and break up long pages of text. For example, duplicating a long list of requisition numbers issued last month will not mean much to report readers. Only the people working with those numbers will know what they mean—surely management readers will not. A chart showing how requisition issues are proceeding against the plan will be much more effective.

Preparing the monthly progress report offers an excellent opportunity to practice my favorite communication tool of placing myself in the intended readers shoes. Read the draft report from the reader's viewpoint and knowledge of the project. Polish up any areas that are unclear from that viewpoint and you will have a *better informed* reader. Good progress reports can be real point winner toward meeting your career goals. That opportunity comes around every month, so don't pass it up!

Progress report contents

The monthly progress report is an important document, so reviewing its contents section by section is worthwhile. The title page or cover sheet is self-explanatory. Using custom-printed covers is not that expensive now that the new printing processes are available. Also, using a decimal numbering system makes the table of contents easy to follow.

Section 2.0, the Management Summary, is a key section of the report, and it's perhaps the most difficult to write. This section should give upper management a thorough review of the project in no more than one or two pages. If upper management people are reviewing many projects, that may be the only section they have time to read.

Creating this section for a large project should tax your ability to write clearly and concisely and still cover the ground. You will find it difficult to include details in the executive summary and still keep it short. If details are necessary to clarify a situation, refer to the detailed section of the report.

The summary need not follow the same format as the complete report. Present the higher-priority items up front in case the reader does not finish the summary. Using a news-reporting style is perhaps the closest analogy to describe the writing style needed here.

Section 3.0, Overview of Project Operations, is a narrative report of continuing operations in the various project disciplines. Discipline leaders should prepare these sections for the project manager in a format suitable for use in the final report. As project manager, you should issue the discipline leaders a standard format for reporting their progress. Using standardized formats will save a lot of editing time to get the various sections into a consistent format.

Collecting data from each discipline group gives you an opportunity to review group performance for the month. Go over each report with the discipline leader as part of your MBO goals review system before incorporating them in the progress report. Concurrently, you may collect a few news items for the executive summary.

Section 3.0 is where we go into more detail about each discipline's actual progress and what is needed to carry out the mission. Remember not to overload the section with boring statistics, just accent the positive and discuss the problem areas. The tone of this section will evolve over the life cycle of the project, starting with process and basic design and then shifting into construction. Don't spend a lot of time on the areas that have moved into a passive follow-up mode.

Section 4.0, Progress Curves, discusses project progress using the monthly updating of the bell and S curves. The cost and schedule curves come from the detailed control reports for use in this section of the monthly progress report. The curves shown in Figures 10.2 to 10.7 represent a typical example of a set of curves.

Discuss the monthly and total progress to date for each curve in the narrative portion of Section 4.0. Analyze and discuss any deviation of actual vs. planned performance. Give particular attention to clarifying the areas that fall below the planned performance curve. Explain the reasons for not meeting the plan and the corrective actions taken or recommended. Also, discuss any serious performance lapses later in section 7.0, Problem Areas.

For example, in Figure 10.2, Engineering Progress, actual progress has fallen below that planned for September. Until September, progress was above planned, which corresponds with the front-end personnel loading bell curve in Figure 10.3. Since the personnel loading is on target, the problem must be elsewhere. One possibility is that our vendor data did not arrive on time to support the design schedule. If that is the case, we must take positive actions to reverse the adverse trend in design progress. Explain the situation in the progress report and continue to monitor that area closely.

Construction progress also is covered in this section. The sudden burst of construction personnel in the past month, shown in Figure 10.4, needs a comment. It has the desired effect of getting construction progress above the planned curve shown in Figure 10.5.

The procurement commitments shown in Figure 10.6 are starting to fall below the planned progress at about the 50 percent complete

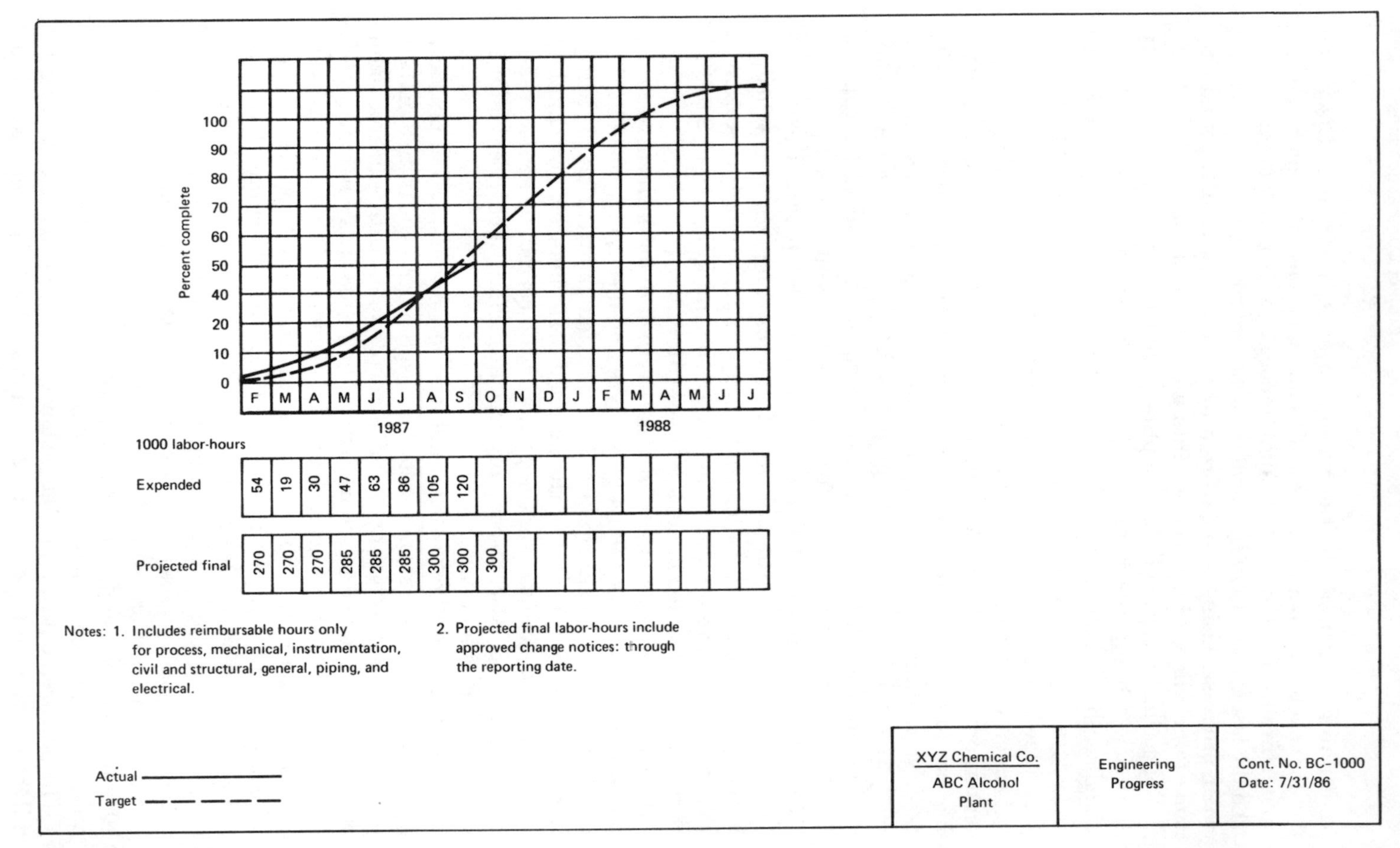

Figure 10.2 Engineering progress.

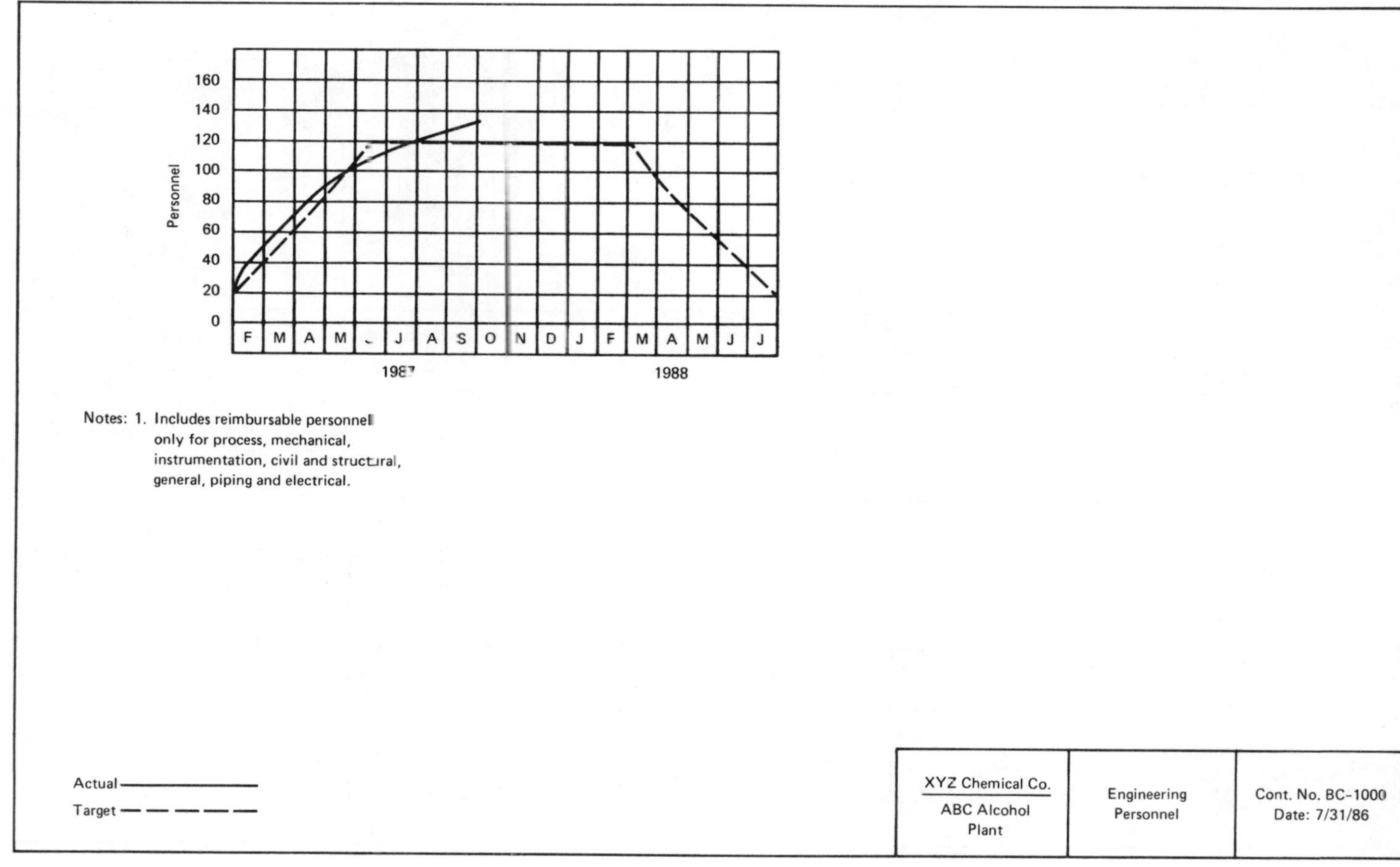

Figure 10.3 Engineering personnel loading.

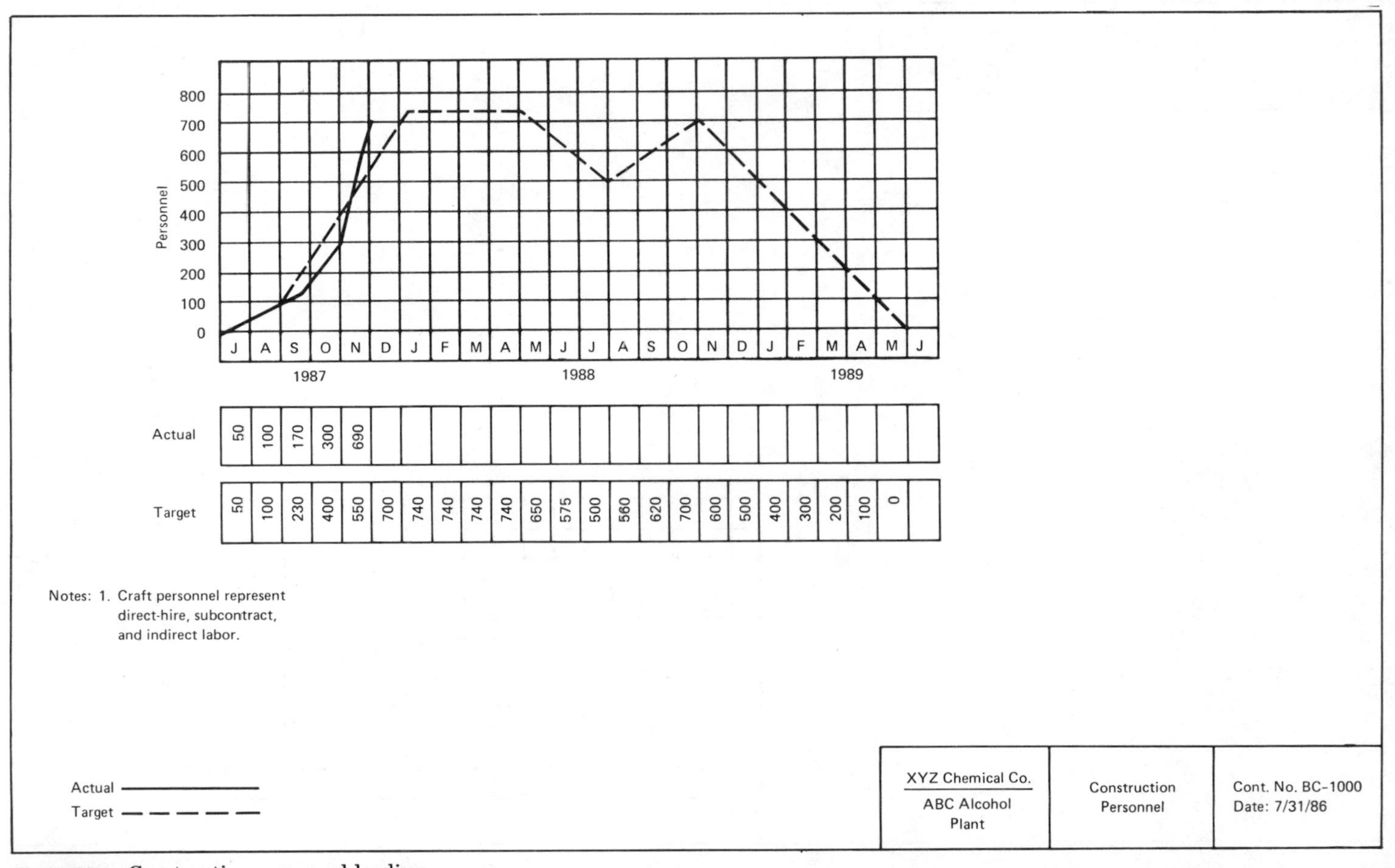

Figure 10.4 Construction personnel loading.

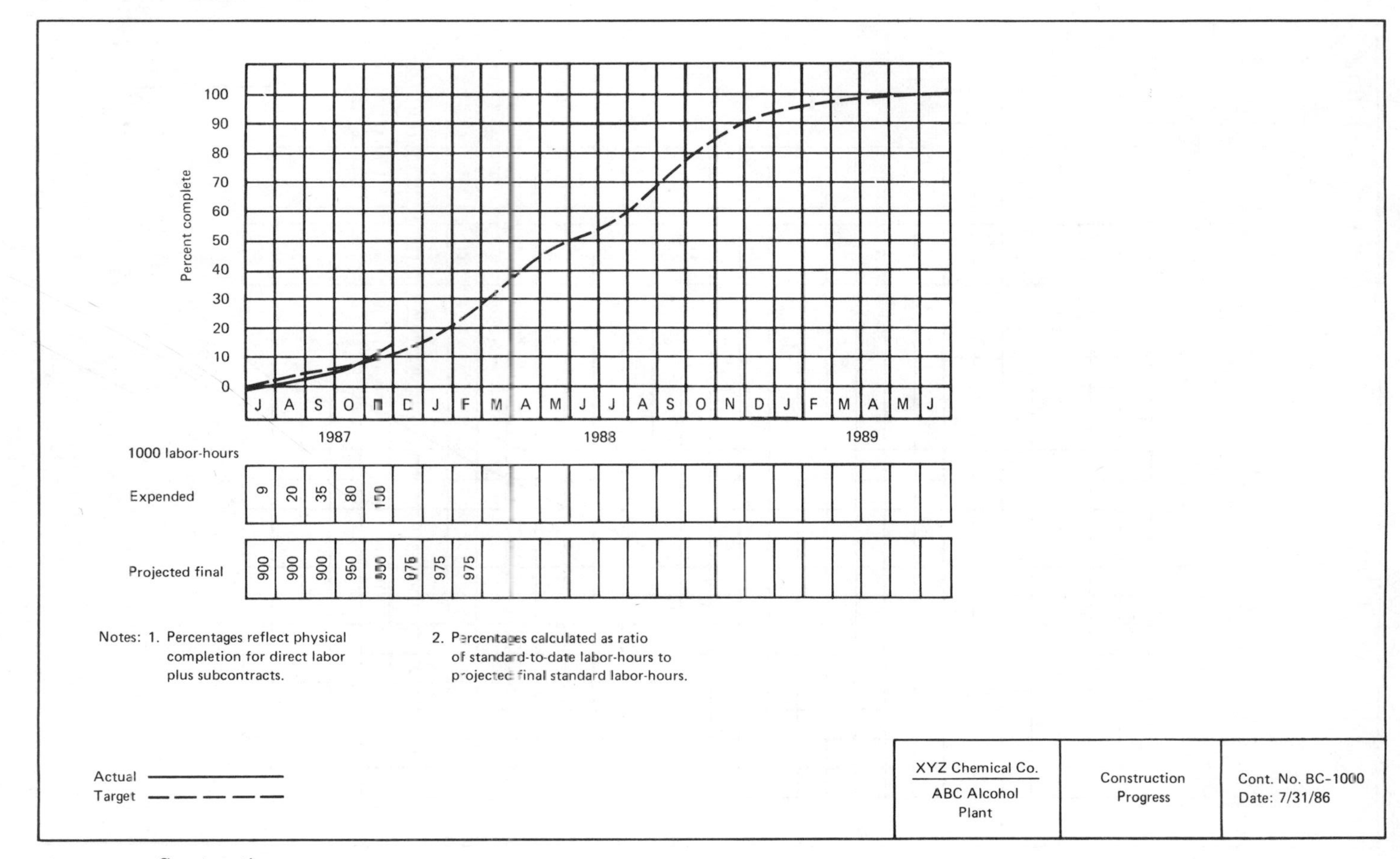

Figure 10.5 Construction progress.

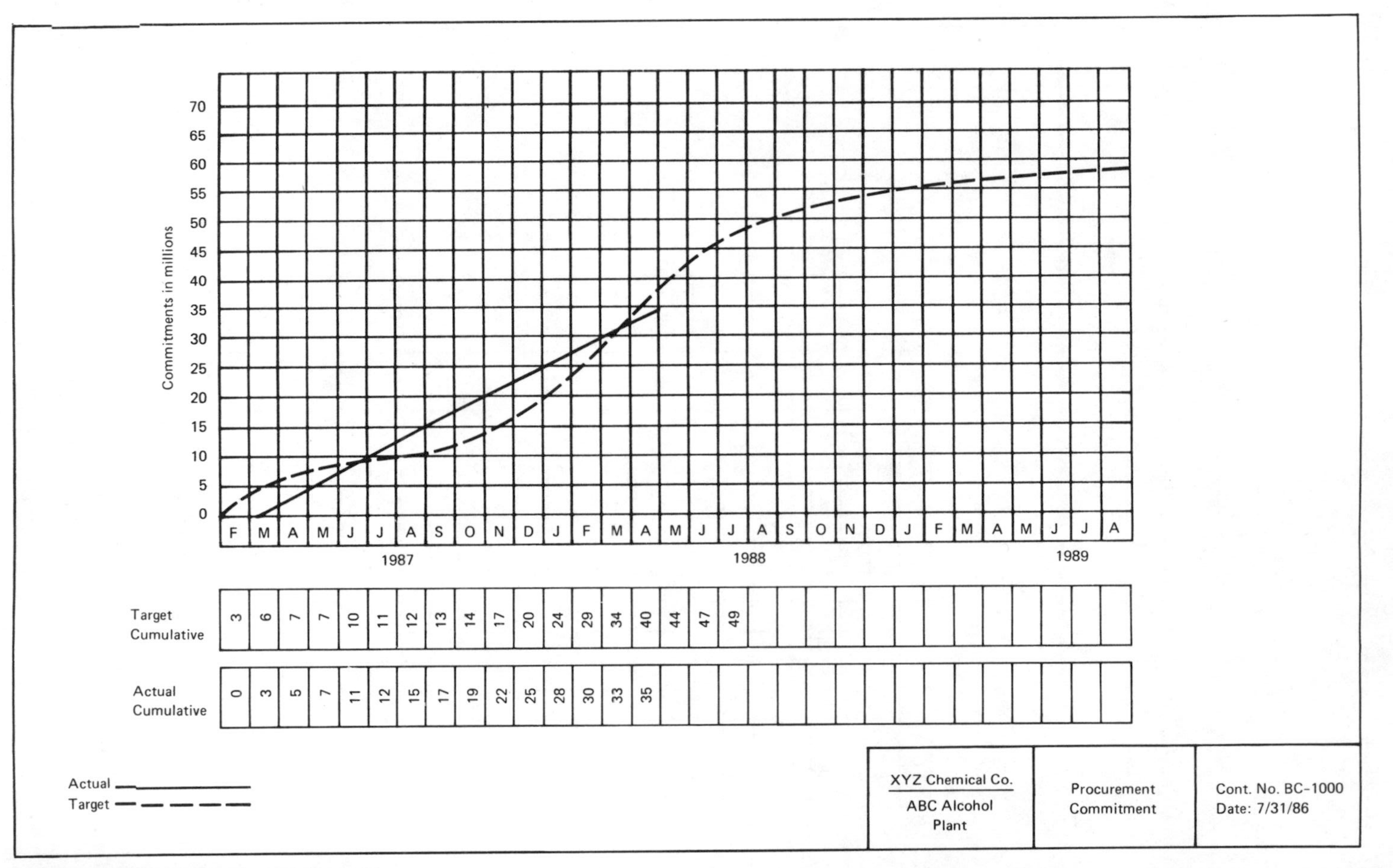

Figure 10.6 Procurement commitments.

point. That adverse trend must be addressed in the report. The purchasing is now in the bulk materials stage, and that is critical to maintaining construction progress.

The cash flow shown in Figure 10.7 has been tracking the projected curve fairly well up to now. This month it appears to be heading off the chart. An explanation of that condition and a short-term idea of what to expect next month are now in order.

Section 5.0, Summary of Project Control Reports, summarizes the detailed control reports which were too bulky to include in full. The schedule report has been discussed in Section 4.0, so this section should highlight cost, material, and quality control matters. Most of the information for this section comes from the executive summary sections of the control reports. The change order log and the project cash flow projections are other good items to include in this section. Also, this section should relate to how well we are achieving our overall and specific project goals. Are we still likely to meet our goal of finishing the project *as specified, on time, and within budget?*

In Section 6.0, Work to be Done Next Month, each discipline lays out its work activities and goals for the succeeding month. Discuss the key milestone dates set for next month. Here are some examples of milestone dates: complete the ordering of long-delivery equipment; finish the steel erection, and issue all general specs.

Be sure that you can meet most of the goals called for, because the items not completed will carry over to next month. Missing monthly goals will require an explanation.

Section 7.0, Problem Areas, is another key section of the report. It requires careful preparation to avoid panic in the reader. Summarize the problem areas mentioned earlier in the main body of the report. For example, we usually have a project form, called an information needs list, that shows all the items needed to complete certain project tasks. If the answers are slow in coming, discuss how the lack of information and decisions is delaying the design program. The adverse effects of that condition will in turn jeopardize schedule throughout the project. Go into the ramifications that that delay will have on construction progress, meeting the strategic end date, and project costs.

Above all, remember to propose some solutions to the problems you have raised. The top-management people reading the report are not going to have the answers. They expect the project manager to produce them. The project sponsors, however, may have some top-management input to help solve the problem.

Leadership, conflict resolution, and problem solving are the project manager's forte. That is a fertile area to practice the fine art of project management!

Sections 8.0, Construction Photos, and 9.0, Drawing and Specifica-

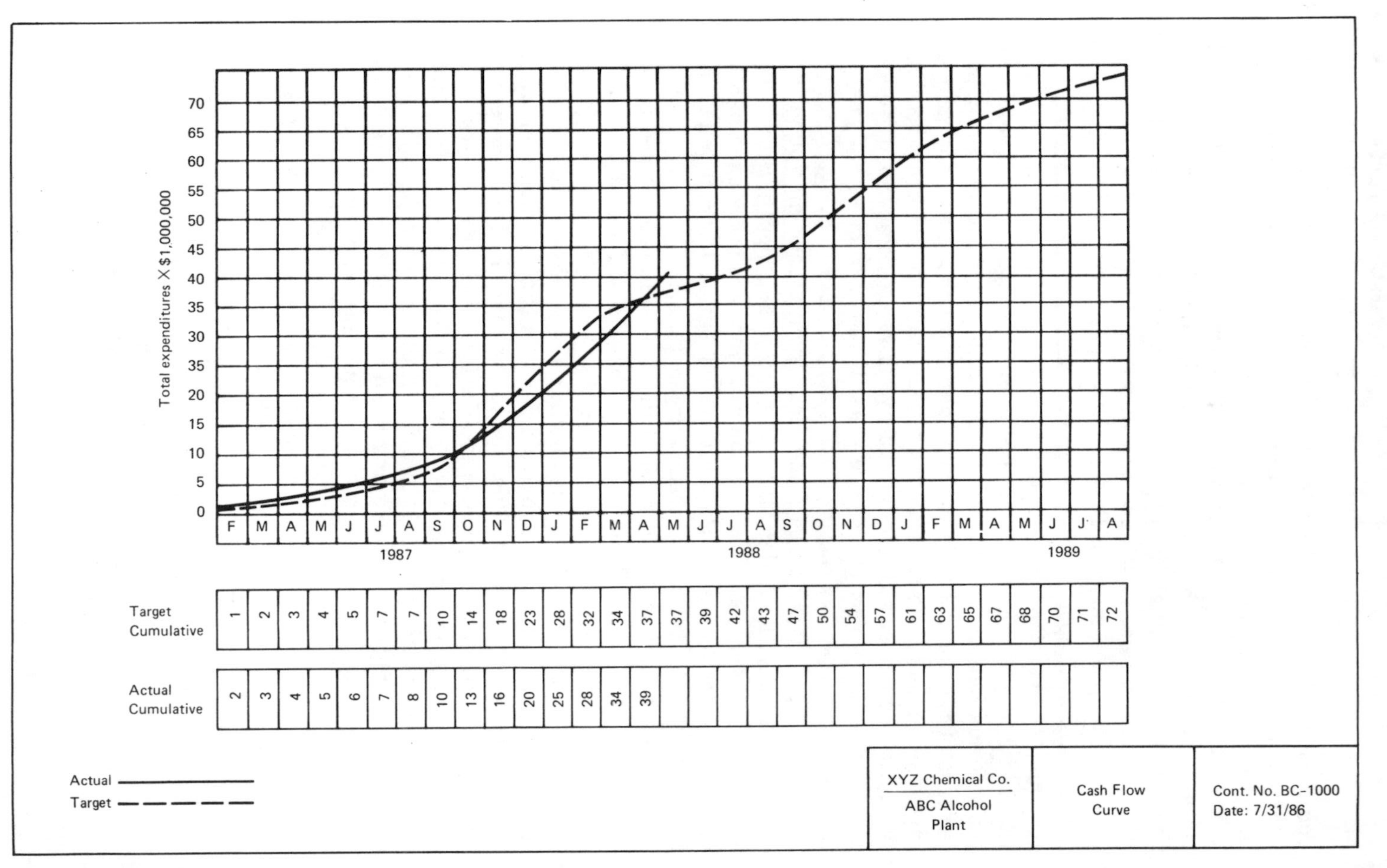

Figure 10.7 Cash flow curve.

tion Lists are optional. Construction photos are good if the project site is out of town or difficult to get to. Home office people are interested in seeing how their work turned out. Construction photos can serve as a good project morale booster for those not able to visit the site.

Drawing and specification lists tend to be bulky and dull so I don't usually include them. Bulky reports cause busy readers to lose interest. If these documents serve no useful purpose, leave them out.

You should consider my suggested table of contents in Figure 10.1 as minimal requirement for a good progress report. Certain project or business conditions may require you to alter or add to it. That is no problem as long as you maintain the overall goals of good organization, standardization, and readability.

Project meetings

Ineffective meetings are major time-wasters in project operations. The project team spends a large part of its time in meetings. The range is from working meetings, planning sessions, problem solving, information gathering, and management review meetings to major presentations.

My goal in this section is to make the large blocks of time spent in meetings as productive as possible. As project managers, we are directly responsible for handling most of the meetings on our projects. Here are some specific goals in improving our meeting results:

- Increase productivity
- Improve quality of results
- Cut wasted time and money
- Build our personal and company images

Our goals in improving meeting results are certainly in line with our total project management goals of *completing the project as specified, on time, and within budget.*

Establish need

"Is a meeting really necessary?" should be the first question asked when considering a meeting. Unnecessary meetings are a form of overcommunication, which is a wasteful practice. If you can settle the matter with a conference call instead of a meeting, you can save time and money.

When group action is needed to exchange information, resolve conflict, or develop ideas or for organizational purposes, a meeting is useful. If it does become necessary, keep the group as small and compact as possible. Six to eight is a reasonable number for a working meeting and twelve is an absolute maximum. Larger groups in information transfer meetings can be effective because the role of the participants is more passive.

I have worked on projects where weekly meetings of thirty or more were commonplace. As expected, the results were minimal. Meeting costs approached $1 per minute per participant. That meant the cost was $30 per minute, $1800 per hour and $3600 per normal 2-hour meeting. That added up to $15,600 per month or $93,600 over a 6-month project. Negative morale resulting from those ineffective meetings easily offset any minor meeting accomplishments, so money spent on the meetings was completely wasted.

Estimating the cost per minute is a good way to keep your meeting size under control. It also allows you to estimate the actual cost/benefit ratio and productivity of your meetings.

Meeting execution philosophy

Since meetings are similar to small projects, they are subject to my Golden Rule of Project Management. *Plan, organize, and control* your meetings to make them productive. That philosophy applies to all meetings, large and small. Obviously, larger, more complex meetings require a more formalized approach. The meeting checklist in the following section also applies to small meetings but on a less formal basis.

I recommend that you follow the meeting checklist for all types of meetings whether they are routine or one of a kind. The easiest *wasted meeting trap* to fall into is the routine weekly meeting, which can become ineffective through slovenly practices. Don't let familiarity breed contempt in routine meetings.

A condensed meeting checklist

I have developed the following meeting checklist to serve as a guide for meeting preparations.

Before the meeting:

- Explore alternatives to having a meeting.
- Set your meeting goals.
- Prepare a detailed meeting plan.
- Make an agenda and visual aids.
- Control attendance (calculate the cost per meeting minute).
- Check key attendee availability.
- Set the meeting time and time limit.
- Distribute the agenda and any reference materials before the meeting.

- Select a proper location and facilities.
- Arrange seating and refreshments if required.
- Assign a recorder and timekeeper.

During the meeting:

- Start on time.
- Follow the agenda.
- Control interruptions.
- Keep your goals in mind.
- Record decisions reached and actions needed.
- Keep participants on the subject.
- Don't waste time.
- Be tactful.
- Rotate part-timers in and out.
- Adjourn on time.

After the meeting:

- Issue minutes of meeting on time.
- Critique your performance.
- Clear up any unfinished items.
- Follow up on action items.

Reviewing that condensed checklist will give you a good idea of just how involved planning, organizing and controlling a meeting can be.

Planning the meeting

Thorough planning is essential if you are to meet your goal of a successful, productive meeting. Proper planning is second in importance only to the ability of the meeting manager to steer the meeting.

Meeting plans should be in writing. The amount of writing will be directly proportional to the size and complexity of the meeting. Even small meetings should rate a written outline.

The first step in the meeting plan is to set the goals you wish to accomplish. What problems are to be solved? What information is to be exchanged? In how much detail? Setting the goals should go a long way toward getting the desired results and having a productive meeting.

The backbone of any meeting plan is the agenda. As prime mover for most project meetings, the project manager must insist that there is an agenda for every project meeting. Failure to have an agenda guarantees a meeting will be unproductive even before it starts.

Assign a priority and a time allotment to each agenda item to tailor the agenda into the meeting plan. The agenda should follow a logical flow pattern. For example, in a problem-solving meeting, state the problem. Next, give the needed background material before starting on the possible solutions. The agenda logic for project review meetings can be more flexible in presenting the subject matter. One universal rule for agendas is to place less-critical items at the end in case time runs out before completing the agenda.

State each agenda item as briefly as possible within the limits of clarity. Each participant in the agenda should have a clear understanding of the matters for discussion under that topic. That is especially applicable to the person who presents the subject in the meeting. Actually, the physical size of a properly drawn agenda has no bearing on the time needed for discussion. A well-prepared agenda serves to shorten the meeting time by giving the chair an effective tool to control the discussion.

After properly preparing and approving the agenda, issue it to the meeting attendees before the meeting. Allow attendees enough time to plan and prepare their contributions to the meeting. Be sure to include any written backup material for attendees to read before the meeting. Passing out reading material during the meeting destroys the tempo of the meeting and wastes valuable time.

With the agenda in hand, you can now complete the list of attendees. Limit attendance to those who are involved in or contribute to the meeting. Attendance is limited because each attendee adds communication channels to the group and increases the meeting control problems. For example, two people have two channels, four have twelve, eight people have fifty-six, and so on. Controlling these additional communications channels adds to the difficulty of controlling the meeting effectively.

The meeting plan must also include any visual aids needed to communicate the message you want to deliver in the meeting. They can be flip charts, overhead transparencies, slides, or handouts. Use a medium that meets the specific needs for presenting the subject matter to the group. Don't go overboard on visual aids just for showmanship. A chalkboard for simple messages can be as effective as an overhead transparency at less cost.

The meeting starting time and duration are important elements in the plan. Calculate the total elapsed time for the meeting from the time values assigned to the agenda items. Add some contingency based on the type of meeting and prior experience while keeping in mind the productivity goals. Using the elapsed time, back off from a natural cutoff event like lunch or quitting time to start the meeting. A natural time barrier makes it easier for the meeting manager to keep the discussion closer to the meeting goals and agenda. Attendees who are inclined to wander from the meeting subject seem to be especially sensitive to natural cutoff times.

The window of time for the meeting must also be accessible by the major participants if you are to have a quorum. A brief telephone poll of key participants by the project secretary can ensure schedule compatibility. Having a couple of options for time and date may be necessary to suit the most critical attendees. Of course, qualified alternate attendees are also usually acceptable.

Certain situations may require a dry run for fine tuning the agenda and presentation. That situation occurs regularly in project review meetings and major management presentations. Include the factor in the meeting plan to allow time for final polishing of the agenda before issuing it.

Organizing the meeting

Organizing the meeting involves making the necessary physical and administrative arrangements. These are the key items in that area:

- Arrange a suitable meeting site.
- Prepare visual aids, handouts, and so on as required.
- Issue agenda and venue information.
- Assign a recorder and timekeeper as required.
- Arrange for refreshments.*
- Make seating arrangements, if necessary.
- Provide note pads and pencils.*
- Hold dry run, if required.*

The starred items are for more formal management-oriented meetings only.

Select the meeting site based on size, creature comforts, visual aid equipment, proper lighting, and so on. The physical requirements are important to maintain the attendees' undivided attention throughout the meeting. Sometimes it is even necessary to get the people away from outside interruptions to have a successful meeting.

Visual aids are important communication tools for information transfer meetings. Make sure that the visual aids are clear and legible to all participants, or they will do more harm than good. If the visual aids are complex, bring in a graphics consultant to prepare them and operate the equipment. B. Y. Auger's book *How to Run Better Business Meetings*[1] is an excellent reference for preparing visual aids.

A working meeting requires a recorder to take and publish the minutes. Issuing meeting minutes is essential to confirm the results to the participants. Publishing the minutes also informs those nonattending project team members of the meeting results.

The recorder can also be the timekeeper to advise the meeting leader on the timing of the agenda items. The meeting leader should not handle the recording and timekeeping duties because they detract from the main duty of leading the meeting. The recorder should be a good communicator and technically qualified in the subject matter. That usually precludes using the project secretary in technically oriented meetings.

It is pointless to record everything said during the discussions leading up to the decision. That would make the meeting minutes too wordy and lead people to ignore them. The meeting leader should sum up the decisions or action required as briefly as possible for the recorder.

It is vital to assign the necessary follow-up action on matters decided in the meeting as part of the recording process. That can be done by providing an action column at the right-hand margin of the minutes of meeting format. That makes it easy for the meeting leader to follow up on the action items after the meeting.

It may be desirable to serve refreshments for larger management-oriented meetings lasting longer than 2 hours. The serving should be at the scheduled break in the agenda or at the start of an early morning meeting. I don't feel that serving refreshments adds anything to working project meetings unless the meetings are unusually long.

Prearranged seating assignments are usually used only in major management-oriented meetings. If necessary, group people from the same company or department together for easy communication. It's a good idea to arrange management people in the normal pecking order to prevent any hurt feelings. That factor is especially important when dealing with international groups having strict cultural habits in that regard. Place upper management attendees near the meeting leader.

Sometimes, the meeting may require only part-time attendance by certain contributors. Schedule the part-timers in and out as required to save their time and keep the meeting within reasonable limits. That gives the meeting leader another incentive to keep the meeting on schedule.

The organization part of meeting preparation is one area the meeting leader can effectively delegate to others. A good project secretary or administrative assistant can arrange much of that detail based on the meeting plan.

Controlling the meeting

The meeting leader's ability to control the meeting is the most important requirement for the meeting's success. Even a well-planned meeting will fall apart if it is not properly controlled from start to finish.

The meeting leader must always keep the meeting goals in mind while controlling the meeting. Writing the meeting goals makes them clearly understood and available for reference during the meeting.

Loss of meeting control is a prime cause for failed meetings. Although steering the meeting requires a firm hand, it must also be done with tact and diplomacy to permit valuable input from all attendees. Autocratic control often causes possible contributors to withdraw mentally from the meeting.

As the first act of meeting control, the leader must start the meeting on schedule. That is a good habit to set up since it encourages perennial latecomers to improve their schedules. Starting late also places the meeting goals in jeopardy even before the meeting starts.

After a few opening remarks and any necessary introductions, the leader should get right to the agenda and follow it closely. Check a few of the milestone targets as the meeting proceeds to ensure that you are maintaining schedule. Some agenda items will finish early and some late so the schedule should even out. Make agenda changes only if there is a good chance of improving the schedule.

Occasionally, some part of the meeting plan may break down and force revisions to the agenda. Handle the changes very carefully to protect the overall meeting goals. Quick thinking here can prevent scheduling a follow-up meeting.

Meeting leaders spend most of their steering efforts controlling the "problem people" in the meeting. Those people fall into these four general groups:

- Those who talk too much
- Those who talk too little
- Those who hold side conversations
- Those who want to usurp the leadership of the meeting

Fortunately, most people attending capital project meetings are technical types who are not garrulous by nature. When a participant becomes too wordy, the meeting leader must restate the sense of urgency required to keep the meeting on track. If that does not work, the chair must find another means to cut off the harangue. You may have to double-team the offender with the cooperation of the recorder or other participants.

Drawing out those who say too little can be even more difficult than controlling those who are wordy. Quite often the silent types make a strong contribution to a successful meeting. Encourage active participation by the reticent by positive recognition when they do open up. Build active contact by drawing out their suggestions and opinions.

Encouraging the clash of differing opinions also opens the way for those who are careful about expressing an opinion.

Control of side conversations is a must to avoid having two meetings at the same time. Side conversations are upsetting to those who are trying to make a contribution and to the meeting manager. If there are any worthwhile ideas in the side conversations, bring them into the full meeting.

Squelching those who would usurp your meeting is the only way to prevent anarchy. Naturally, you should use extra tact if the usurper happens to be your boss. Occasionally, visiting executives do take over meetings, but they will normally back off when gently reminded of their actions.

Meeting control involves many human relations factors in a live performance. A good way to build your skills in that area is to study the human relations involved and practice various ideas in real situations. Use those that work and discard those that don't. Rate your performance after every meeting to see where you can improve. Antony Jay's article from the *Harvard Business Review*[2] presents some new ideas for that type of information.

Another useful meeting control device is the tension breaker. Quite often tensions will develop during meetings when conflicts are being resolved or sticky points are being negotiated. Perhaps the most effective tension breaker is humor. It can relax tensions, but use it carefully. It can backfire if it embarrasses someone.

A prompt adjournment should occur when you clear the agenda or the allowed meeting time has expired. If some minor agenda items remain, try to settle them without convening another meeting. If they are important enough, however, schedule a later meeting.

Postmeeting activities

Issue the meeting minutes promptly to remind people to get started on their "action items." Don't give them an excuse to forget about their assignments. Make every effort to issue the minutes no later than 24 hours after the meeting.

Some international meetings require signing a *protocol* before the participants leave the meeting site. That puts extreme pressure on preparing and issuing the minutes to avoid tying up the meeting participants. In that case the documentation of the minutes must begin before the meeting is over.

In addition to listing the action items, the minutes serve as a record of the actions taken during the meeting. They are major contributors to project documentation. Place a copy of them in the project file as a part of the project record.

Shortly after the meeting, the leader should review the meeting results with key staff members to measure overall performance. Some searching questions to ask *and answer* are the following:

- Did we meet our goals?
- Were we properly prepared?
- Did we control the meeting?
- Did we finish the agenda?
- What were the good points?
- What were the bad points?
- Where can we improve next time?

If you answered the first four questions no, you need to work on those areas to improve your meeting performance. Also, remember the main purpose of the meeting. Follow up on the action items to be sure that the meeting time was spent effectively.

Approaching meetings with a proactive stance as I have put forward here should improve your meeting performance. Naturally, you will have to shape these ideas into your own operating environment. Effective project meetings are too important a communications tool to waste, so the quicker you improve performance in that area the better!

Be a good listener

Most of our discussion to this point has involved outward messages, which comprise only half of the communication equation. Receiving the messages is equally important, but often it does not get the same emphasis.

Much project management communication deals with conflict resolution. Resolving conflicts effectively requires PMs to listen to both sides of the conflict. Since project morale is a key factor in successful project execution, PMs often play the project chaplain role. The most difficult disputes and decisions are the ones that wind up in the PM's office for final settlement.

I don't mean to imply that PMs have unlimited time to spend on these matters. They must gain good listening habits that allow them to separate the wheat from the chaff to give timely and sound decisions.

To be a good listener requires being patient. Many PMs lack patience because of the continual sense of urgency in project management. Therefore, both listening ability and patience are traits most project managers need to develop.

I know of no training given in the area of patience, so patience must be self-taught. However, there are several good courses for improving listening ability. I think the speed of improved listening comprehension from such training courses will pleasantly surprise you. A refresher course every few years is also effective in keeping that skill sharp.

Summary

Effective communication in every form is a potent weapon in the project manager's arsenal of skills. That vital ingredient is essential for successful project execution. Human relations form the basis for sound communications. The project manager must be an effective speaker, writer, presenter, meeting manager, and listener. A PM must train and practice to become proficient in all those areas.

References

1. B. Y. Auger, *How to Run Better Business Meetings,* Minnesota Mining and Manufacturing Company, Minneapolis, Minn., 1979.
2. Antony Jay, "How to Run a Meeting," *Harvard Business Review,* March–April 1976.

Case Study Instructions

1. As project manager of your selected project, prepare the communications section of your Project Procedure Manual. Include all items such as correspondence (and logs), project reports, meetings, minutes of meetings, telephone conversations, etc.
2. Write a project progress report for your selected project at the 50 percent design completion point. Construction is 20 percent complete. Assume that design productivity is below standard and you are 3 weeks behind schedule. The delay in design is also starting to affect field progress.
3. The chairman of the board and three key directors are touring the company's major capital projects. They are due to review your project in 2 weeks. The plan is to review the design, which is 75 percent complete, and then visit the site, where the work is 35 percent complete. Your management has placed you in charge of presenting the home office part of the program as well as planning and executing the field portion of the trip. Funding for the next project similar to yours depends on how well your project is going. Present your detailed plan for managing this major presentation.
4. The owner (or client) has requested an in-depth project review meeting before taking lump sum construction bids. The design includes a model and design drawings, along with full construction specifications and an engineer's estimate. The owner's team is

available for one day from 9:00 AM to 5:00 PM. Prepare your detailed meeting plan and agenda for the meeting. Include all physical arrangements, visual aids, and so on that you plan to use.

5. As construction manager, give a detailed plan of how you propose to run the weekly field planning meeting on your selected project. The work is about 60 percent direct hire and 40 percent subcontracted. The home office has provided a PC-generated CPM schedule.

Chapter

11

Human Factors in Project Management

If there were no people involved, project management would be duck soup! Whenever I talk to project managers about problem areas, people problems always head the list. Perhaps it is because we spend so much time working on methods and procedures and so little time on human factors.

In-depth study of human factors in the business world has taken place only in the past two decades. An example of that situation shows up in Fredrick Herzberg's work cited in earlier chapters.[1] Since the middle 1960s, considerable research and data on the humanities have been published in management journals.

The general management community has eagerly accepted the work on human factors. Also, literature on human factors in project management has increased somewhat, but it does not seem to have affected actual practice. If it had, project managers would not continue rating human factors as their number 1 problem area.

I believe proper handling of the human factors is the critical third leg of the stool leading to successful projects. It rates right up there with my Golden Rule of Project Management and project communications as a requirement for successful projects. Even though we plan, organize, and control our projects perfectly, the projects can fail if we mishandle the human factors. The human element in project management involves human relations, personality traits, leadership and career development. Strength in those areas can help us do a better job of managing projects, as well as improve the chances of meeting our personal goals. Years ago, good project managers with a natural flair for human relations consistently had better projects than those without it. With the research and literature in the humanities available now, we no longer have to depend on chance. All we need is the right attitude, study, training, and practice.

Obviously, I am not a professional in human relations. I base my ideas in this chapter on practical observations made in managing capital projects for many years. This chapter cannot possibly cover all of the human factors involved in project work since the field is so broad. The areas I have selected for discussion are the key ones for improving your project performance.

Qualities for a Successful Project Manager

Let's start by taking inventory of the most important personal traits necessary to manage projects successfully. You can then study the list to see how to shape your own qualifications to meet the requirements. A periodic review of your performance in these key areas can be a valuable tool for keeping the necessary skills up to date. My list of key qualities for managing projects is as follows:

Effective manager/administrator	Resourceful
Strong leadership capability	Problem solver
High standards of ethics and integrity	Responsive
People-oriented	Multidiscipline capability
Decisive	Knowledgeable in the business
Personal drive	Creative and imaginative
Physical stamina and mental toughness	Patient
Strong communicator	Common sense

That is an impressive list of characteristics. It further confirms how broadly based project management work really is. Most of the traits involve human factors in one form or another. We have discussed many of the traits in earlier sections of the book. This chapter concentrates on the traits not previously covered.

Personality

In looking over the list, it becomes readily apparent that a person matching the requirements must have a dominant personality. Clearly, a person satisfying the qualifications must be a self-starter with enough drive to stay the course.

We also know that capital projects are not successfully completed by brute force alone. Because of the people-oriented nature of project work, a lot of persuasion is also necessary. Combining the two personality traits results in the dominant-persuasive personality which is ideally suited to project management.

The largest number of project management professionals come from the technical side of the business. Understanding the technical nature of the work is a prerequisite for managing capital projects. Technology seems to appeal to people with introverted personalities because complete concentration and hard work are needed to meet the degree requirements. The intensity of the technical course work virtually excludes study in the humanities, which could broaden the technical person's personality.

Given that type of background, most of us are still on the introverted side by the time we reach the project management level. If your personality tests on the introverted side, it is time to do something about it.

Much psychological research shows that we are products of our forebears and environment. That research also shows that our personalities are formed by the time we reach age 7. I find it difficult to accept those theories at face value because of my personal situation. My memories at age 7 are no longer very vivid. However, I do remember being very shy, introverted, and lacking in leadership capabilities all the way through high school. Without the benefit of formal humanities training, I expanded my personality, became more outgoing and dominant, and had a successful career in project management. If I could do it, anyone can do it! All it takes is desire, training, and practice, practice, practice.

The training for curing introversion is not so difficult. I recommend one of the personality-expanding courses such as offered by the Dale Carnegie group. That course has proved itself over the past 50 years with millions of graduates. Following the course with some work in the Toastmasters' Club should start you on the road to generating an enthusiastic, more outgoing personality. These basics will also open the way for further study and practice in the human element of project management.

There are a few people on the extroverted end of the personality spectrum who make it to the project management level. Extroversion is a bigger problem, because I know of no training for lowering high levels of it except the school of life. If you have a problem with extroversion, work on it gradually without killing off your enthusiasm in the process.

Ethics and integrity

As in any profession, high standards of ethics and integrity are important in project management. Project managers often control large sums of money that don't belong to them. Usually, accountants thoroughly audit the funds, but there is still a lot of room for possible conflict of interest. A conflict of interest can arise on either side of the client-contractor equation. Both parties must conduct project activities on a businesslike basis and show no favoritism to vendors, suppliers, or subcontractors. That includes accepting favors, kick-backs, and gifts. Today most companies have a statement of policy covering all employees who handle company business.

Be certain that you stay within those guidelines in all your dealings. You cannot build a professional reputation in this business if you don't adhere to the rules. There are a lot of insidious ways to circumvent the rules, so be alert for any traps. Some firms vying for the project business may not be as scrupulous as you are about bending the conflict-of-interest rules. As leader, maintaining the ethical standards of the project team rests on you. Set a good example yourself, and make it known to the others where you stand on ethics and personal integrity.

Controlling that situation is fairly easy in the United States, but it can be difficult in countries with cultures that condone (and even promote) conflict of interest. I seldom recommend going against management policy, but in that situation I do. If your management tries to place you in a conflict of interest situation, don't accept it. Trading your professional reputation for a job is not a good bargain. If you find yourself working for that sort of company, you are better off without it!

Personal drive

Project management is a demanding profession that often generates long working hours and extensive travel. Conflict resolution and problem solving add stress to the job. All of that is taking place within an environment of change further compounding the pressure. Many people won't consider these working conditions as attributes to accepting the job. Only those people who look on that working environment as a challenge and have a strong personal drive to excel in it, need apply.

On the plus side, a certain amount of authority, power, and respect come with the job. The pay is usually attractive, and the personal satisfaction of a job well done is invigorating. Effective project managers are always in short supply so job security is good. (A notable exception to that statement occurred during the 1982 recession.)

The mental side of the job demands a sound physical side to support it. Project managers should keep in shape to handle their demanding schedules. That means having a regular exercise program and good dietary habits throughout your career. The conditioning program and diet should be sensible and matched to your individual needs. Jogging, for example, is not an absolute necessity.

Relaxation is an equally important part of a project manager's personal schedule. No one can live in a pressure cooker forever. Relaxing when time permits is necessary, but don't play as hard as you work. Some people have tried that unsuccessfully. Performing with a hangover the next morning does not improve project management!

Multidiscipline capability

Project people with experience across several disciplines are best suited for the job. Most of us who became project managers moved up from the technical ranks by designing and building capital projects. If you plan to get into project management, seek assignments that broaden your technical scope. That includes such areas as cost engineering, scheduling, procurement, and construction.

The most difficult cross-discipline transition is the one from nonprocess to process projects. The key here is the chemical engineering expertise required in the process environment. If you are making such a transition, a course in chemical engineering for nonchemical engineers would be helpful. Obviously, it will not make you a chemical engineer, but it will give you the necessary basics. The same advice holds for training in any other discipline in which overcoming a weakness will help your career.

The major weakness of most project managers is a lack of skills in the business side of running projects. Earlier, I mentioned the missing humanity subjects in most technical curricula. Also missing are business courses. The educational institutions leave it to us to get the business training necessary for technical management. We can do that in several ways. The first is through on-the-job training while working as a project engineer under a project manager/mentor for a few years. That approach can be effective, but it is time-consuming and costly because of the mistakes made during the learning process.

The second is by taking an MBA degree. Preferably that is while working as an engineer rather than right after you earn your bachelor degree. That way you will get a better grasp of what the MBA work is about because you can base it on your work experience. The MBA is an added advantage in strengthening your résumé in addition to gaining knowledge.

If you are past the age for standing up to the MBA course work, there is still hope. General management training is also available through courses such as the Alexander Hamilton Institute's *Modern Business Program*. That is a correspondence course requiring 2 years of study on either an individual or group training program. I used the course successfully as an individual and also the group training course for technical managers. In addition, I recommend a continuing program of training seminars and professional meetings to broaden and maintain your skills. Frequent self-analysis of your personal and business performance is valuable in evaluating your skills inventory. Be honest with yourself and cover both on and off the job activities.

Human Relations

The area of human relations in project management is broad because of the many interpersonal contacts involved in a largely conflict-oriented environment. In addition to the external human relations, there are the personal ones concerning the project manager's career. Of course, the two are closely related. The rest of this chapter covers those crucial human factors and picks up the traits not covered in the preceding list. I have found the following areas to be the most critical to effective project management.

Client-contractor relations	Leadership
Contract administration	Common sense
Project relations	Keeping your cool
Public relations	Negotiating ability
Labor relations	Patience

Client-contractor relations

When contractors' project managers are polled about problem areas, client relations seem to come out on top every time. Ten years of leading project management seminars has brought me in contact with about 500 client people who feel the same way about contractors. Obviously, there is something wrong when these key project groups have trouble working as a team.

In this discussion, I refer to the client-contractor relationship in its broadest sense. That includes project managers who serve clients within their own companies as well as outside contractors and subcontractors. Some examples are central engineering departments and plant engineering groups that serve operating departments and divisions of their own companies. In each case it is difficult to get around the them-and-us syndrome.

Contractor-client relations involve two very important aspects in the business of meeting project goals. They are too valuable an asset to be sacrificed in a breakdown of human relations. Contractor-client relations can make or break:

1. The success of the current project or, at the very least, meeting the project goals.
2. A contractor's opportunities for future work with the client.

These are two powerful incentives for making the client-contractor relationship work throughout the project. The project managers can give up meeting their personal goals, but they don't have the authority to give away their company's goals. They accepted those goals when they took on the project assignment.

It is impossible to write a scenario in which everyone has successful relations with a client or contractor. After all, a lot of personal chemistry is involved. Much can happen over the complex course of an average capital project. However, there are some things I can tell you that will improve the odds considerably. The primary necessity in any client-contractor scenario is having both parties realize that the relationship is essential to project success. Neither project manager can afford to have the project branded a failure, regardless of the personalities involved. Poor contractor-client relations can completely wipe out an otherwise outstanding performance. That makes an effective working relationship the number 1 priority throughout the project, both for your company and for your personal reputation in the business.

No two people will handle a given situation in the same way. You must handle human relations in a natural, unforced way to make any solution work. You will have to try various techniques with each contractor (or client) to see which one works for the given situation. Some key areas in the client-contractor relationship are discussed in the following sections.

Ethical conduct

Any failure in the key area of ethical conduct by either party will seriously undermine the relationship from the start. Neither party can condone unethical practices in the other. Fortunately, unethical conduct does not occur up very often. When it does, squelch it immediately. PMs have a fiduciary responsibility to their respective managements to spend the project budget wisely. That does not leave any room for showing favoritism or accepting kickbacks from vendors, suppliers, or subcontractors. Most companies have a code of conduct for handling company business which covers these points. Be sure that everyone on your project has a copy of and abides by the code of conduct. If your company does not have a code of ethics, write one for the project to define your ground rules.

Responsiveness

Contractors must be responsive to the client's needs first. That is a typical buyer/seller arrangement that places the customer in a preferred position. It does not, however, give the customer an automatic right to make unreasonable demands on the contractor or supplier.

Responsiveness is also a two-way street. Clients must be equally responsive to the contractor's needs for proper execution of the work. It is vital for clients to make their inputs and approvals in a professional and timely manner as defined in the contract. A strong effort by both project managers on this point will result in a highly successful working relationship.

Mental toughness

Weakness is seldom respected in any culture, not even by bullies. To keep control of their projects, project managers must resist domination by their counterparts. The buyer/seller relationship is not grounds for an uneven playing field. The contract should define the client-contractor relationship if there is any doubt in your mind. I recommend that you gently, but firmly, cut off any one-upsmanship activity whenever it occurs on your project. People who are prone to that type of activity will usually get back into line when pressed on the matter.

Contract administration

Contract administration is an area which sometimes causes friction in the client-contractor relationship. Even the best-crafted contract requires some interpretation from time to time. It is important that both parties take a proactive stance in meeting contractual requirements. That avoids friction over the need for one party to constantly remind the other of the contract requirements.

Also, there are areas of give and take in any contract which can make a contractual relationship run smoother. However, neither party can give away the store. Likewise, the other party should not ask for it. This area requires a lot of judgment for proper handling, so approach it with caution. Take a few smaller calculated risks until you get the feel of it.

Relations of the project staffs

The project managers must also monitor the client-contractor relations throughout their project staffs. They must resolve any interstaff relations not resolved at the local level. The number of personality problems seems to be directly proportional to the size of the staff. The best way to reduce problems in this area is for the project managers to set good examples as role models for the rest of the staff. Also, be sure that all members of the project staff know your policies on how you expect them to handle their contractor (or client) relations.

Those are a few areas that I feel are critical to ensuring good contractor-client relations. If you learn how to handle them well, any minor areas should fall into place as well. Remember, try different approaches to suit different situations until you find the combination that works for you. Also, nobody bats a thousand in the area of contractor-client relations. There will probably be a time where nothing seems to work, so making a change is inevitable. If it happens more than once or twice, there is cause for concern. You must review your performance and correct the basic cause of the problem.

Project relations

The next most important human relations problem area is the project team's working relationships with the rest of the company. Eventually, you will have business contacts with everyone in the company from general management down to the mail room.

The most frequent contacts are those with the department heads and managers of the design, procurement, construction, and project control groups. They are the people who will be supplying the members of your project team and having a technical input to the work. To a lesser degree, you will have contact with business development, corporate legal, accounting, and operating management. Certainly, the first four items we covered under client-contractor relations also apply to project relations. Those areas were ethical conduct, responsiveness, mental toughness, and holding others to their commitments. The one ingredient missing from the earlier discussion is the buyer-seller relationship, since we are now all working for the same organization.

That makes it more important to establish yourself as a knowledgeable manager with a mature outlook and respect for other people. It is essential to command respect and cooperation from the people who must perform if your project is to succeed. That reputation is not won by putting on your Superman Suit every morning before leaving for work. We build our respected reputations gradually, brick by brick and stone by stone, much as we build our projects. Throwing the rocks and bricks at the other guys, even when they seem to be asking for it, will not build a solid reputation.

My philosophy is that it takes everybody pulling for you, in addition to a good performance on your part, to have a successful project. Having a few people waiting around to pull the rug out from under you increases the chances for failure.

You should also remember that project relations work in all directions—upward, downward, and sideways. Most managers are attentive to their upward relations as a matter of personal survival. That is from where both the rewards and retribution stem. We certainly must maintain those relations if we are to meet our project and personal goals. Relations downward are sometimes another matter. Here, we hold power and authority over the project staff. How we dispense that power and authority is a major factor in maintaining project morale. Good project morale is the intangible factor that has a strong bearing on a successful performance. You are not likely to meet the project goals of quality, schedule, and budget without it.

Public relations

The project manager usually acts as the public relations (PR) representative for the project. PMs should mesh their project PR responsibilities into the overall corporate policies set by top management. It is natural to want your project presented in the most favorable light. A proactive approach to project PR is the best way to accomplish that. Bad PR can get in the way of meeting the project goals. PMs should be alert to spot newsworthy events on their projects which could interest the business or the public. That is one of those subjects not covered in our technical educations, so developing a feel for recognizing newsworthy items is another self-taught skill.

It is even more important to take a proactive stance on items that are likely to be detrimental to your project or corporate image. Try to look ahead for any negative PR concerning the project that might be developing in the community. Make sure that any adverse publicity gets a fair counter presentation the first time around. If an adverse or erroneous item gets into the media, it is virtually impossible to get a retraction later.

Establish the initial contacts with local government and community officials before opening the site. Tell them about your plans to handle such problem areas as traffic, dust, noise, and fumes which could adversely impact the area. To keep the public on your side, stress the positive effects of the project on the community. The project manager often acts as the project's technical representative in speaking to civic and government groups about major environmental matters. The presentations require careful preparation, including clearance from top management. Be especially careful in handling the news media. Handing out a well-written and well-checked news release is safer than an impromptu presentation.

Labor relations

Labor relations are a prime responsibility of the construction department or the contractor. However, they impinge heavily on the success of the project, so project managers should understand basic labor relations policy. They need to know enough to make knowledgeable choices when making project level labor decisions.

The main goal of any labor policy is to maintain high productivity and reduce work stoppages which hamper job progress. Local labor practice and usage play a large role in labor relations and make it difficult to develop specific guidelines covering all situations. The relatively recent arrival of the open-shop construction concept has further complicated labor policy. The open-shop concept is most prevalent in the right-to-work states of the Sunbelt. The introduction of open-shop operations has created some competition for the construction labor supply once monopolized by the labor unions. That has forced the unions to be more productive and cooperative on working rules and labor rates.

The basis for labor relations is the union contract or the open-shop agreement which sets the working rules for the field labor. Sometimes a supplemental agreement, called a site agreement, is used to define the special conditions in effect for the duration of that project. The purpose of the contract or site agreement is to lay down a detailed set of rules governing the use of construction labor on the project. The agreement sets out working hours, pay scales, fringe benefits, grievance procedure, and so on, along with various management and labor policies.

It is the nonroutine matters which cause the most problems on construction projects. Chief among them are jurisdictional disputes. Such a dispute is the result of two or more craft unions claiming the right to perform a certain type of work. That does not occur in open-shop work, where construction management decides which craft will do the work based on the skills required.

Jurisdictional disputes are reduced by holding a prejob conference. The various craft business agents meet with construction management to resolve their differences before the work starts. They discuss each major work activity in detail to agree on who will handle it. The agreements reached are in writing and are signed by all participants. If you cannot head off disputes before construction starts, you must eliminate their ill effects. Make an all-out effort to keep the rest of the work going while arguing any fine points of interpretation. Take all possible legal steps within the agreement to avoid setting up picket lines.

The contract usually includes strikes in the force majeur clause excusing any delays they may cause. As the owner's project manager, it's in your best interest to get the work restarted as quickly as possible. Therefore, you should offer any help your organization can give to the contractor in settling the dispute. However, owners should never approach labor representatives directly; they should always go through the contractor.

Leadership

Leadership is an area which touches on most of the other human factor areas in project management. It is a crucial requirement for effective total project management. Building leadership skills will greatly enhance your chances of success in that field.

The following list of eight areas of importance in management illustrates my thoughts about the nature of leadership in project management. It is a condensation of the material given by Henry Mintzberg in his book, *The Nature of Managerial Work.*[2]

1. Peer skills—the ability to set up and maintain a network of contacts with equals.
2. Leadership skills—the ability to deal with subordinates and the kinds of complications created by power, authority, and dependence.
3. Conflict-resolution skills—the ability to mediate conflict and handle disturbances under psychological stress.
4. Information-processing skills—the ability to build networks, extract and validate information, and transmit it effectively.
5. Skills in unstructured decision making—the ability to find problems and solutions when alternatives, information, and objectives are ambiguous.
6. Resource allocation skills—the ability to decide among alternative uses of time and other scarce resources.
7. Entrepreneurial skills—the ability to take sensible risks and implement innovation.
8. Skills of introspection—the ability to understand the position of a leader and the leader's impact on the organization.

Each one of those areas has a direct bearing on the practice of project management. Now let me tie them into our discussion.

Peer skills

We discussed this point earlier in connection with setting up effective working networks with department heads and other operating groups.

The contacts give you the necessary outside support to execute the project.

Leadership skills

Management delegates a lot of authority to project managers to execute the project. Knowing how to use that authority to build and motivate an effective project team is essential to successful project management. That is the skill most involved with human relations!

Conflict-resolution skills

Project management abounds with stressful conflict-resolution situations. Project managers must learn how to cope with those emotionally charged situations quickly, calmly, and fairly without damaging working relationships and project morale. Try to present decisions as win-win solutions to the parties involved. That will allow participants in the decision to keep their self-esteem and maintain enthusiasm.

Information-processing skills

Virtually all project communication passes over the PM's desk. In addition to assimilating that information, PMs must get out into the trenches to find out what is going on. That requires every communication skill: speaking, listening, reading, writing, and presenting information. This area is crucial to maintaining the good client-contractor and project relations mentioned earlier.

Skills in unstructured decision making

Positive leadership leaves no room for shilly-shallying. Decisiveness is one of the character traits listed earlier for successful project managers. "Shooting from the hip" on every decision is often hazardous, so I don't recommend that either. Take the time available for making the decision, but don't drag it out unnecessarily.

Resource allocation skills

Project managers oversee the disposition of all the project resources. That includes time, money, people, material, equipment, and systems. Each of them makes a contribution toward meeting the project goals, so allocate it wisely. That skill interacts closely the decision making just discussed.

Entrepreneurial skills

This is my favorite! The project is your business (profit center) to run. Some projects require a Mom-and-Pop approach; others are big business. In either case, you should take sound calculated risks when the payout looks good. By internalizing your project numbers, you should know what to expect when the project reports arrive. Entrepreneurs must also be creative, so use some imagination in running your business.

Skills of introspection

In addition to understanding your position as a leader, skills of introspection mean periodic self-analysis of your project performance. Is everything possible being done to reach the project goals? Does your performance measure up to the personal standards you have set? Are your leadership skills getting effective results?

Applying introspection to your job environment is one of the best teaching tools available to you. Management schools and training courses can only point you in the right direction. Introspective practice of management theory is the quickest way to learn what really works for you!

As you reviewed the list of eight areas of importance to management, I am sure you thought of several specific examples of recent project situations applicable to each one. Run through those examples in your own mind and rate yourself on how you actually handled them. If the answers are not good, try to mold your leadership skills to improve the outcome next time. Making a frank appraisal of your performance against the checklist a couple of times a year can improve your leadership skills immensely. Improvement in that area will also help you mold the dominant-persuasive personality necessary for total project management.

Common sense

Common sense is one of those intangible attributes that some of us were lucky enough to be born with. Others, not so lucky, must acquire it. The dictionary defines common sense as "sound practical judgment that is independent of specialized knowledge, training or the like; normal native intelligence." To me, the operative words here are "sound practical judgment." As a start, I recommend using sound practical judgment in areas where you do have specialized knowledge and training, such as project management.

Areas where you don't have specialized training and knowledge also call for common sense. You need common sense when people seek to promote the use of impractical ideas on your projects.

Suppose, for example, a department head is trying to impose using an unproven, overdetailed, costly scheduling technique on your small project. That is when common sense should tell you to ask some pertinent questions. Do we really need it on that type of job? Will it really work as well as you say? Can we afford to experiment with it on that small job? Can we stand the extra cost?

The project management system is really nothing but the application of common sense. The application of sound practical judgment in practicing total project management is what I have been talking about throughout this book. One learns common sense by observation, practice, and experience.

Keeping your cool

Keeping your cool is another trait we can do something about. Project managers cannot survive by pushing the panic button. They must learn to deal with panic situations calmly to avoid becoming nervous wrecks.

It is better to reserve your energies for clear thinking on how to solve the problem. If the project leader is running around in a state of panic, the panic will spread to the rest of the project team. No serious problem was ever solved by creating a state of panic.

Panic is a symptom of extreme worry. If you have a tendency to worry, you are also likely to panic in difficult situations. One way I have used to overcome extreme worry is to analyze the problem to see what is the worst that could happen. Many times it turns out to be less catastrophic than first imagined—you can live with it. Starting from that premise, coolly explore ways to improve on the worst-case scenario. Anything salvaged over the worst case is an improvement. In many cases, you can turn the panic situation around canceling out most of the adverse affects.

Furthermore, having successfully handled the panic coolly makes you look more professional in the eyes of your management and peers. That builds the desired reputation as a mature, knowledgeable, and respected practitioner of total project management.

Negotiating ability

Project managers routinely find themselves in negotiating situations. Such situations arise with peers in almost every aspect of executing the project. They include areas like staffing, estimating, and scheduling in addition to such normal areas as contract negotiation, purchasing, and change orders.

Negotiation is one of the arts of project management that you can learn through training and practice. There are several good courses offered in the management training marketplace.

Remember, negotiation is a special form of meeting, so a detailed meeting plan and agenda are critical to success. Before going into the negotiation, plan your basic strategy. That includes studying your short- and long-term goals, a profit strategy, and selection of a negotiating team. Next, set your information-gathering and processing systems and determine who makes the decisions. You should then be ready to study and implement your tactical approach and finally arrive at a satisfactory close.

It is also important to control the negotiating meeting to your advantage, even if you are the seller or plaintiff. It may surprise you to find how often that can be done when your preparation is better than the opposition's. Review the section of meetings in Chapter 10.

Patience

Patience is a virtue—most of the time. However, don't confuse patience with allowing an existing problem area to continue hoping that it will go away.

Patience is something most of us develop naturally as we mature; we base it on common sense and practical judgment. That is why we often see effective project teams made up of mixtures of seasoned hands and young Turks to get patience and push.

Patience is akin to controlling one's temper in difficult situations. When you lose your temper, you lose the outcome of the situation. Patience is also a must in the client-contractor relationship discussed earlier. That is especially true in international projects when cultural differences are involved.

If you must have one weakness in your project management makeup, lack of patience is the most acceptable. I have found it to be a useful tool when undergoing annual appraisals. According to good management practice, the reviewer must ask what you think your weaknesses are. After a few moments of thought, admit to being a little impatient. The reviewer will immediately transpose your admitted weakness into a strength and go on to the next subject. Thus, you have neatly dodged the mine field of admitting your weaknesses. You are bound to increase your score by a couple of points!

Personal Human Factors

To wrap up human factors in project management, there are a few items that affect the project manager personally. The following tips can improve your performance and promote your career.

- Selling the organization
- Bending the rules

- Knowledge of the business
- Personal public relations

Selling the organization

Selling the organization is an important area that can relate to contractor-client or project relations. Never poor-mouth your company or any part of it in front of clients, contractors, or management. To do so is a common failing, especially after you have developed a close personal relationship with the other person. That does not mean you should try to convince anyone that a shoddy performance is great. When you run into a poor performance in the organization, get it corrected and make sure the right people hear about the corrections. Don't reinforce the ill effects by saying, "I always knew that was a lousy department."

On the positive side, make sure people hear about a good performance in an unpretentious way. When that is done properly, one builds the image of the project team and improves morale simultaneously. Proper handling in that area can make an average performance look good and a good one look excellent. It is more difficult to get a poor performance up to average, but it has been done.

Bending the rules

Bending the rules has to do with the art of project management. I have spent most of the space in this volume laying down strict rules for you to follow in practicing total project management. Now I want to talk about bending some rules. What's going on here?

Rules are necessary to keep organization in and chaos out. However, rules do offer the disadvantage of stifling creativity and reasonable risk taking. Creativity and risks can often pay large dividends if they succeed. We hate to miss out on any benefits like that, so occasionally it pays to bend the rules.

The problem is knowing when and how much bending is safe. When you spot an opportunity, analyze the situation carefully from all angles including a sound fall-back position. Justify taking the risk by calculating the potential gains and losses to determine if the bending is really worthwhile. List the pros and cons and weigh each one to determine the chances of success. Make sure you have all the information necessary to make the decision and review the plan with your key peers. If the risk looks sound, give it a try.

Learning when and where to bend the rules can come only with time and experimentation. I recommend you start in a small way and try to build on your experience. The cardinal rule is this: Make surethe potential rewards warrant the risk.

Knowledge of the business

You must have a thorough knowledge of the project management business and the fields in which you are practicing. Keep current on general economic conditions and how they may affect your business. How are interest rates going? What is the outlook on inflation? What current events are likely to affect your project or business?

You should take time to read your industry trade journals so you keep up with business trends. Keep current on what the competition is doing, what big projects are breaking, and so on. All of that background is necessary to participate intelligently in top-level conversations with management, clients, contractors, and peers. Being knowledgeable about your business builds your image as a competent and professional project manager. Membership in professional societies and outside reading are key sources for that general background information. When you plan your time, leave some room for that sort of mind-expanding activity.

Personal public relations

Your personal PR program is something you have to do yourself—assuming you have not hired a press agent. One area of effort is getting letters of commendation at the end of the project. Naturally, you should expect them only after a good or outstanding performance. People often volunteer such letters, which is fine. However, if they don't volunteer, some timely hints may help.

Another personal PR activity is submitting newsworthy articles about unusual features of your project for house organs and trade journals. Such articles are image builders. With some imagination, you should come up with a few more.

References

1. Fredrick Herzberg, *Work and the Nature of Man,* Thomas Y. Crowell, New York, 1966.
2. Henry Mintzberg, *The Nature of Managerial Work,* Harper & Row, New York, 1973.
3. Chester L. Karrass, *Give and Take,* Thomas Y. Crowell, New York, 1974.
4. Edward Levin, *Negotiating Tactics,* Fawcett Columbine, New York, 1980.
5. Robert J. Graham, *Project Management, Combining Technical and Behavioral Approaches for Effective Implementation,* Van Nostrand Reinhold, New York, 1985.

Case Study Instructions

1. Your corporate public relations department has asked for a list of project factors likely to impact on the local area. It plans to use the information to indoctrinate local citizens groups, public officials, and labor leaders on the effect the new project will have on their

community. Make a list of potential positive and negative effects your project will have on the community and the state.

2. After the 25 percent design review, you are convinced that the contractor's project manager is not performing to your satisfaction. How would you approach correcting the situation?
3. Perform a self-analysis on your personality as it relates to becoming a successful project manager. Include checking the inventory of the eight areas of importance in management.
4. As the contractor's project manager, you are starting to have problems with your client relations. The owner is not living up to the contract requirements in regard to information supply and drawing approvals. You already have requested improvement in the progress report, but it doesn't seem to be working. How would you approach that problem to ensure meeting your project goals?

Appendix

Project Manager's Job Description

1.0 Concept

The project manager is responsible to company management for execution of the project to ensure completion of the work as specified, on time, and within budget. The PM directs all aspects of project execution required by the project scope. That includes such key areas as engineering, design, procurement, construction, and facility start-up. The PM supervises all contractual and financial matters and client relations.

The project manager is the focal point for all project activities from start to finish. He or she shall also play a lead role in the proposal stage and serve as the company's technical representative in contract negotiation. Company management shall issue a project management charter giving the PM broad authority to manage the project execution within the guidelines set up in this job description.

2.0 Foreword

Project managers must establish a reputation as knowledgeable managers who know the company, its organization, and the work to be done. They must prove their ability to manage, coordinate, and motivate people to maintain high standards and to produce successful projects.

Planning, organization, and control are key functions of the PM's work. He or she must plan the work and organize the available people and methods to permit routine matters to proceed with minimum effort and delay while responding quickly and effectively to unusual or emergency needs.

The PM sets the project goals and priorities early in the project and frequently reviews them with the project staff and management. To avoid being enmeshed in technical problems at the expense of losing managerial control, PMs must be true managers and not simply doers or supervisors. They must also be *effective* delegators to project their expertise and project philosophy throughout the project organization. Training and developing new project managers are byproducts of effective delegation.

3.0 Duties and Responsibilities

Following each planning, organizing, and controlling duty is a number in parentheses that denotes one of the following decision levels:

(1) Project manager has complete authority to act.
(2) Project manager must first advise company management before taking action.
(3) Actions that require prior company management approval.

3.1 Planning

a. Become completely familiar with all documents and special project and client (or contractor) requirements, and prepare and distribute pertinent project information to those having a "need to know." (1)
b. Develop the master plan for executing and controlling the project in conjunction with Company Operating Management. (2)
c. Prepare and issue written project procedures to govern all project activities consistent with the approved project plan, company standards and client (or contract) requirements. (2)
d. Oversee the preparation and approval of the project budget and establish the procedures for controlling and reporting project costs for all phases of the work. (1)
e. Oversee the preparation and approval of the project schedule and establish procedures for monitoring, controlling, and reporting project progress throughout its life. (1)
f. Establish specific project goals and set priorities for all facets of the project and issue them to the project staff. Incorporate these into a *management by objectives* system to ensure meeting the goals. (1)
g. Oversee the development of basic project design criteria and general specifications and establish quality control procedures to ensure quality products, services, and performance throughout the project. (1)
h. Oversee the project purchasing agent's preparation of a project procurement plan defining the procurement activities required to ensure timely delivery of physical resources to the project. (1)

i. Review project plans and procedures regularly and revise and update them as necessary to keep the project current. (1)
j. Establish and monitor all channels of communication (including meetings) among all key participants in the project. (1)

3.2 Organizing

a. Develop a project organization chart showing the lines of authority and interrelationships of key personnel and activities on the project. (3)
b. Prepare job descriptions detailing the duties, responsibilities and project objectives for key project team personnel. (1)
c. Initiate and participate in the selection of key project supervisors and establish an MBO system for the project. (2)
d. Finalize and implement the overall project personnel plan in conjunction with key project supervisors and applicable department heads in accordance with project timing. (1)
e. Formulate procedures for assignment and/or transfer of project personnel to fill positions at the jobsite or satellite offices. (3)
f. Participate with company management and applicable corporate groups to organize special project requirements such as taxes, insurance, legal services, lease or purchase of space or equipment, and vehicles. (3)
g. Continually review the project organization and make adjustments to suit actual project needs, especially near the end of the project. (1)
h. Arrange the internal and client (or contractor) project kickoff meetings to initiate the execution of the project in an orderly manner. (2)
i. Issue the project procedure manual within the first 30 days of the project. (Use "Holds" where necessary.) Use the PPM as the basis for project staff orientation sessions. (1)
j. Keep the contracting parties informed of changes in the project organizational structure and manning, and secure contractual approvals if required. (1)

3.3 Controlling

a. Closely monitor project activities for conformity to contract scope requirements and establish a change order procedure for scope revisions. Monitor contractual requirements and recommend adjustments when required. (1)
b. Administer and enforce compliance with the terms of the

contract, the project master plan, the project procedures, and management directives paying particular attention to guarantee and warranty requirements. (1)

c. Regularly monitor the systems for controlling project costs, schedule, and quality to see that they are working effectively in meeting project objectives. The control systems should forecast activities to project completion. (1)
d. Maintain effective communications with client, contractors and key project participants by: (1)
   - Conducting project review meetings on a regular basis.
   - Issuing high-quality project reports covering the status of physical progress vs. schedule, cost vs. budgets, material status, and so on with an explanation for any off-target items.
   - Discussing any problem areas along with your recommendations for solving them.
   - Documenting the project with minutes of meetings, correspondence, telephone call confirmations, daily diaries, and so on to build working files and a project history.
e. Review project personnel requirements regularly to ensure that the human resources are properly matched to the workload and schedule. (1)
f. Monitor the flow of design data, vendor data, and project information to ensure that all parts of the project are progressing smoothly. (1)
g. Maximize use of existing company standards, methods, and procedures for maximum efficiency. (1)
h. Monitor all project invoices and payments to ensure adherence to the cash flow plan. Also, periodically review the escalation and contingency accounts to see that they are being properly allocated. (1)
i. Promptly inform company management of any unusual project events or problems so they are kept current on unforeseen happenings. (1)
j. Review and approve all outside communications to ensure that the project and company images are being presented fairly. (1)
k. Practice control by exception and give immediate attention and corrective action to off-target items. (1)

4.0 Authority

In order to have strong and effective project management control of a project, the Project Manager's authority must be established in writing and supported by top-management policy and actions.

The project manager shall have the authority to:

1. Communicate across departmental lines on any matters relating to the project.
2. Participate in the selection of personnel who will be assigned to the project and shall be consulted prior to any proposed changes in assignment of project personnel.
3. Request the presence of any departmental personnel whose services are required to serve the project.
4. To arbitrate interdepartmental and interdiscipline differences on matters pertaining to the project. Occasionally upper management may be required to approve the decision.
5. Approve all expenditures and commitments for the project within any limits which may be set by upper management.

5.0 Working Relationships

To successfully complete a project, the project manager must have the full cooperation of all departments within the company in addition to the authority granted by top management. To gain that cooperation, the project manager must maintain good working relationships within and across all organizational lines of the company. As a minimum the project manager must:

1. Cooperate with corporate staff members, department heads, and other management personnel in matters relating to their assigned areas of responsibility.
2. Cooperate with other operating units, management centers and/or affiliates so that the best interests of the company and the project are served at all times.
3. Keep company operating management and department heads current on all project matters which can affect their operations.
4. Be responsive to requests for information and services from both company and project operating groups.
5. Provide routine and/or special reports required by the company procedures, operating management, or the client.

6.0 Leadership Qualities

A successful project manager must have strong leadership qualities to ensure that the project team is performing at top efficiency at all times.

To develop into a true leader the project manager must:

1. Direct the project work to meet the project and company contractual obligations at all times while maintaining high project team morale.
2. Develop and maintain a system of decision making within the project team whereby decisions are made at the proper level per the project procedures, position descriptions, and company policies.
3. Promote the development and growth of key project supervisors and encourage them to do likewise with their people.
4. Establish written project objectives and performance goals for all key members of the project team and review them periodically via an MBO system.
5. Promote an atmosphere of team spirit with the project staff and client (or contractor) staff.
6. Project leaders shall conduct themselves at all times in an exemplary manner to set a good example for all team members to follow.
7. Be a good listener and fairly resolve any problems or differences between project personnel, clients, department heads, contractors, and so on, which may arise.
8. Anticipate and minimize potential problems before they arise by maintaining a frequent contact and current knowledge of all project activities, project status, client and contractor attitudes, and outside factors which might affect the project.
9. Maintain a positive attitude toward the project staff, clients, contractors, management and peers at all times.
10. Attack problem areas quickly no matter how distasteful they may be. All problems must be brought into the open and resolved as soon as possible.

Appendix

B

Project Services Checklist

PROJECT SERVICES CHECK LIST

Project No.:______________________

Project Name:______________________

Client:______________________

Project Manager:______________________

Type of Contract:______________________

Scope of Services:______________________

Schedule Dates

Start:______________

Finish:______________

Issued By:______________

Date:______________

SERVICES REQUIRED	BY CONTRACTOR	BY OWNER	BY OTHERS (SPECIFY)	NOT REQUIRED	REMARKS
PROCESS AND BASIC DESIGN					
Laboratory Testing					
Field Measurement and/or Testing					
Site Plans					
Equipment Layouts					
Process Design Basis					
Process Description					
Process Flow Diagrams					
Heat and Material Balances					
Utility Summary					
Equipment List					
Equipment Data Sheets					
Process P&ID's					
Utility P&ID's					
Control Logic Diagrams					
Instrumentation					
Instrument List					
Flow Instrument Data Sheets					
Control Valve Data Sheets					
Safety Valve Data Sheets					
Rupture Disc Data Sheets					
Analyzer Data Sheets					
Control Computer Logic Diag.					
Thermal Rating of Heat Exchangers					
Catalyst and Chemical Summary					
Motor Horsepower Estimates					
Operating Manuals					
Start-up Assistance					
Process Studies (List)					
____________________					
____________________					

SERVICES REQUIRED	BY CONTRACTOR	BY OWNER	BY OTHERS (SPECIFY)	NOT REQUIRED	REMARKS
MECHANICAL EQUIPMENT					
Equipment Specifications					
Equipment Requisitions					
Technical Bid Analysis					
Review Vendor Drawings & Data					
Review Vendor Calculations					
Update Requisitions as Purchased					
Prepare Lubrication Schedule					
Prepare Noise Specifications					
Perform Noise Surveys					
Spare Parts Specification					
Spare Parts Requisition					
Vendor Coordination Trips					
Field Trips					
Equipment Studies (List)					
______________________________					
______________________________					
CIVIL ENGINEERING					
Soils Investigation & Report					
Foundation Design Analysis					
Site Development					
Grading Plan					
Drainage Plan					
Roads					
Railroads					
Fencing (Temporary/Permanent)					
Topographic Survey					

SERVICES REQUIRED	BY CONTRACTOR	BY OWNER	BY OTHERS (SPECIFY)	NOT REQUIRED	REMARKS
CIVIL ENGINEERING Cont'd.					
Property Survey					
Foundation Plans					
Foundation Design					
Buildings (Layout)					
Buildings Design					
Structural Design					
Concrete					
Steel					
Platforms and Ladders					
Paving					
Review Vendor Drawings					
Review Vendor Calculations					
Field Trips					
Civil Studies (List)					
______________________					
______________________					
PIPING					
Specifications					
Materials					
Fabrication & Erection					
Testing					
Flow Diagram Drafting					
P&ID Drafting					
Plot Plans					
Equipment Layouts & Elevations					
Orthographic Drawings (Specify Areas)					

SERVICES REQUIRED	BY CONTRACTOR	BY OWNER	BY OTHERS (SPECIFY)	NOT REQUIRED	REMARKS
PIPING Cont'd.					
Equipment Orientation Drawings					
Platform Layouts					
Standard Piping Details					
Piping Design Models (Specify Areas)					
Model Shipping					
Isometric Drawings (Manual or Computer)					
Material Take-off (Computer or Manual)					
Underground Piping					
Steam Tracing					
Pipe Supports and Hangers					
Specialty Items					
Spare Parts					
Underground Piping Design					
Sewers					
Process Lines					
Water Systems					
Firewater					
Piping Requisitions					
Piping System Stress Analysis					
Analog Studies					
Review Vendor Drawings					
Review Shop Drawings					
Steam Tracing					
Field Trips					
Piping Studies (List)					

SERVICES REQUIRED	BY CONTRACTOR	BY OWNER	BY OTHERS (SPECIFY)	NOT REQUIRED	REMARKS
VESSELS AND EXCHANGERS					
Equipment Specifications					
Equipment Requisitions					
Technical Bid Analysis					
Review Vendor Shop Drawings					
Review Vendor Calculations					
Calculate Equipment Weights					
Update Requisitions as Purchased					
Spare Parts Requisitions					
Mechanical Design of Exchangers					
Vessel Drawings					
Shop Coordination Trips					
Field Trips					
Vessel Studies (List)					
________________					
________________					
ELECTRICAL					
Power Source: ________________					
________________					
Main Substation					
Specifications					
General					
Major Equipment					
Power System Design					
Major Load Study					
Short Circuit Analysis					

SERVICES REQUIRED	BY CONTRACTOR	BY OWNER	BY OTHERS (SPECIFY)	NOT REQUIRED	REMARKS
ELECTRICAL Cont'd.					
Relay Coordination Study					
Main Single Line Diagram					
Medium Voltage Single Line					
Low Voltage Single Line					
Area Classification Drawing					
Logic Diagrams (Interlocks)					
Motor Schematics & Wiring Diagrams					
Substation Layouts					
Conduit Plans					
Lighting Plans					
Communication Plans					
Equipment Requisitions					
Technical Bid Analysis					
Review Vendors Drawings					
Prepare as Built Drawings					
Bulk Material Take-offs					
Subcontract Bid Package					
Vendor Trips					
Field Trips					
Spare Parts Requisitions					
Electrical Studies (List)					
______________________________					
______________________________					

SERVICES REQUIRED	BY CONTRACTOR	BY OWNER	BY OTHERS (SPECIFY)	NOT REQUIRED	REMARKS
INSTRUMENTATION					
General Specifications					
Instrument List					
Instrument Standard Details					
Panel Layout Drawings					
Control Room Layout Drawings					
Computer Hardware					
Computer Software					
Control Schematics					
Logic Diagrams					
Instrument Loop Drawings					
Instrument Piping Drawings					
Instrument Requisitions					
Technical Bid Analysis					
Update Requisitions to as Purchased					
Instrument Bulk Material Take-Offs					
Instrument Bulk Material Requisitions					
Vendor Drawing Review					
Subcontract Bid Package					
Vendor Visits					
Spare Parts Requisitions					
Field Trips					
Instrumentation Studies (List)					
____________________					
____________________					

SERVICES REQUIRED	BY CONTRACTOR	BY OWNER	BY OTHERS (SPECIFY)	NOT REQUIRED	REMARKS
BUILDINGS					
Architectural Layouts					
Specifications					
Architectural Details					
Structural Design and Drawings					
Electrical Design and Drawings					
Plumbing Design and Drawings					
Heating Ventilating & Air Conditioning Design and Drawings					
Subcontract Bid Packages					
Subcontractor Coordination					
Field Inspection					
Review Vendor Drawings					
Architectural Studies (List)					
______________________					
______________________					
DUST COLLECTION SYSTEMS					
Review Applicable Codes					
Specifications					
Flow Diagrams					
Equipment Requisitions					
Subcontract Bid Package					
Technical Bid Analysis					
Vendor Drawing Review					
Review Vendor Calculations					
Field Trips					
Dust Collection Studies (List)					
______________________					
______________________					

SERVICES REQUIRED	BY CONTRACTOR	BY OWNER	BY OTHERS (SPECIFY)	NOT REQUIRED	REMARKS
Materials Handling					
Specifications					
Equipment Layout					
Flow Diagrams-Material Balance					
Chute & Hopper Drawings					
Equipment Requisitions					
Interlock Logic Diagram					
Technical Bid Analysis					
Subcontract Bid Package					
Review Vendor Drawings					
Spare Parts Requisition					
Vendor Trips					
Field Trips					
Material Handling Studies (List)					
______________________					
______________________					
GENERAL ENGINEERING					
Insulation Economic Analysis					
Insulation Specification					
Insulation Take-Offs					
Painting Specification					
Painting Take-Off					
Subcontract Bid Packages					
Technical Bid Analysis					
Refractory Specifications					
Refractory Requisitions					
Field Trips					

SERVICES REQUIRED	BY CONTRACTOR	BY OWNER	BY OTHERS (SPECIFY)	NOT REQUIRED	REMARKS
GENERAL ENGINEERING Cont'd.					
Welding and Metallurgy					
Material Selection					
Welding Specifications					
Review Welding Specs (Vendors)					
Field Trips					
Studies (List)					
________________					
________________					
ENGINEERING ADMINISTRATION					
Engineering Schedule					
Drafting Room Schedule					
Physical Progress Measurement					
Spec/Req/Drawing Status Report					
PROJECT SCHEDULING					
Project Schedule Type:________					
Project Schedule Reporting & Update Cycle________					
Key Activity Sorts (CPM)					
Interim Schedule					
COST CONTROL					
Home Office Cost Report					
Code of Accounts					
Field Cost Report Period:________					
Project Close-Out Report					
Cash Flow Report					
Special Reports (List)					
________________					
________________					

SERVICES REQUIRED	BY CONTRACTOR	BY OWNER	BY OTHERS (SPECIFY)	NOT REQUIRED	REMARKS
ESTIMATING					
Feasibility Estimate					
Capital Cost Estimate					
Definitive Estimate					
Estimate Trending					
Prepare Definitive Budget					
Estimate Change Orders					
Cost Studies (List)					
____________________					
____________________					
PROCUREMENT SERVICES					
Prepare Vendors List					
Maintain Inquiry Status Report					
Issue Inquiries					
Prepare Bid Tabs					
Review Commercial Terms					
Issue Purchase Orders					
Issue Change Orders					
Approve Vendor Invoices					
Expedite Vendors					
Phone Contact					
Shop Visits					
Prepare Expediting Reports					
Expedite Vendor Data					
Expedite Sub-vendors					
Issue Delivery Status Reports					
Establish Quality Control Standards					

SERVICES REQUIRED	BY CONTRACTOR	BY OWNER	BY OTHERS (SPECIFY)	NOT REQUIRED	REMARKS
PROCUREMENT Cont'd.					
Qualify Welding Procedures					
Perform Pre-award Shop Survey					
Shop Inspection List					
Witness Equipment Run Tests					
Third Party Inspection Required?					
HOME OFFICE CONSTRUCTION SERVICES					
Construction Scope:______________					
______________________________					
Pre-construction Planning					
Field Staffing					
Field Procedures					
Heavy Lift Studies					
Heavy Equipment Routing Plan					
Area Labor Survey					
Pre-job Conference					
Project Labor Agreement					
Subcontract Procurement					
Subcontract Administration					
Field Scheduling System					
Special Requirements (List)					
______________________________					
______________________________					

Appendix

C

Case Studies

These case studies are hypothetical projects which have been created for use with the Case Study Instructions given at the end of each chapter. A broad spectrum of project types is offered to cover the various types of working environments as shown in Figure 1.1. In the event that you do not find a project on that list, you can create a do-it-yourself project by using the generic format given at the end of the list.

Since it is not possible to give you every detail of a project in this limited space, you will have to make assumptions for any of the missing details you will need to create a viable solution to the case study.

You may choose to assume the position of project manager for either the owner or contractor as best suits your working environment or career goals.

Project 1

A major U.S. insurance company is developing a series of combination office and light industrial parks at three U.S. locations. The project will have a staged development of the three locations starting on the west coast, in the midwest, and on the east coast, in that order. Site development and infrastructure have been estimated at about $8 million at each site exclusive of land, design fees and overhead. One design firm will handle all three projects to ensure consistency of design and project execution.

Each site will start with the construction of a speculative office building and a light-manufacturing building for leasing purposes by the owner. These will be 50,000 and 100,000 sq ft, respectively. Your firm will be responsible to provide services for site selection, preliminary and final design, construction contractor selection, and field inspection of the work on all sites. Construction is planned to be done with lump sum construction contracts with the first unit to be opened in 12 months from design contract award, and subsequent units will come on stream at 6-month intervals.

Project 2

Your firm is going to build a grass roots paper mill in northern British Columbia, Canada. The preliminary cost estimate is $400 million for the initial plant with one paper machine being installed now and more machines added later as sales warrant. The site is remote with limited port facilities, an indigenous population of Indians, and little or no infrastructure. The provincial government has agreed to furnish some of the infrastructure to encourage development in the area.

The schedule for the project is 48 months from conceptual design approval to plant start-up. The owner plans that the work will be performed on a turnkey basis by a single major engineering-construction contractor selected from a worldwide slate of contracting firms. The owner has prepurchased the paper machine because of the long delivery and will turn the order over to the successful contractor for completion of the delivery. A continuous digester has been purchased as an installed package from a Swedish firm. A paper-coating process has been licensed by the owner from a firm in Finland.

Financing will be from internal funding and government loans, so cash flow and budgets will be very tight. Cost and schedule overruns will be catastrophic for the financial plan for the project; therefore, good project control is vital to the success of the project. You are to start the project at the proposal stage and assume you win the award.

Project 3

You are a project engineer in a medium-size manufacturing plant responsible for plant modifications and process improvement projects in the $25,000 to $750,000 range through an in-house team of 15 technical people. A plant addition for the production of a new product has been delegated to your group in addition to your normal workload of plant projects.

The process for the new plant has been developed in the corporate R&D center in another city, and your group will be responsible for the final process design and selection of the production machinery. Your management is expecting your team to deliver a working plant within 18 months of project release. Your group has the responsibility for cost and schedule control and you are permitted to use outside help if you feel that it is required to meet the project goals.

Project 4

The Department of Energy (DOE) has been chosen as the agency to construct a new supercollider for the National Science Foundation on a site not yet selected in the United States. The project has a preliminary price tag of $4.5 billion and is urgently required to reestablish the preeminence of the United States in superconductivity research.

The supercollider will take 7 years to build and will contain the following major parts: a 53-mile oval particle accelerator tunnel, a ground level 350 acre campus of 15 buildings, 4 lab buildings on the oval, and smaller support buildings every 5 miles on the tunnel. The interior of the oval not occupied by buildings will be left in its native state. Access roads and a security system will be required to service the total facility.

Financing is going to be furnished by the federal and state governments as well as foreign countries interested in participating in the venture. It is estimated that about 50,000 contractors and suppliers will be participating in the construction of the facility and its equipment. Your group will be responsible only for the construction of the capital facilities portion of the project.

You may select the position of project director for the owner, the DOE, or a major contractor on the project in your area of expertise.

Project 5

This is a *generic project* format which you may use to construct a do-it-yourself sample to suit your particular area of expertise in the capital project field. Assume you start the project at the proposal stage and carry it through start-up and turnover.

OWNER:	___ government, ___ industry, ___ developer, ________________ other
FACILITY:	___ institution, ___ plant, ___ building, ___ laboratory, ___ commercial, ___ study, ________________ your choice
COMPLEXITY:	___ addition, ___ green-field site, ___ revamp, ___ hi-tech, ________________ other
SIZE:	___ sq ft, ___ capacity, ___ stories, $____________ estimated cost, ___ labor-hours, ________________ other
LOCATION:	___ country, ___ urban, ___ remote, ___ developed country, ___ third world, ________________ other

INFRASTRUCTURE: ___ existing, ___ nonexisting,
___ partial, __________ transportation,
___ utilities, ___ labor supply,
________________ other

SERVICES: ___ project development, ___ design,
___ procurement, ___ construction,
___ inspection, ___ start-up, ___ extras,
________________ other

FINANCING: ___ internal, ___ construction loan,
___ participative, ___ government loan,
________________ other

SCHEDULE: ___ loose, ___ tight, ___ medium,
___ fast-track, ___ impossible,
________________ other

BUDGET: ___ liberal, ___ tight, ___ average,
___ improbable, ________________ other

DESIGN BASIS: ___ well-defined, ___ preliminary,
___ loose, ___ by owner ___ by contractor,
___ environmental permits,
________________other

EXECUTION PLAN: ___ in-house, ___ prime contract,
___ separate subcontracts, ___ mixed basis,
________________ other

LONG-DELIVERY ITEMS: equipment____________________________
materials____________________________
other factors____________________________

OTHER FACTORS: ____________________________________
____
____
____
____
____
____
____
____
____
____
____
____

Appendix

D

Typical Project File Index

0.0 File Index
1.0 Project Proposal
2.0 Contract and Legal
    2.1 Copy of contract with amendments
    2.2 Change notices
        2.2.1 Change orders
        2.2.2 Change order log
        2.2.3 Internal change orders
    2.3 Licensing agreements
    2.4 Secrecy agreements
    2.5 Legal correspondence
    2.6 Subcontracts (a),(b),(c), etc.
        2.6.1 Subcontract documents
        2.6.2 Subcontract correspondence
    2.7 Insurance
        2.7.1 Certificates
        2.7.2 Claims
3.0 Project Procedures
    3.1 Project procedure manual
    3.2 Company policies
    3.3 Others as required
4.0 Project Correspondence

- 4.1 To client (or contractor)
- 4.2 From client (or contractor)
- 4.3 Project memorandums
- 4.4 Correspondence logs
- 4.5 Minutes of meetings
- 4.6 Field correspondence
- 4.7 Subcontractor correspondence
- 4.8 Others as required

5.0 Project Reports
- 5.1 Monthly progress report
- 5.2 Cost reports
- 5.3 Scheduling reports
- 5.4 Material status reports
- 5.5 Design office progress report
- 5.6 Labor budget reports
- 5.7 Field reports
- 5.8 Others as required

6.0 Process Design
- 6.1 General correspondence
- 6.2 Process design basis
- 6.3 Process description
- 6.4 Process flow diagrams

7.0 Architectural Design
- 7.1 General correspondence
- 7.2 Project design basis
- 7.3 Preliminary phase submission
- 7.4 Approved preliminary design
- 7.5 Others as required

8.0 Civil-Structural Design
- 8.1 General correspondence
- 8.2 Civil design basis
- 8.3 Structural design basis
- 8.4 Codes and zoning correspondence
- 8.5 Others as required

9.0 Electrical Design
- 9.1 General correspondence
- 9.2 Electrical design basis
- 9.3 Main power source development
- 9.4 Others as required

10.0 Mechanical Equipment, Vessels, etc.
- 10.1 General correspondence
- 10.2 Equipment list
- 10.3 Spare parts policy

11.0 Piping Design
- 11.1 General correspondence
- 11.2 Piping design basis
- 11.3 Plot plans
- 11.4 Equipment layouts
- 11.5 Piping material control
- 11.6 Others as required

12.0 Instrumentation
- 12.1 General correspondence
- 12.2 Process control philosophy
- 12.3 Instrument list
- 12.4 Others as required

13.0 Design Office Services
- 13.1 General correspondence
- 13.2 Reproduction
- 13.3 Computer services
- 13.4 Facilities
- 13.5 Others as required

14.0 Procurement
- 14.1 Procurement correspondence
- 14.2 Procurement plan
- 14.3 Expediting reports
- 14.4 Procurement status reports

15.0 Project Scheduling
- 15.1 General correspondence
- 15.2 Project scheduling policy
- 15.3 Original project schedule
- 15.4 Others as required

16.0 Project Estimating and Cost Control
- 16.1 General correspondence
- 16.2 Estimating plan
- 16.3 Original project estimate
- 16.4 Cost control correspondence
- 16.5 Others as required

17.0 Document Control (Central Files)
- 17.1 Document control procedure
- 17.2 Document transmittal file
- 17.3 Master drawing files
- 17.4 Vendor document control and files
- 17.5 Master specification files
- 17.6 Master requisition files
- 17.7 Others as required by project

Notes:

1. The project secretary controls files 0.0 through 16.0
2. The document control administrator controls 17.0
3. Each design group maintains a working file for the discipline.

Index

Acts of God, contract coverage for, 57
Acts of man, contract coverage for, 57
American Arbitration Association, 58–59
American Institute of Architects (AIA), short-form contracts of 33, 43, 92
Applicable law clause in contract, 58
Appropriation estimates, 129–130
Arbitration, 58–59
Architect/engineer or central engineering group, 10
Architectural and civil works projects, 68
Architectural design group , 81
Architectural and engineering firm, 88, 94
Architectural and engineering projects, 146–147, 164–165, 265
Architectural group, 81
Architecturally driven project, 81
Arrow diagramming method (ADM), 106–110
Audio-visual presentations, 292–293
Average project hourly rate, 225

Bar chart schedules, advantages and disadvantages, 117
Bar charts, 103–105
Basic network diagramming, 106–107
Beneficial occupancy, 93, 266
Bidding, 239–240
Bonus, 12
Bonus-penalty contracts, 40, 43, 45
Budget(s), 120, 145–151, 221, 225–226, 235, 268
Bulk material control, 261–262

Capital cost estimates, 130
Capital project, 4–5, 68–84, 245–250, 262, 357–360
Case studies, 22–23, 357–360
Cash flow, 14, 151–152, 245–250
Central engineering group, 10–11
Change orders, 53–54, 77, 185, 208, 229, 240, 245
Civil, architectural, and structural (CAS) design, 74–75, 81, 130
Civil works project, 81
Client-contractor relations, 322–323
Client relations, 285
Code of accounts, 147–148, 217
Commercial terms of contract, 50–59
Communications, 36–37, 284, 289–314, 326, 334
Compartmentalization syndrome, 184
Computer(s), 2, 99, 117–118, 134–135, 150, 208, 235, 292
Computer-aided design (CAD), 99, 141–142
Computer-aided engineering (CAE), 142
Conceptual phase, 70–71, 80
Conflict resolution, 313, 322, 328
Constructibility analysis, 170, 242
Construction all risks insurance, 57
Construction bidding/project follow-up in nonprocess design, 81, 91–92
Construction costs, controlling, 241–244
Construction documents in nonprocess design, 90
Construction group, 77, 83, 184, 241–244
Construction interface, 283–284
Construction labor, 225–226, 254–255
Construction personnel planning, 161–162
Construction phase, 77, 83–84, 90
Construction progress, 257–260
Construction project manager, 92, 183
Construction in project procedure manual, 208–209
Contingency:
  budgeting for, 148–149
  estimating for, 124, 137–138, 143
Contingency account, release of, 149
Contract(s), 33, 37–59, 92, 201
Contract administration, 59–61, 324
Contract tickle file, 60
Contracting capital projects, 37–46
Contractor, 25, 33, 53
Contractor selection, 27–36
Control by exception, 268–269

Controlling material costs, 228–233
Convertible contract, 42, 43, 44
Correspondence log, 292
Cost control, 120, 214–245
 (*See also* Project control)
Cost-effectiveness ratio, estimating, 132–133
Cost engineering, 215
Cost engineers, 77
Cost-escalator clauses, 54–57
Cost estimate, 124, 151–152
Cost plus fixed fee contract, 29, 34–35, 39–40, 43, 44, 51, 52, 54
Cost reduction, 215
Cost reporting, 215
Cost trending, 131–132
Critical path, analysis of, 109–110
CPM/Critical path method, 2, 98, 105–117, 119, 121, 163, 259, 268
Curve, estimating, 128

Definitive estimates, 130–131
Delay claims, calculating, 54
Departmental functions, 193–195
Design area and cost control, 236–237
Design data coordinator, 280
Design development in nonprocess design, 90
Design document schedule, 99
Design estimate and budget, 139–144, 276–277
Design execution in nonprocess capital project, 70–84
Design organization chart, 179–182
Design phase, scheduling of, 95–99
Design procedures in project procedure manual, 205–206
Design project manager, 92
Design schedule input, 277–278
Design subcontracting, problem areas, 164–165
Design team, mobilizing, 278–279
Detailed design in capital project, 74
Document approval procedures, 205
Dummy activity, 109

Earned-value system, 206, 251–252
Effective delegation, 189, 338
Electrical design:
 in capital project, 75–76
 of nonprocess capital project, 82
Engineering costs, 19, 78
Engineering schedule, 95–99
Engineering warranties, 49
Engineer's estimate, 130
Entrepeneurial skills, 216, 329
Equipment guarantees, 49
Equipment and material control, 260–262

Escalation, 135–137, 143, 148–149
Estimate(s), 123–145, 207
Estimate to completion, 99, 221, 224, 232
Estimating, cost of, 132–133
Estimating methods, 128, 133–135
Ethics, 239–240, 319–320, 323

Feasibility study, 80
Field change orders, 185
Field cost control group, 185, 243, 257–259
Field cost engineers, 243
Field craft labor planning, 162–163
Field engineering group, 77, 185–186, 265–266
Field labor productivity, 226–228, 243, 258
Field labor rates, 226
Field labor survey, 77
Field project manager (FPM), 83, 183–184
Field-required date, 100
Field scheduling group, 185, 258–259
Field superintendent, 183, 184
Field supervisory staff, 161, 243
Financial resources, 171
Financial rewards, 12
Force majeure clause, 57, 327
Foreign currency fluctuations, 138–139

Gantt chart, 103
Goal-oriented project groups, 9
Guarantees, 48–49

Herzberg, Frederick, 317
Home office services, 139–144, 156–161, 237–238, 255
Human factors, 284–286, 317–334
Human resources planning, 155–163

Icarus Corporation, 134
Incentive contract, 45, 51
Inflation projections, 135–136
In-house operating mode, 26
Insurance clauses in contract, 57–58
Integrated design-and-build projects, 92
International projects, 52, 58, 136–137, 138, 165, 227

Job descriptions, 15–16, 188, 337–342
Jurisdictional disputes, settling, 327

Kiss pricinple, 13, 27, 200

Labor cost estimating, 138, 221–226, 228
Labor hours, underrunning, 227–228
Labor rates, 138, 225, 226
Labor relations, 326–328

Lead project engineer, 274–276
Leadership, 286, 328
Letters of intent, 60–61, 92
Liability, limits of, in contract, 53
Licensing arrangements, 48
Life cycle curve, 4–7, 72, 78
Line and staff functions, 175–176, 181
Linear Particle Accelerator project, 19
Liquidated damages, 45
Listening skills, 313–314
Logic-based schedules, 103, 105–106
Logic diagram terminology, 107
Long-delivery equipment, 93, 97, 100, 168–169
Lump sum contract, 41, 43–45, 53–54, 149
Lump sum projects, 51, 52, 247, 249
Lump sum proposal, 29

Management-by-exception techniques, 17, 114, 268
Management by objectives, 16, 188–189
Management sponsor, 181
Manual budget, 150, 235
Manual scheduling methods, 117–118
Master project execution plan, 65—67
Material resource costs, 228–245
Material resources planning, 167–171
Material takeoffs (MTO), 261–262
Matrix system, 191, 193, 195, 198–200
MBO (management by objectives), 16, 188–189
Mechanical completion, 266
Mechanical design, 74, 82
Megaprojects, handling, 19–21
Multiproject environments, 274

National Society of Professional Engineers (NSPE), 43
Negotiating skills, 331–332
Nonprocess capital project, 5, 78–84, 88–94

Office manager, construction, 186
Open-book cost estimate, 44
Open-ended orders, 45, 240
Operational planning, 65–66
Organizational design, 174–176
Organization charts, 177, 186, 190–191
Outside estimating services, 133–134
Overtime work, 254–255
Owner-client goals, 9–10
Owner-contractor operating mode, 27

Panic, control of, 131
Patience, 332
Payment terms, 51–53
Percent accuracy, estimating, 124
Performance guarantees, 48–49
Performance test, 83, 267
Personnel planning, 157–167
Personnel loading curves, 156–161
PERT (program evaluation and review technique), 105, 108
Physical progress, 250–254
Piping design in capital project, 75
Piping and instrumentation diagrams (P&IDs), 5, 73, 75, 76, 130
Planning and scheduling, 63
Postmeeting activities, 312–313
Precedence diagramming method, 110–112
Prejob conference, 327
Process design package, 73
Process group, functions of, 73, 95–98
Process plant, 68–78, 267
Process project, 5, 94–102, 178–179
Process schedule, 95–98
Procurement, 76, 91, 99–100, 177, 207, 219–220, 238–241, 256–257, 282–283
Productivity, 226–227, 243, 254, 258
Pro forma contract, 33
Programming, 81, 89, 236
Progress report, 293–305
Project, 4, 118
Project architects, duties of, 273–274
Project cash flow curve, 246–247
Project cash flow plan, 151–152
Project change notices, 53–54, 219, 244–245
Project closeout, 89, 266–267
Project control, 14–15, 77–78, 82–83, 166–167, 207–208, 213–271, 281–282, 339–340
Project coordination, 6, 204–205
Project correspondence, 204, 292
Project cost estimating, 123–145
Project cost report, 150–151, 228–235
Project definition, 44, 125–128
Project director, role of, 19, 183
Project engineer, 273
  assistant, 275
  and client relations, 285
  and communications, 284
  and construction interface, 283–284
  as deputy project manager, 279
  as design data coordinator, 280
  as design manager, 275–279, 280–283
  duties of, 273–284
  lead, 274–276
  monitoring design scope, 280–281
  in multiproject environment, 274
  and procurement interface, 282–283
  and project manager, 285
  and project relations, 286
  and quality control, 281–282
Project engineering, 273–274
  administrative side of, 275–279

Project engineering (*Cont.*):
environment for, 274–275
human factors in, 284–286
leadership, 286
technical side of, 280–284
Project environment, 1
Project execution approach, 25–26, 66–67, 102
Project execution plan, 13–14, 34, 66–67, 102
Project file index, 205, 361–364
Project functions, 195–196
Project goals, 8–12
Project history, 267
Project invoicing, 249
Project life cycle, 4–7, 71, 72, 78
Project management:
communications in, 289–314
control in, 14–15, 213–271
cost control in, 244–245
defining, 4–6
functions of, 7–8
goals of, 8–9
Golden Rule of, 12–13, 306, 317
history of, 1–3
human factors in, 317–335
megaprojects in, 19–21
organization in, 14, 173–210
philosophy of, 12–13
planning in, 13–14, 63–84
project goals in, 8–12
project manager in, 15–16
project size in, 16–19
project variables in, 4–8, 16–19
resources planning in, 155–172
scheduling in, 87–120
Project manager, 7
analysis of owner's goals by, 36
audio-visual presentations of, 292–293
authority of, 340–341
bending of rules by, 333
and client contractor relations, 322–325
common sense of, 330–331
in conceptual phase, 71
construction, 92
in construction phase, 83–84, 92, 182–183
and contract administration, 46, 60, 324
and cost controls, 216, 217, 244–245
and cost engineering, 123
creed of, 8, 9
deputy, 279
and design subcontracting, 164–167
development of organization chart, 36, 177–182
ethics of, 239–240, 319–320, 323
and field activity, 260
handling of project acceptance by, 267
Project manager (*Cont.*):
and human relations, 322–330
importance of reputation of, 12
interest of, in contract, 37–38, 46–50
job description of, 15–16, 337–342
knowledge of business by, 334
and labor relations, 326–328
and management sponsor, 181
patience of, 332
personal drive of, 320–321
and personal human factors, 332–333
personal public relations of, 332–334
personality of, 318–319
and preparation of live presentation, 36–37
and process design, 73
and procurement, 76–77, 91, 256–257
and project control, 78, 207–208
and project correspondence, 292
in project master plan, 67
and project relations, 325–326
and project scheduling, 87, 102
and public relations, 326
qualities needed for successful, 318
in quality control, 264, 266
and project engineer, 285
and project staff, 324–325
and reporting, 207–208
responsibilities of, 52–53, 183, 293–305, 338–340
responsiveness of, 323–324
in selecting contractor, 28
selling organization by, 333, 334
skills of, 36–37, 59, 291–292, 313–314, 321–322, 324, 328–331, 341–342
and working relations, 341
Project master plan, 66–67
Project meetings, 305–313
agenda for, 307–308
checklist for, 306–307
controlling, 310–312
establishing need for, 305–306
execution philosophy, 306
minutes for, 309–310
organizing, 309–310
planning, 307–309
and postmeeting activities, 312–313
visual aids in, 309
Project mobilization, 189–190
Project money plan, 14, 67, 123
budgeting, 145–151
controlling, 214–235
and currency fluctuations, 138–139
project cash flow plan, 151–152
project cost estimating, 123–145
Project organization, 14, 173–210
design chart, 179–182
integrated project, 186
line and staff functions, 175–176

Project organization (*Cont.*):
matrix system, 191, 193, 198
organization charts for, 176–186, 190–191
organizational design, 174–175
organizational procedures, 200–210
process project, 178–179
project modes, task force vs. matrix, 191–197
project procedure manual (PPM), 200–209
project team in, 186–189
role of project manager in, 339, 340, 341
satellite office procedures, 210
task force, 191–193, 198–199
Project planning, 13–14, 63, 172, 338–339
Project procedure manual, 200–209
Project relations, 285–286, 325–336
Project reporting, 293–305
Project resources planning, 67, 155–172
construction subcontracting, 167
design subcontracting, 164–167
financial, 171
human resources planning, 155–163
material resources planning, 167–171
and subcontracting, 163–167
Project scheduling, 87–120
construction, 100–101
for nonprocess project, 88–94
Project secretary, 292
Project services checklist, 47, 276–277, 343–356
Project size, 16–21
Project team, 65
and cost control, 220, 244–245
goals of, 11–13
motivating, 187–190
selection of, 186–187, 278–279
Project variables, identifying, 4–8, 16–19
Promotion, 12
Proposal effort, managing, 36
Proposal goals, 36–37
Proposal phase:
of capital project, 72
of nonprocess capital project, 80
Proposal schedule, 29, 102
Proposals, evaluation of, 34–36
Public relations, 326, 334
Punchlist phase, 266
Purchasing commitment curve, 256, 257

Quality control, 166, 262–268
in capital projects arena, 262
in construction, 265–266
design, 263–264
of equipment and materials, 264–265
and project engineers, 281–282
Quality control (*Cont.*):
in project procedure manual, 206
Quality and material standards, 196

Recorder in project meetings, 309–310
Request for proposal (RFP), 27, 31–34
Requirement for gtood estimate, 144–145
Research and development (R&D), 70–71
Responsibility of owner and contractor, 49
Retainage, contractor's, 53
Richardson Engineering, 134
Risk management, 57–58

Satellite office procedures, 210
Schedule control, 250–260
Scheduling budget, 120
Scheduling systems, 103–120
Schematic design in nonprocess design, 89–90
Selecting the project team, 186–187
Secrecy agreements, 48
Site development, 21
Software development, 76
Span of control, 174
Split-contract A&E project, 44, 47–48, 92, 241
Split design-and-build project, 92
Staff functions, 181, 219
Standard labor-hours, 138, 226
Start-up sequence, 95, 97, 100
Statement of interest, 30
Strategic planning, 65, 71
Structural design of nonprocess capital project, 81–82
Subcontracting, 59, 164–167
Substantial completion, 266
Suspension and termination clauses, 56–57

Task force approach, 191–200
Technical proposal, 34
Technical scope of work, 46–48, 280–281
Temporary construction facilities, 242
Third-party constructors, 94, 178–179
Time control, 250
Time-and-material contract, 42, 45
Time plan, 14, 67, 250
Total installed cost (TIC), 4, 19
Turnkey project, 67

Unit price contracts, 41, 45–46
Upset price, 45

Value engineering, 215
Vendor, data, 100, 206, 282
Vendor list, suppliers on, 207, 239
Visual aids in project meetings, 292, 309

Wage-price inflation, 135

Warranties, engineering, 49
Work activities, 68, 78, 107–109
Workload curves, 31, 35
Work, scope of, 47–48, 54, 280–281

Zero balance bank accounts, 52, 246

About the Author

George Ritz is an expert in project management, with 40 years experience in executing projects in the United States and abroad. He has had a distinguished career with several well-known consulting and construction firms including his present association with CRS Sirrine in Greenville, S.C. His assignments have included a broad spectrum of projects from schools for the handicapped to world-class petrochemical plants.

Mr. Ritz holds a BSChE degree (with honors) from the University of Illinois and is a graduate of the U.S. Merchant Marine Academy at Kings Point with a major in marine engineering. He is a Registered Professional Engineer, a member of the American Academy of Forensic Sciences, and a seminar leader for the *Chemical Engineering* Seminars and Conference Group.